The Economics and Econometrics of the Energy-Growth Nexus

T0311928

The Economics and Econometrics of the Energy-Growth Nexus

Edited by

Angeliki N. Menegaki

TEI Stereas Elladas, University of Applied Sciences, Lamia, Greece
Hellenic Open University, Patras, Greece

ACADEMIC PRESS

An imprint of Elsevier

Academic Press is an imprint of Elsevier
125 London Wall, London EC2Y 5AS, United Kingdom
525 B Street, Suite 1800, San Diego, CA 92101-4495, United States
50 Hampshire Street, 5th Floor, Cambridge, MA 02139, United States
The Boulevard, Langford Lane, Kidlington, Oxford OX5 1GB, United Kingdom

Notices
Knowledge and best practice in this field are constantly changing. As new research and experience broaden our understanding, changes in research methods, professional practices, or medical treatment may become necessary.

Practitioners and researchers must always rely on their own experience and knowledge in evaluating and using any information, methods, compounds, or experiments described herein. In using such information or methods they should be mindful of their own safety and the safety of others, including parties for whom they have a professional responsibility.

To the fullest extent of the law, neither the Publisher nor the authors, contributors, or editors, assume any liability for any injury and/or damage to persons or property as a matter of products liability, negligence or otherwise, or from any use or operation of any methods, products, instructions, or ideas contained in the material herein.

Library of Congress Cataloging-in-Publication Data
A catalog record for this book is available from the Library of Congress

British Library Cataloguing-in-Publication Data
A catalogue record for this book is available from the British Library

ISBN: 978-0-12-812746-9

For information on all Academic Press publications visit our website at
https://www.elsevier.com/books-and-journals

Working together
to grow libraries in
developing countries

www.elsevier.com • www.bookaid.org

Publisher: Candice Janco
Acquisition Editor: J. Scott Bentley
Editorial Project Manager: Susan Ikeda
Production Project Manager: Priya Kumaraguruparan
Designer: Vicky Pearson Esser

Typeset by Thomson Digital

Contents

Chapter 5: Critical Issues to Be Answered in the Energy-Growth Nexus
              (EGN) Research Field...**141**

Angeliki N. Menegaki, Stella Tsani

PART 2: The Econometrics of the Energy-Growth Nexus **185**

Chapter 6: Practical Issues on Energy-Growth Nexus Data
              and Variable Selection With Bayesian Analysis **187**

Aviral K. Tiwari, Anabel Forte, Gonzalo Garcia-Donato, Angeliki N. Menegaki

Chapter 7: Current Issues in Time-Series Analysis for the Energy-Growth Nexus
     (EGN); Asymmetries and Nonlinearities Case Study: Pakistan..............229

Muhammad Shahbaz

Chapter 8: Panel Data Analysis in the Energy-Growth Nexus (EGN) 255

Can T. Tugcu

List of Contributors

Mohammad Al-Saidi   Institute for Technology and Resource Management in the Tropics and Sub-tropics (ITT), University of Applied Sciences, Cologne, Germany

Nicholas Apergis   University of Piraeus, Piraeus, Greece

Heli Arminen   Lappeenranta University of Technology, Lappeenranta, Finland

Alper Aslan   Nevşehir Hacı Bektaş Veli University, Nevşehir, Turkey

Daniel Balsalobre-Lorente   University of Castilla-La Mancha, Ciudad Real, Spain

Anabel Forte   University of Valencia, Valencia, Spain

José A. Fuinhas   University of Beira Interior and NECE-UBI, Covilhã, Portugal

Gonzalo Garcia-Donato   University of Castilla-La Mancha, Albacete, Spain

Vladimír Hajko   Mendel University in Brno, Brno, Czech Republic

Roula Inglesi-Lotz   University of Pretoria, Pretoria, South Africa

Arthur Kraft   Chapman University, Orange, CA, United States

John Kraft   University of Florida, Gainesville, FL, United States

Antonio C. Marques   University of Beira Interior and NECE-UBI, Covilhã, Portugal

Angeliki N. Menegaki   TEI STEREAS ELLADAS, University of Applied Sciences, Greece; Hellenic Open University, Greece

Maamar Sebri   University of Sousse, Sousse, Tunisia

Muhammad Shahbaz   Montpellier Business School, Montpellier, France

Aviral K. Tiwari   Montpellier Business School, Montpellier Cedex 4, France

Ebru Topcu   Nevşehir Hacı Bektaş Veli University, Nevşehir, Turkey

Stella Tsani   Athens University of Economics and Business; International Centre for Research on the Environment and the Economy, Athens, Greece

Can T. Tugcu   Akdeniz University, Antalya, Turkey

Foreword

Arthur Kraft*, John Kraft**

*Chapman University, Orange, CA, United States; **University of Florida, Gainesville, FL, United States

It has been over 5 decades since the founding of the Organization of the Petroleum Exporting Countries (OPEC) in 1960, over 4 decades since the oil embargo of 1973, and almost 4 decades since we published our paper on the link between energy and gross national product (GNP) (Kraft and Kraft, 1978). Never did we believe this simple bivariate analysis of causation would be the spark igniting a stream of research on the energy-growth nexus (EGN). We are pleased to be part of something that has lasted so long and has had such an impact.

The chapters in this book represent a significant contribution to this body of work. *The Economics and Econometrics of the Energy-Growth Nexus* is an excellent summary of the efforts by a significant number of leading global scholars on this important topic. We applaud the authors for presenting a broad array of topics surrounding the EGN.

Since the exogenous shock of 1973 some interesting dynamics have emerged. Many oil-producing countries have themselves become significant energy consumers and have implemented aggressive infrastructure spending. A recent Wall Street Journal article shows that the current price of $52.52 per barrel is less than the price per barrel needed by 11 OPEC members to balance their national budgets (Sider, 2017). In addition to this, the stated objective by several countries to achieve 100% electric vehicle production in 2 decades will have an impact on the demand for fossil fuels and the global automobile industry. On the supply side, new technology has had an impact on shale oil and shale gas production. These events, as well as climate issues, will impact future research on the EGN.

In the unfolding of quality and sophisticated future research exploring the link between energy and growth, this book represents a significant first step.

References

Kraft, J., Kraft, A., 1978. On the relationship between energy and GNP. J. Energy Dev., 401–403.
Sider, A., 2017. Oil prices settle above $50. Wall Street Journal, (July 31 Issue).

Introduction

Angeliki N. Menegaki[*][**]

[*]*TEI STEREAS ELLADAS, University of Applied Sciences, Lamia, Greece;* [**]*Hellenic Open University, Patras, Greece*

Energy is feeding growth, because it is an input to every kind of good or service. Particularly, after the industrial revolution, growth has become inextricably woven with energy use, together with other important inputs, such as capital and labor. While the primitive man used wood for the production of the required energy in his everyday life, the man progressively passed to other energy vehicles, such as coal, oil, and other fossil fuel types, with the latter two being the mostly used and consumed fuel types since 1820 as shown in Fig. 1 (note that $1 \text{ EJ} = 10^{18} \text{ J}$).

The Evolution of the Consumption of Different Energy Sources

While Fig. 1 enables an overview of the consumption of the different fuel types from as early as 1820, but only up to 2000; Figs. 2–4 enable a more detailed and recent overview (up to 2012), separately for fossil fuels, nuclear energy, and renewable energies. After

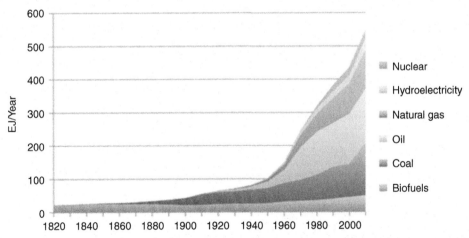

Figure 1: World Energy Consumption by Source, Based on Vaclav Smil Estimates From Energy Transitions: History, Requirements, and Prospects Together With BP Statistical Data for 1965 and Thereafter.
The Oil Drum, 2012. World Energy Consumption Since 1820 in charts. Available from: http://www.theoildrum.com/pdf/theoildrum_9023.pdf.

1950, there is a tremendous increase in energy consumption, which accompanies mass industrialization, urbanization, and transportation. Industrialization generated an increased variety and volume of manufactured goods and an improved standard of living for the countries that experienced it. An urban life promises better education and improved public health access, as well as a safe and uninterrupted provision of water, food, and shelter. Urban people have access to the aforementioned goods by paying for them and they can devote their own time to more creative ways of achieving their socioeconomic potential. Based on Figs. 2–5, which follow, both high and low income countries reveal a generally descending trend in fossil fuel energy consumption, but each one of them for different reasons. High-income countries have become more sustainable and seek ways to become more energy secure in front of oil price fluctuations. Conversely, low-income countries, confronted with less energy and finance options, just go about with less energy consumption and become trapped in their vicious cycle of poverty. Noteworthy is that about 1.3 billion people in Sub-Saharan Africa and Asia do not have access to electricity (IEA 2011). This reveals the existent margin for additional energy consumption in those countries. The rest of the middle-income groups follow a slightly ascending trend in their fossil energy consumption.

Traditionally, countries with huge fossil reserves play an important geopolitical role. Energy prices have been a very important factor for the development of all economies in the world. Price shocks and energy supply turbulences can wreck economies. However, energy alone does not suffice for the development of a country, as there are countries with huge energy

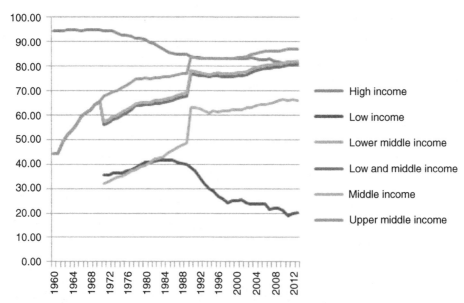

Figure 2: Fossil Fuel Energy Consumption (as % of Total).
Wordbank, 2017. Wordbank development Indicators. Available from: data.worldbank.org/data-catalog/world-development-indicators.

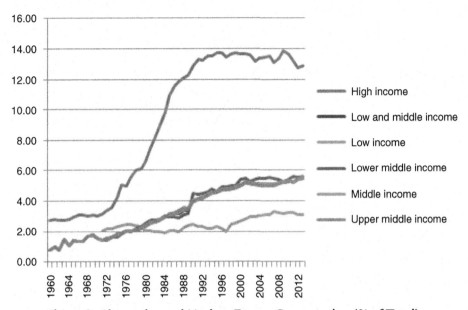

Figure 3: Alternative and Nuclear Energy Consumption (% of Total).
Wordbank, 2017. Wordbank development Indicators. Available from: data.worldbank.org/data-catalog/world-development-indicators.

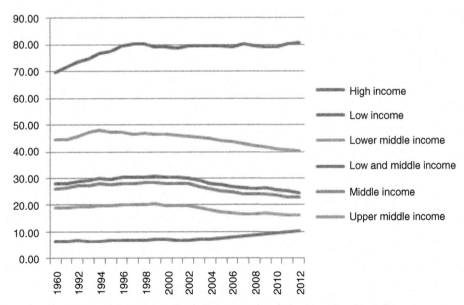

Figure 4: Renewable Energy Consumption (as % of Total).
Wordbank, 2017. Wordbank development Indicators. Available from: data.worldbank.org/data-catalog/world-development-indicators.

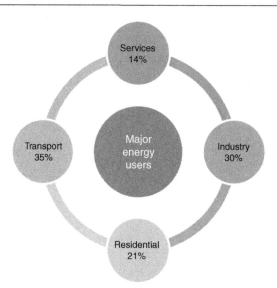

Figure 5: The Distribution of Energy Consumption Across Major Economic Sectors and Activities.
International Energy Agency (IEA), 2013. Key world energy statistics 2017.
Available from: www.iea.org/publications.

reserves; however, those have reached little economic development, due to their economic and political structure. Last, fossil fuel resources are gradually driven to exhaustion and this threatens global economies.

Based on Fig. 3, alternative and nuclear energy follows an increasing trend in all countries, particularly in high-income countries. This energy type is responsible for the production of 11% of global electricity needs (World Nuclear Association 2017). However, this rise has more or less stabilized itself from 1990 up to date for most income classes of countries. Fourteen European Union Member States are among the nuclear energy producers. Germany has decreased its production by 60,682 GWh, followed by Lithuania at 17,033 GWh, which also has ceased operation of its nuclear facilities in 2009, Belgium by 16,619 GWh and Sweden by 11,837 GWh in the period from 1990 to 2015 (European Commission 2017). Nuclear energy produces about 20% of electricity in the United State, however, it is not expected to increase any further due to the lower cost of other alternatives and the safer technologies, as well as the safety concerns generated, particularly after the Fukushima nuclear disaster, in Japan, in 2010.

As far as the evolution of renewable energy consumption is concerned, this follows a downward path for most income groups of countries and an increasing trend for high-income countries, which have the means to finance their costly infrastructure. Also, low-income countries appear to be stable in their renewable energy consumption. Fig. 5 depicts the distribution of energy consumption across major economic sector and activities, and show that transport and industry are the largest consumers of energy.

Energy Consumption and Economic Growth

Mainstream economics still base their theories in the neoclassical growth (Solow) model in which labor and capital are the protagonists in economic growth. Thus, the latter are regarded as the primary inputs, while energy is treated as any other input, which will exist perpetually in cheap and large quantities, namely, as an intermediate factor (Stern 2004). However, energy is exhaustible and nonreproducible; and for this reason it receives another theoretical treatment from ecological economics. The close relationship and the high correlation between energy consumption and economic growth are observed in Figs. 6 and 7.

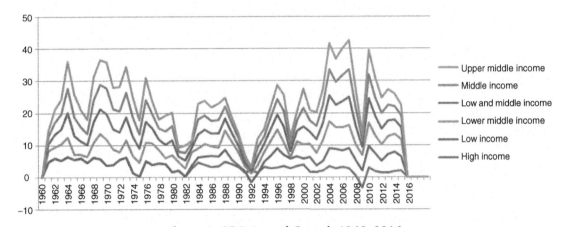

Figure 6: GDP Annual Growth 1960–2016.
Wordbank, 2017. Wordbank development Indicators. Available from: data.worldbank.org/data-catalog/world-development-indicators.

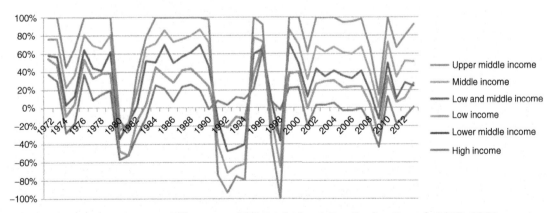

Figure 7: Energy Use (Kilogram of Oil Equivalent) Per Capita Growth 1972–2012.
Wordbank, 2017. Wordbank development Indicators. Available from: data.worldbank.org/data-catalog/world-development-indicators.

This correlation reveals a strong connection, but it does not give information on the direction of this connection and the relationship generation. Energy use has fuelled economic growth, and economic growth generates the need for using additional energy. The two magnitudes appear to have shared a lot of comovement.

Besides the aforementioned, the energy sector also contributes to employment, as a significant amount of jobs is offered therein, for example, in the United State this percentage is 5% of the total employment (Energy.Gov, 2017). The energy sector bears impacts on resource efficiency and the environment. Energy is heavily interlinked with food and water. To manage this relationship, a holistic approach is necessary that will take account all these new aspects too. The current book proposes this approach in various occasions, particularly in Chapters 1 and 5.

When a country is using its energy judiciously, it becomes less dependent on imports and its CO_2 emissions are lessened. Renovation and expansion of energy grids is important for the unhindered provision of energy. Europe aspires to reduce its greenhouse gas emissions at 20% of 1990 levels by 2020, increase by 20% its share of RES and 20% its energy savings. European funds channel investment into technologies. The United Nations Climate Change Conference in Paris entailed commitments with respect to the prevention of temperature increase. However, it is difficult and costly for countries to embark on renewable energy infrastructure.

The increase in fossil fuel consumption has generated emissions. The latter have caused pollution and the subsequent environmental damage with their repercussions. Melting of glaciers and the rising sea levels, the increasing temperature and the existence of extreme weather conditions are only some of the most important aspects of climate change. The evolution of CO_2 emission is shown in Fig. 8.

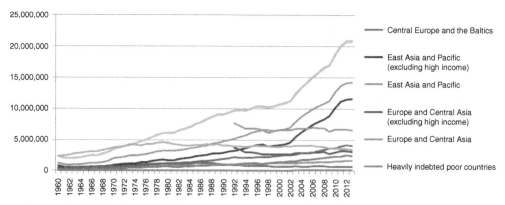

Figure 8: CO^2 Emissions (Kilogram Per 2010 US Dollar of GDP) From 1960–2012.
Wordbank, 2017. Wordbank development Indicators. Available from: data.worldbank.org/data-catalog/world-development-indicators.

Despite the continuous decline in energy intensity, more effort is needed to keep temperature rise to 2°C (IEA-International Energy Agency, 2016b). Decoupling emissions from energy use is vital and this can take place through the adoption of renewable energy and the establishment of more energy thrift technologies, as well as eliminating coal-fired power plants, phasing out the fossil fuel subsidies, and reduce methane emissions, which constitute another greenhouse gas that traps significant amounts of solar radiation in the atmosphere. Overall, there is a decoupling trend between energy consumption and GDP with the latter increasing by 90% and primary energy only at 56% with a huge variation in energy efficiency across sectors (IEA-International Energy Agency, 2016a,b).

Energy use is bound to increase in the forthcoming years due to the economic development and population growth. This phenomenon will be more likely to take place in developing countries rather than the already developed and industrialized ones, which have reached a stage of economic maturity. Until 2013, the bulk of energy production stemmed from fossils (GreenFacts, 2001–17). However, given the economic deceleration throughout the world because of the economic crisis and the reduced demand, energy consumption, even in China, has not increased as much as in the past. Technological developments toward energy efficiency and the increased availability of alternative energy sources do not put the same high pressure to energy demand as in the past (BP, 2016).

The Energy-Economic Growth Nexus

This is an important field of energy economics due to the importance of the two concepts it deals with. Energy sources are scarce and most of them generate emissions, which contribute to climate change. Different countries have stipulated different agreements and have set up various plans for the reduction of emissions and for refraining from using nonrenewable energy sources. Up to date, the energy-growth nexus (EGN) research field has generated ample papers with a lot of contradicting results, which makes it hard for the policy makers to find out how growth will be affected when energy conservation programs will be applied in a single country, groups of countries, or certain sectors. Therefore, the current book entitled as *The Economics and Econometrics of the Energy-Growth Nexus* attempts to collect and systemize all knowledge around the nexus with the hope to make current and future researchers and practitioners more able to drive the field into convergence.

The book is a constructive synthesis of knowledge and experience from 20 experts who are internationally recognized for their work and contribution into the EGN. They hope they will give useful insights and will put new order in the field, which demands much more research to be done in the future, so that conclusive and convergent evidence is reached. The book consists of two parts. Part 1, covering Chapters 1–5, deals with the economics of the EGN; while Part 2, covering Chapters 6–11, deals with the econometrics of the EGN.

Part 1: The Economics of the Energy-Growth Nexus
Chapter 1—The Energy-Growth Nexus: History, Development, and New Challenges

Profs. Vladimír Hajko, Maamar Sebri, Mohammad Al-Saidi, and Daniel Balsalobre-Lorente contribute to this chapter. Their contribution introduces readers to the origins of the EGN, the main routes it has followed up to date and to the routes that need to be taken in the future. The chapter recognizes the need for the EGN to reach consensus and it places great emphasis on the directions, which are spawned from the "Hajko critique." The problems the "Hajko critique" underlines are additionally recognized and further expanded in Chapter 5. Chapter 1 also outlines the history of the EGN, places emphasis on some other branches of the EGN, such as the environmental Kuznets curve and last emphasized on the new perspectives that gradually take place in the EGN, such as the food-energy-water nexus or the energy-water nexus fields.

Chapter 2—Disaggregation in the Energy-Growth Nexus: An Indicative Literature Review

Prof. Alper Aslan and Dr. Ebru Topcu have written about disaggregate energy consumption. They distinguish between energy type disaggregation and sectoral disaggregation and explain why this is important in the EGN. They provide a selective and representative recent literature review with examples and results from every sector together with a case study.

Chapter 3—On the Dynamics of Renewable Energy Consumption (Aggregated and Disaggregated) and Economic Growth: An Approach by Energy Sources

Profs. Antonio C. Marques and José A. Fuinhas have written about renewable energy consumption and growth. Given the particular importance of renewable energies and their importance for sustainable development, devoting a separate chapter on them was absolutely worthy. This chapter analyses the electricity-growth nexus in a context of energy transition from conventional to renewable sources. That transition requires making options in what concerns the composition of the electricity mix, namely, by choosing the contribution from each source to the mix. As such, this chapter assessed the interactions of sources and the consequences of that diversification on the economic growth, by focusing on Germany, which is in the phase-out of the nuclear base load source.

Chapter 4—The Role of Potential Factors/Actors and Regime Switching Modeling

Prof. Roula Inglesi-Lotz casts the foundations on the alternative variables, which can be used in the EGN because they too affect or shape the production function characterizing an economy. Among them, the analysis includes production factors (capital, labor, and technological progress), international trade, and financial development, as well as less discussed ones, such as militarization and tourism development. Also, alternative considerations of the growth measurement will be presented. Regime switching modeling is mostly concerned with political economy variables that create a different cadre within which an economy works.

Chapter 5—Critical Issues to be Answered in the Energy-Growth Nexus Research Field

Dr. Angeliki N. Menegaki and Stella Tsani review the international environmental, climate, and energy agreements that a lot of countries have embarked on, as well as the sustainability concerns raised by them. The urging need to reach the stipulated targets demands policy makers to reduce energy consumption. Thus, policy makers are urged to know if and by how much energy use reduction will retard economic growth. This chapter also discusses the ways with which energy efficiency can be hosted in the EGN framework. Another issue covered is the difference of results between single countries studies versus panel data studies with respect to the interpretation of their results in policy making.

Part 2: The Econometrics of the Energy-Growth Nexus

Chapter 6—Practical Issues on Energy-Growth Nexus Data and Variable Selection With Bayesian Analysis

Profs. Aviral Kumar Tiwari, Anabell Forte, Gonzalo Garcia-Donato, and Angeliki N. Menegaki coauthor the chapter and guide readers to a number of practical topics and technical topics. Profs. Tiwari and Menegaki provide a brief tour on international databases, and practical advice on the ways to select data, transform data, use and interpret results in the EGN. Profs. Forte and Garcia-Donato provide a new insight for variable selection in the EGN, while introducing Bayesian analysis in the EGN.

Chapter 7—Current Issues in Time-Series Analysis for the Energy-Growth Nexus; Asymmetries and Nonlinearities Case Study: Pakistan

Prof. Muhammad Shahbaz introduces the reader with new trends in the time-series analysis in the EGN. The acknowledgement of asymmetries and nonlinearities, both in stationarity and cointegration, as well as in causality (but this topic is covered in Chapter 9 too), produces a more accurate representation of these concepts; and thus more detailed suggestions for policy making in the long- and short-run. Besides presenting the reason of the generation of these new concepts and giving a succinct and down-to-earth overview of them, the chapter applies these concepts in a new case study for Pakistan.

Chapter 8—Panel Data Analysis in the Energy-Growth Nexus

Prof. Can T. Tugcu writes about panel data analysis in the EGN. Data constraints in energy economics encourage the researchers to employ panel data methodologies rather than others. However, there are quite many panel approaches to be utilized and ample statistical failures to be solved. Thus, research may get stuck in a confusing atmosphere in terms of investigating the correct diagnostics, following the right analysis sequence, selecting the appropriate estimating methodology, and interpreting the results. Under such a framework, this chapter sheds light on the aforementioned issues and provides either theoretical or empirical knowledge about panel data analysis in the EGN.

Chapter 9—Testing for Causality: A Survey of the Current Literature

Prof. Nicholas Apergis reviews the most recently developed causality tests. Special emphasis is placed on certain causality methodologies, such as bivariate causality tests in time series, multivariate causality tests in time series, alternative causality test approaches in time series, asymmetric causality, linear panel causality, and nonlinear and nonparametric causality. The chapter also and foremost contains rich varieties of case studies.

Chapter 10—Simultaneous Equations Modeling in the Energy-Growth Nexus

Prof. Heli Arminen presents a simplified simultaneous equations model that can be used in the EGN. Key concepts are illustrated, which are related to simultaneous equations modeling (SEM). SEM estimation methods with cross-sectional, time-series, and panel data are also covered; and the framework is used to demonstrate the methods in practice. The chapter concludes with recommendations for future research.

Chapter 11—The Energy-Growth Nexus Checklist for Authors

Stella Tsani and Dr. Angeliki N. Menegaki discuss the research on EGN from a prospective author's and a reviewer's point of view. The analysis summarizes the questions that have received much attention to date and those that are less addressed. The chapter pays attention to the common methodological approaches employed in the existing studies, and it highlights possible paths that future research in the field could consider. This chapter also discusses the ways to address and ensure transferability of the results beyond the energy-related scientific community reaching for policy design and implementation. Last, this chapter discusses the technical and practical requirements and the necessary steps to be taken from prospective authors and researchers in the field, while preparing their manuscripts for publication.

The contributors of this book truly believe that his book will change the way researchers, practitioners, and students think about EGN. Their aim is to put the pillars for new directions in the EGN and lead this field into convergent evidence, eloquent enough, for policy making.

References

BP, 2016. BP Statistical Review of World Energy. Available from: https://www.bp.com/content/dam/bp/pdf/energy-economics/statistical-review-2016/bp-statistical-review-of-world-energy-2016-full-report.pdf.

Energy.Gov, 2017. 2017 US Energy and Employment Report. Available from: https://energy.gov/downloads/2017-us-energy-employment-report.

European Commission, 2017. Nuclear Energy Statistics. Available from: http://ec.europa.eu/eurostat/statistics-explained/index.php/Nuclear_energy_statistics.

GreenFacts, (2001–2017). Forests and Energy. Available from: http://www.greenfacts.org/en/forests-energy/l-3/2-prospects-energy-supply.htm.

International Energy Agency (IEA), 2016a. Energy Efficiency Indicator Highlights. Available from: https://www.iea.org/publications/freepublications/publication/energy-efficiency-indicators-highlights-2016.html.

International Energy Agency (IEA), 2016b. Energy Efficiency Market Report 2016. Available from: https://www.iea.org/eemr16/files/medium-term-energy-efficiency-2016_WEB.PDF.

International Energy Agency (IEA), 2011. Energy for all. Available from: https://www.iea.org/publications/freepublications/publication/weo2011_energy_for_all.pdf.

Stern, D.I., 2004. Economic growth and energy. Encyclopedia of Energy 2. Available from: http://sterndavidi.com/Publications/Growth.pdf.

The Oil Drum, 2012. World Energy Consumption Since 1820 in charts. Available from: http://www.theoildrum.com/pdf/theoildrum_9023.pdf.

World Nuclear Association, 2017. Nuclear Power in the world today. Available from: http://www.world-nuclear.org/information-library/current-and-future-generation/nuclear-power-in-the-world-today.aspx.

The Economics of the Energy-Growth Nexus

CHAPTER 1

The Energy-Growth Nexus: History, Development, and New Challenges

Vladimír Hajko*, Maamar Sebri**, Mohammad Al-Saidi†,
Daniel Balsalobre-Lorente‡

*Mendel University in Brno, Brno, Czech Republic; **University of Sousse, Sousse, Tunisia; †Institute for Technology and Resource Management in the Tropics and Sub-tropics (ITT), University of Applied Sciences, Cologne, Germany; ‡University of Castilla-La Mancha, Ciudad Real, Spain

Chapter Outline

1 Introduction: What is the Energy-Growth Nexus?

The Energy-Growth Nexus (EGN, sometimes also referred as Energy-Economy Nexus) is a broad collection of empirical literature investigating the causal relationship between energy consumption and economic growth. As the name suggests, the very core of the topic is represented by two main variables of interest. The results are essentially represented by a description of how these two variables influence each other. This is a relatively simple definition, but given the economic context, it is inherently much more complicated.

The EGN developed gradually over time, with multiple occasions when new methods were introduced or flaws in previous estimation procedures were identified. However, the topic emerged in response to world oil crises and the associated reduction in energy supply (and consequently in energy consumption). In this period a number of discussion papers, projects, and reports related to the nature of energy in the national economy and the changes in foreseeable future emerged (Allen et al., 1976; Hitch, 1978; Khazzoom, 1976; Long and Schipper, 1978). It is notable that even at that time, several authors argued for what is today known as decoupling of energy consumption and economic growth. Decoupling is the direct opposite of main thesis of EGN, that is there exists a relatively stable "fundamental" relationship between energy and output and as such it can be used for forecasting. If decoupling holds, then not only any potential "fundamental EGN relationship" is subject to substantial changes over time (rendering it moot), but previous instances of causally linked development of energy consumption and economic growth cannot contribute to the policy agenda in the future. Mielnik and Goldemberg (2002) and Steinberger and Roberts (2010) provide further discussion of decoupling.

1.1 The Basic Causal Relationships

There are four commonly recognized outcomes of the identification of Granger causality between the two main variables of interest, that is, energy and economic output. These four basic outcomes are: (1) no causality (the neutrality hypothesis), (2) unidirectional causality from energy consumption to economic growth (the growth hypothesis), (3) unidirectional causality from economic growth to energy consumption (the conservation hypothesis), and (4) bidirectional causality (the feedback hypothesis).

The neutrality hypothesis implies that neither energy conservation nor energy supply expansion policies will affect the economic growth and vice versa. It might also be the case that such influence is so small, that it is beyond our detection ability.[1] This hypothesis naturally serves as the statistical null hypothesis in the empirical investigations.

Bidirectional causality or the feedback hypothesis represents the very opposite point of view. This hypothesis presumes mutual interdependence of energy and growth and their joint determination. However, it is unclear how to describe such an influence, because should we account for all potential variants, it also splits into multiple cases, distinguished by individual causality signs, i.e. +/+, +/−, −/+, and −/−. It is therefore obvious that the notion of bidirectional causality alone provides no specific policy guidelines.

---

[1] This is in line with yet another energy economics strand of research, focusing on the influence of energy prices on economic growth, stemming from the so-called oil shocks literature. A strong argument for its validity is the observation that energy costs represent only a small proportion of the overall GDP - for instance, Rotemberg and Woodford (1996) reported the value of oil inputs in the US economy to be less than 4% of the total value added. Even rather significant changes in the energy prices and consequent changes in the energy consumption will likely have rather limited capability to influence overall GDP, where these effects will be effectively overshadowed by other factors.

The major point of interest is represented by the unidirectional causality types, that is, either from economic growth to energy (the conservation hypothesis) or from energy to economic growth (the growth hypothesis). The typical implication of the evidence for the conservation hypothesis is that energy conservation plans may be implemented without any impact on economic growth, that is, it lends support for the implementation of energy conservation policies motivated by global warming mitigation efforts. Beaudreau (2010) argues that non-economists usually believe that causality runs from energy consumption to output (i.e., they consider energy as a production factor), while economists think output causes energy consumption growth (i.e. they consider energy as an intermediate good and the demand for energy is mainly regarded as derived demand).

The growth hypothesis is based on the assumption that energy is a necessary production factor. A reduction in the energy supply (and consequently in energy consumption) will therefore cause an economic slowdown. However, Beaudreau (2010) stresses that there is no physical link between energy consumption at time t that might cause a change in GDP in time $t + i$, where $i > 1$. The major problem associated with this definition is closely associated with the definition of Granger causality. Granger causality simply states that a variable X is said to Granger-cause Y, if Y can be better predicted using the historical development of both X and Y, compared to the predictive capability when using the own lagged observations of Y alone.

That definition does not provide any explanation on why this is so. An observation lagged by one period can have a different coefficient sign than the observation lagged by two periods, with both coefficients being statistically significant and Granger-causing the development of the other variable of interest.

The vast majority of the EGN articles ignore such nuances of their results, and simply report the results in such a fashion, that when X Granger causes Y, it means that a decrease in X causes a decrease in Y (this description is labeled as a positive sign of causality in this text). Such an assumption of the positive sign of the causality is rather strong. It is evident that two other situations can arise. One is that a negative sign of the causality is observed, that is, when a cumulated effect of an increase in X causes a decrease in Y, or a decrease in X causes an increase in Y (such a situation reverses the typical policy recommendations aforementioned). Another situation is when the coefficients of the lagged variables of X have alternating signs. For example, when the effect of a positive change in X lagged by two periods causes Y to decrease, but a positive change in X lagged by one period causes Y to increase. This is more complicated and does not lend itself well to policy recommendations.

Few papers actually attempted to handle the aforementioned problem. Among the first one is probably Sari and Soytas (2007) who used the generalized impulse response technique to analyze the dynamics between energy and economic growth, using capital and labor as control variables; and Bowden and Payne (2009) who calculated the sum of the lagged coefficients in multivariate vector autoregression (VAR) with capital and labor proxies.

1.2 Why did the EGN Emerge?

Worldwide energy consumption has seen a huge ramp-up especially since the second half of the 20th century; and according to projections, it will continue to follow the same pattern over the next decades (EIA, 2016). Indisputably, the combined effect of expanded population, fast economic growth, and significant structural changes that have occurred in the world, is likely the primary responsible of this sharp increase in energy use. The relationship between energy consumption and economic growth has been the subject of an extensive debate. This debate is still far from over on both theoretical and empirical strands. First, from a theoretical strand, the role of energy in economic growth is contradictory; in the way it is emphasized by ecological economists (Ayres and Warr, 2005, 2009; Cleveland et al., 1984; Hall et al., 1986, 2001; Murphy and Hall, 2010) and neoclassical economists (Aghion and Howitt, 1998; Barro and Sala-i-Martin, 2003; Mankiw, 2006; Solow, 1956). Thus, while from a resource and ecological perspective, energy is considered as a key factor of production; and therefore economic growth, for the mainstream neoclassical school of economics, energy does not really represent a constraint or the reason for economic growth (Stern, 2011). Second, from the empirical strand, the EGN has evoked much discussion and has built quite a bulk of literature. This field of research has been gaining an empirical momentum especially due to new data and the development of more sophisticated econometric tools.

The EGN came to existence during the 1970s energy crisis (tightly linked to the so-called oil shocks in 1973 and 1979). The situation in the world suggested a changing framework with respect to the availability of relatively cheap and abundant energy sources. It was widely believed that the conventional energy supply would remain limited and would have its repercussions on the continuously rising energy prices.

This belief was boosted by various Malthusian visions, such as the oil peak debate[2] continually reappearing since the 1950s, with controversial empirical evidence from estimates of peak that always "moved forward" in time (Hirsch et al., 2005; Caruso, 2005), or the Club of Rome and their influential report (Club de Rome et al., 1972). The latter amplified the bleak visions of the future concerned with overpopulation and the running out of depletable resources. Both these strands hint heavily at Hotelling (1931) rule, leading to the predictions of increasing energy commodity prices (a situation during the oil crises lend these approaches a notable voice in the public discussions).

Nevertheless, there was also a substantial technological progress affecting the energy consumption in consumer electronics and appliances, the fuel efficiency in transportation, the residential energy consumption (e.g., air conditioning), and most of all the industrial

---

[2] Peak oil denotes a point of time when the rate of oil extraction is at its maximum, that is, extraction rates go down after this peak.

energy efficiency. In other words, the changing energy efficiency leads to a situation when increased economic output can be achieved while energy consumption remains the same or even decreases. This is particularly apparent in the industrial energy consumption in the post-transformation countries in Europe, but a similar, even though much less pronounced trend can be observed in all developed countries. However, recall that these trends are observed in aggregated energy consumption. For a discussion on the importance of energy quality and the decoupling from growth, see Warr and Ayres (2010); Stern (2010); Bithas and Kalimeris (2013); Fiorito (2013); and Cleveland et al. (1984).

1.3 The Global Warming Link

Today, one of the main energy related policy topics is the global warming. The vast majority of the developed countries have implemented a certain amount of policies aimed at mitigating global warming. As can be shown with Kaya identity (which we explain later in the text), this in practice translates into energy conservation policies. In short, global warming mitigation is directly or indirectly projected into energy conservation policies. The former takes place through goals of lower energy consumption and the latter through terms of higher energy efficiency (Berkhout et al., 2000; Saunders, 1992),[3] which can, however, be outweighed by the so-called rebound effect.

Anthropogenic global warming is closely associated with carbon dioxide emissions. Scientific consensus is that substantial future warming will occur if no abatement policies are implemented (Nordhaus, 2010) and that anthropogenic activities influence the global warming (Cook et al., 2013, 2016). In turn, apart from country-specific policies, there are supranational efforts, such as the Kyoto protocol and the post-Kyoto negotiations led by the United Nations Framework Convention on Climate Change (UNFCCC).

Since the main energy source today are fossil fuels and their combustion, significant reduction of carbon dioxide emissions is unavoidably linked to either lower *overall* consumption, higher energy efficiency, lower carbon intensity of energy or lower *energy* consumption. Lower overall consumption can be achieved through radically changing the consumption patterns of the majority of people or reversing the population trend—either of these is highly unlikely to happen. Other options to achieve reduction of carbon emissions are measures promoting higher energy efficiency (leading to lower primary energy intensity) or lower carbon intensity, which in turn require dramatic technological improvements without the occurrence of a significant rebound effect. Once again rather difficult objectives, leaving reduction of energy

---

[3] Essentially a reformulation of Jevons Paradox (Jevons, 1865). Given the increased efficiency, the costs expended on a certain goods or activity go down, which in turn boosts the consumption; so the overall amount of energy might not decrease as much, remain the same, or even increase; however, see also Gillingham et al. (2013, 2015) for a strong criticism of its importance.

consumption as the most prolific option. The links between these variables are best illustrated with the so-called Kaya identity as follows:

$$C \equiv N \times \frac{GNP}{N} \times \frac{E}{GNP} \times \frac{C}{E}$$

where C stands for the carbon emissions, N for population, GDP/N is the economic level (GDP per capita), E/GDP, the primary energy intensity, and C/E the carbon intensity of energy (Hoffert and Caldeira, 2004).

European Union's (EU) main energy document, "Energy Policy for Europe" defines obligatory goals for energy policies of the EU member countries. This European supranational energy policy is also known as the 20-20-20 strategy. This name refers to the three goals that should be reached by the year 2020; that is, 20% reduction in the carbon dioxide emissions, 20% share of renewable in the energy mix, and 20% reduction in the primary energy consumption (EU strategy update, Roadmap To 2050 has further extended these goals to 40% GHG reduction by 2030 and 80% GHG reduction by 2050).

While it seems there is a general consensus on the necessity to adjust the emissions, the EGN basically provides an answer to the "but what if?" question for the main method of carbon reduction, that is, energy consumption reduction.

The main issue raised throughout the whole historical development of the EGN is the worry that a limitation of energy supply might cause economic slowdown (and with that will come additional economic costs associated with energy conservation).

This worry is inherently associated with *two* strong assumptions. First, that energy is essentially a production factor, and as such there is likely an influential causality running from energy consumption to economic growth. Second, there is no abrupt change in the role that energy plays in the economy. That is, there are likely no dramatic technological changes or energy use changes that can also change the relationship abruptly. Note, how this is in direct contradiction with the empirical observation of the so-called decoupling, that is, the disassociation between energy consumption and economic output.

The fact that such worries are definitely present and is evident from the discussion regarding, for example, the prices of energy commodities preceding the 2008 crisis. Particularly, recall the record-breaking oil prices, which also refueled the peak oil debate. Both increasing energy prices and the peak oil debate are closely tied with the problem of energy security (i.e., security of sufficient energy supply, and energy commodities supply). Energy security is a significant issue, evidenced particularly in huge risks and potential damages during the electricity blackouts. For example, the consulting company ICF estimated the costs of a 3-day long blackout in 2003 in the northeast United States and central Canada to range from US$ 6.8–10.3 billion (ICF, 2003).

Another concern stems from the geopolitical influence of certain large-scale producers and exporters, particularly Middle East countries and Russia. Furthermore, China is building its influence in Africa to secure the African energy supplies, such as coal. The Russian Federation has used its position as a large supplier of natural gas for the EU to develop an influence on multiple occasions. Such legitimate concerns show that energy supply plays a significant role in the economy. The question is, whether and how the EGN is able to best capture it.

1.4 The Environmental Kuznets Curve (EKC): The Not-So-Distant Cousin of the EGN

The environmental Kuznets curve (EKC) literature investigates the empirical relationship between economic growth and the environmental pollution, often proxied by carbon dioxide (CO_2) emissions. One should note the close similarity of the EKC research to that of EGN, particularly given the fact that energy production today is very closely tied to fossil fuels, and hence CO_2 generation in their burning. Usually the EKC aims to explain environmental pollution as a polynomial (often quadratic) of economic level (e.g., GDP per capita)—with the expectation that pollution rises only to a certain point of the economic growth.

The economic growth literature has analyzed the relationship between income and environmental degradation (Grossman and Helpman, 1991; Grossman and Krueger, 1995; Panayotou, 1993; Selden and Song, 1994). These primary findings showed the existence of an inverted U-shaped relationship between economic growth and environmental degradation, defined as the EKC.[4] The underlying argument behind the EKC theoretical model is that environmental pollution is an increasing function of the level of economic output, until a critical income level -turning point- is reached (apart from the evidence of the presence or absence of a particular EKC relationship, the identification of the turning point itself is often the objective of interest). Initially, Grossman and Helpman (1991) proposed an inverted-U relationship between environmental degradation and income level (Fig. 1.1).

The inverted-U EKC model (Fig. 1.1) depicts the relationship between the level of income and environmental degradation (Ang 2007; Apergis and Payne, 2009; Halicioglu, 2009; Jalil and Mahmud, 2009; Saboori et al., 2012; Shahbaz et al., 2012, 2014a,b, 2015a; Tang and Tan 2015). This behavior suggests that in the early stages of economic growth, environmental pollution levels rise until a certain turning pointlet's say $X(1)$ is reached, beyond which economies find a correction in pollution levels.

The EKC model reflects on how the economic cycle influences environmental quality through three channels: scale, composition, and technical effects (Grossman and Helpman, 1991; Grossman and Krueger, 1995). The scale effect states that increases in output levels will

---

[4] The EKC was initially developed by Grossman and Helpman (1991). It takes its name from the inverted-U relationship between economic growth and environmental degradation, whose pattern of conduct resembles that studied by Kuznets (1955) in his analysis of inequality and economic growth (Panayotou, 1993).

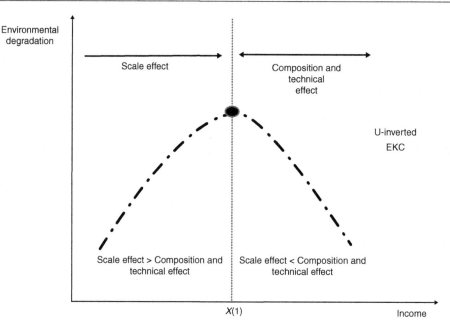

Figure 1.1: The Inverted-U Environmnetal Kuznets Curve (EKC).

decrease the environmental quality. Consequently, economic growth initially has a negative impact on the environment. When economic systems are in a developing stage, with low-income levels and high rates of economic growth, the scale effect overcomes the composition and the technical effect. Torras and Boyce (1998) consider that the scale effect suggests that even if the structure of the economy and the technology of the countries do not change, an increase in production will result in decreased environmental quality.

Second, at high levels of economic growth, the composition effect positively affects environmental quality. The composition effect reflects the transition from a developing stage, with highly polluting production processes, to a developed stage characterized by a production pattern involving less-polluting activities or the evolution from an economic system based on the primary sector to another economic system dominated by the service sector (Hettige et al., 1998).

Finally, when economies are in a developed stage, characterized by high-income level and low economic growth rates, the technical effect captures the productivity advances and the adaptation to cleaner technologies, where the environmental correction process is conditioned by the innovation processes (Aghion et al., 2014). In addition, Andreoni and Levinson (1998, 2001) propose that when income increases, the level of environmental contamination will be corrected, essentially through technological factors.

Originally, the EKC model revealed an inverted-U relationship between income and environmental degradation (Grossman and Helpman, 1991; Shafik and Bandyopadhyay, 1992;

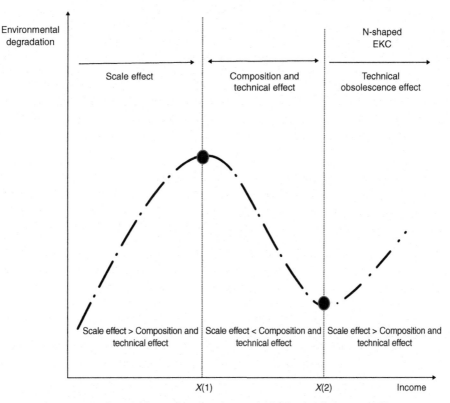

Figure 1.2: The N-Shaped EKC: The Technical Obsolescence Effect.

among others). Another branch of studies recognizes the existence of an N-shaped relationship (Fig. 1.2), to explore the long-run relationship between income and environmental degradation (Balsalobre and Álvarez, 2016; Dinda, 2004; Grossman and Krueger, 1995; Selden and Song, 1994; Torras and Boyce, 1998). Note: some studies, such as that by Balsalobre and Álvarez (2016), also address technical obsolescence as additional effect, acting similar as scale effect, that occurs as a result of deficient or inadequate energy regulation.

Fig. 1.2 reflects an N-shaped link between income and environmental degradation. This behavior occurs when the relationship between pollution and income is initially positive, but it becomes negative once a given income threshold is reached, until, after a certain turning point this relationship becomes positive again. The N-shaped EKC patterns reflect that environmental degradation increases during an economic developing stage, but once it reaches an initial turning point, let's say $X(1)$, environmental degradation decreases; this process continues until a second turning point, let's say $X(2)$, where environmental pollution once again begins to experiment with an increase. The third stage reflects high-income levels with low economic growth rates and inefficient and insufficient environmental innovation measures. This process is justified by diminishing returns in terms of the technological change that reduces the amount of contamination because of the collapse of technical obsolescence (Balsalobre et al., 2015; Torras and Boyce, 1998).

Torras and Boyce (1998) consider that a return to an upward pollution path appears when the margin for successive improvements in the distribution is exhausted; in other words, when there are diminishing returns in terms of technological change in reducing the pollution because of "obsolescence." Therefore, when regulatory measures that are linked with energy innovation are implemented, they delay the scale effect; additionally, a decreasing level of emissions is maintained (Balsalobre and Álvarez, 2016). Álvarez et al. (2017) conclude that when economies are in a developing stage, it is necessary to accelerate energy-innovation measures to reduce the scale effect and achieve a reduction in pollution levels. In addition to this, once economies reach a developed stage, they must continue to pursue energy innovation measures to elude technical obsolescence. When the total effect of the relationship between economic growth and environmental pollution is broken down, the technical effect is considered to be the main factor in the correction of the environmental pollution process.

A brief literature review of the EKC

There is a wide range of literature that assessed the relationship among energy consumption, economic growth, and environmental degradation. The empirical evidence suggests that the EKC model allows us to differentiate between developing and developed economies. The first stage of economic development is based on the overexploitation of fossil sources and high-income rates, where the tertiary sector bears a low weight.

Assuming that, advances in the energy sector have transformed economic systems: the increase in energy efficiency improvements is currently leading to the creation of a sustainable model, one that has less energy intensity and lower dependence on non-renewable sources. To review the role of the energy sector in the income-environmental degradation nexus (Ang, 2007; Farhani et al., 2014; Halicioglu, 2009; Jalil and Mahmud, 2009; Lau et al., 2014; Ozturk and Acaravci, 2013; Ozturk and Al-Mulali, 2015; Pao et al., 2011; Pao and Tsai, 2011b; Saboori et al., 2012; Shahbaz et al., 2012, 2013a, 2014a; Tan et al., 2014; Tang and Tan, 2015), it is necessary to remind ourselves the four known hypotheses (growth, conservation, feedback, and neutrality) that support the idea of interdependence between energy consumption and economic growth.

The literature has developed different methodologies to explore the relationship between economic growth and energy use. We have collected various types of indicators and techniques (Table 1.1) that have been used to show the role that energy consumption plays on the income-environmental degradation relationship. Many studies consider other variables beyond energy consumption, to proxy the energy sector. Shahbaz et al. (2014b) examine the relationship in CO_2 emission, electricity consumption, GDP and GDP square, urbanization,

Table 1.1: Short literature reviews, which support EKC hypothesis and the energy consumption.

Authors/ References	Time Period	Countries/ Regions/ Organizations	Methodologies	Variables used in the Study	Causalities
Ang (2007)	1960–2000	France	ARDL bounds testing and VECM Granger causality	CO_2 emission, GDP, GDP square, and EC	GDP → CO_2
Jalil and Mahmud (2009)	1975–2005	China	ARDL bounds testing and Pairwise Granger causality	CO_2 emission, GDP, GDP square, EC, and trade openness	GDP → CO_2
Halicioglu (2009)	1960–2005	Turkey	ARDL bounds testing and VECM Granger causality	CO_2 emission, GDP, GDP square, EC, and trade openness	GDP ↔ CO_2 EC → CO_2
Apergis and Payne (2009)	1971–2004	Central America	Pedroni cointegration, fully modified OLS, and VECM Granger causality	CO_2 emission, GDP, GDP square, and EC	GDP → CO_2 EC ↔ CO_2
Apergis and Payne (2010)	1992–2004	Commonwealth of independent states	Pedroni cointegration, fully modified OLS, and VECM Granger causality	CO_2 emission, GDP, GDP square, and EC	GDP → CO_2 EC ↔ CO_2
Wang et al. (2011)	1995–2007	China	Pedroni cointegration and VECM Granger causality	CO_2 emission, GDP, GDP square, and EC	GDP → CO_2 EC ↔ CO_2
Pao and Tsai (2011a)	1992–2007	BRIC	Pedroni cointegration and VECM Granger causality	CO_2 emission, GDP, GDP square, EC, and foreign direct investment	GDP → CO_2 EC → CO_2
Shahbaz et al. (2012)	1971–2009	Pakistan	ARDL bounds testing, Gregory–Hansen cointegration test, and VECM Granger causality	CO_2 emission, GDP, GDP square, EC, and trade openness	GDP → CO_2
Saboori et al. (2012)	1980–2009	Malaysia	ARDL bounds testing, Johansen–Jusellius cointegration, and VECM Granger causality	CO_2 emission, GDP, GDP square, and EC	GDP → CO_2
Hamit-Haggar (2016)	1990–2007	Canada	Pedroni cointegration, fully modified OLS, and VECM Granger causality	CO_2 emission, GDP, GDP square, and industrial EC	GDP → CO_2 EC → CO_2
Lau et al. (2014)	1970–2008	Malaysia	ARDL bounds testing and VECM Granger causality	CO_2 emission, foreign direct investment, trade openness, GDP, and GDP square	GDP → CO_2

(Continued)

Table 1.1: Short literature reviews, which support EKC hypothesis and the energy consumption. (*cont.*)

Authors/ References	Time Period	Countries/ Regions/ Organizations	Methodologies	Variables used in the Study	Causalities
Shahbaz et al. (2014a)	1971–2010	Tunisia	ARDL bounds testing and VECM Granger causality	CO_2 emission, GDP, GDP square, EC, and trade openness	GDP → CO_2 EC ↔ CO_2
Farhani et al. (2014)	1971–2008	Tunisia	ARDL bounds testing and VECM Granger causality	CO_2 emission, GDP, GDP square, EC, and trade openness	GDP → CO_2 EC ↔ CO_2
Shahbaz et al. (2014b)	1975–2011	United Arab Emirates	ARDL bounds testing and VECM Granger causality	CO_2 emission, electricity consumption, GDP, and GDP square, urbanization, and exports	GDP → CO_2 EC → CO_2
Tang and Tan (2015)	1976–2009	Vietnam	Johansen cointegration and VECM Granger causality	CO_2 emission, GDP, GDP square, EC, and FDI	GDP ↔ CO_2 EC ↔ CO_2
Kasman and Duman (2015)	1990–2010	European Union	Pedroni cointegration, fully modified OLS, and VECM Granger causality	CO_2 emission, GDP, GDP square, EC, trade openness, and urbanization	GDP ↔ CO_2 EC ↔ CO_2
Shahbaz et al. (2015a,b)	1980–2012	African countries	Pedroni cointegration, fully modified OLS, and VECM Granger causality	CO_2 emission, GDP, GDP square, and energy intensity	GDP ↔ CO_2 EC → CO_2

Notes: (→) indicates the direction of causality. (↔) indicates bidirectional causality. CO_2 emission, GDP, and EC or its components.
ARDL, Aautoregressive distributed lag; EC, energy consumption.

and exports in the United Arab Emirates for the period 1975–2011. The results confirm the existence of the EKC in the country, as well as unilateral causality from GDP to emission and bidirectional causality between energy consumption to emission. Shahbaz et al. (2013b) and Tiwari et al. (2013) employ coal consumption as a proxy for the energy sector and found evidence for the EKC hypothesis in South Africa and India.

Saboori and Sulaiman (2013) consider total energy consumption, coal, gas, electricity, and oil consumption, as a proxy for the energy sector. This study supports the existence of the EKC in Malaysia, where there is bidirectional causality between GDP and emissions, as well as between different energy indicators and emissions. Al-Mulali et al. (2015) used both fossil fuels energy consumption and renewable energy consumption as indicators of the energy sector in Vietnam for the period 1981–2011.

Other studies use panel techniques to examine the validity of the EKC hypothesis, as well as the pattern of causality among the variables that include energy consumption as the sole indicator of the energy sector. For example, Pao and Tsai (2011a) examine the relationship between CO_2 emission, GDP, GDP square, energy consumption, and foreign direct investment in Brazil, Russia, India, and China for the period from 1992 to 2007. The results show that the EKC exists in the country, in addition to the unidirectional causality, from GDP to emissions and energy consumption to emissions. Studies with similar evidence for the EKC within a panel data framework include Apergis and Payne (2009, 2010), Kasman and Duman (2015), and Shahbaz et al. (2015a,b) for Central American, Commonwealth of independent states, EU, and African countries, respectively.

Apparently, the EKC literature suffers from a controversy similar to the one found in the typical EGN studies. See for instance Richmond and Kaufmann (2006), Koirala et al. (2011), or (regarding deforestation EKC) Choumert et al. (2013)—with the remark that reads as follows: "Our main results are the following: Early influential studies favoring the EKCs are counterbalanced by recent estimates that do not corroborate the EKCs for deforestation." This is remarkably similar to the situation regarding the oil-shocks literature and the EGN literature.

López-Menéndez et al. (2014) also document that despite the fact the numerous contradictory empirical results do not offer any ground for consensus about the type of "fundamental" EKC relationship, some evidence for the support of the EKC is provided in the majority of the published studies. Only 11.5% of the studies indicate no evidence of the EKC relationship. This is another close resemblance to the EGN literature. For further discussion the interested reader should read Hajko (2017) and Fanelli (2011); the latter reports on empirical literature in general and documents the share of papers publishing negative results (finding no evidence for the rejection of their respective null hypotheses) has dropped from about 30% in 1990 to approximately 14% in 2007. We should not also forget the ongoing debate on the (low) reproducibility of published results. For instance, a study by Begley and Ellis (2012) managed to reproduce the results of only 6 out of 53 published studies (studies conducted by elite departments and published in top journals).

2 The Methodological Development of the EGN

As aforementioned, the EGN is an empirical exploratory area that needs more concrete theoretical foundations. To develop the latter, we need literature surveys and robust metaanalyses. Both of these, however, rise and fall with the quality of the underlying studies. Careful inspection and handling of known limitations is therefore highly desirable.

2.1 Literature Surveys Regarding the EGN

There are multiple literature surveys available that can document the development of the topic in general (Ozturk, 2010; Payne, 2010b), with a particular focus on electricity studies (Payne, 2010a) or with a focus on country-specific studies (Omri, 2014). However, apart from notable cases of being the first studies to employ a new method, the rest of the studies often provide conflicting results and on their own cannot contribute to the development or enrichment of the theory. This can only be implemented through continuous meta-analysis.

2.1.1 Meta-analysis

The EGN has generated a huge volume of literature that needs to be reviewed, summarized, and classified in order to be useful in future research. The traditional literature review is not without some merit, but not sufficient in giving precise conclusions and describing the potential sources of the outcome heterogeneity. The meta-analysis technique appears to be a beneficial approach in these circumstances. It is defined as a statistical method that allows the cumulative findings of a large collection of well-defined empirical studies to be pulled together. It is intended to draw credible conclusions from a large sample of empirical results after converting them to one or more metrics called effect sizes that are combined across studies (Littell et al., 2008).

Several studies have focused on this systematic review and aggregation of existing estimates. Examples of such studies concerning EGN are Chen et al. (2012), Bouoiyour et al. (2014), Bruns et al. (2014), Menegaki (2014), Kalimeris et al. (2014), Sebri (2015), and Hajko (2017). These studies provide a detailed description of numerous individual results in the EGN papers. In fact, they have systematically encoded the description of the individual results and performed analysis on this type of data.

Chen et al. (2012) have used 174 pairs of causality tests from 39 primary studies and the multinomial logit model to analyze the effect of some sample characteristics and econometric methods on the probability of finding 1 type of relationship out of 4 directions of Granger causality, involving electricity consumption and growth (i.e., no causality, unidirectional causality from growth to electricity consumption, unidirectional causality from electricity consumption to GDP, and bidirectional causality). Empirical results suggest that these four types of relationship are significantly affected by many factors, including the time period, economic development, estimation methods, and greenhouse gas emission reduction plans, including a carbon tax.

Bouoiyour et al. (2014) have employed findings from fourty three empirical studies in the electricity-growth nexus. The authors subdivided the total number of studies into four groups according to the supported hypothesis (growth, conservation, feedback, and neutrality) and followed a five-step procedure (Hunter et al., 1982) to outline the leading factors affecting the causality direction. Main findings of this meta-analysis have confirmed those provided in Chen et al. (2012).

Kalimeris et al. (2014) have tried to explain the general trends in the direction of the EGN causal relationship based on findings from one fifty eight empirical studies. By using the rough set data analysis (RSDA), a classification and filtering technique, and the multinomial logit model estimates, they found evidence that no solid and well-defined fundamental causal relationship could be established.

Menegaki (2014) focused on the variation from study to study of the long run elasticity of the GDP growth with respect to energy consumption. Based on findings extracted from 51 empirical studies, metaregression results suggest that the elasticity of the GDP with respect to energy consumption is significantly affected by the econometric analysis methods and the omitted variable bias.

Bruns et al. (2014) focused on a well-discussed issue in the meta-analysis method: the genuine effect and publication bias. Specifically, they investigated the question of whether there is a genuine causal relationship between energy and growth or the large number of significant results is just related to publication or misspecification bias. They employed a sample of seventy two empirical studies on the EGN, providing 574 pairs of causality tests. Their main findings revealed that a genuine effect existed from growth to energy use; however, only in models that include energy prices as a control variable. There is also a genuine effect from energy use to growth when employment is controlled for and when cointegration is found.

Sebri (2015) carried out a meta-analysis on the causal dynamics between renewable energy and growth. This study is motivated by increasing number of related studies and the absence of a meta-analysis on the renewable EGN. The author used a multinomial logit model following Chen et al. (2012) approach to find that the causality direction between the two variables is attributed to several factors, including model specification, data characteristics, estimation techniques (cointegration methods and causality tests), and development level of the country on which a study was conducted.

Finally, Hajko (2017) applied the machine-learning based classification trees and forests technique to classify a sample of 104 empirical EGN studies to investigate whether a fundamental energy–economy relationship can be found. As in previous metaanalyses, the author found no evidence of any fundamental EGN relationship and documented that the literature suffers from several deficiencies, such as publication bias, variables omission, and study sample characteristics, which contribute to the failure of a genuine energy–growth relationship.

None of these studies managed to find any evidence for the evasive fundamental EGN relationship. It is noteworthy that all meta-analysis papers agree on the fact that the results are mainly influenced by estimation methods. This is a remarkable consensus that should not be overlooked; and it calls for great reconsideration of the way the EGN research is being done. Meta-analysis studies also document that many published studies suffer from serious methodological flaws. Also note that a criticism of nearly all problems identified in the empirical investigation in the EGN methodology already appeared in the earlier methodological critique by Beaudreau (2010).[5]

---

[5] In addition, Beaudreau (2010) argues for the use of *energy availability* instead of energy consumption [inspired by "species–energy" relationship in biology and ecology as discussed in Evans et al. (2006)].

Why does the EGN seem to be a good field of application of meta-analysis technique? The perceived strength in the relationship between energy and growth has motivated scholars to increasingly study this dynamics. In doing so, various econometric techniques, geographic areas, time frames, and model specifications are employed yielding a broad spectrum of outcomes and leading sometimes to conflicting interpretations. Traditional literature review may be subjective (Florax et al., 2002; Stanley, 2001; Stanley and Jarrell, 1989) in the sense that authors generally try to just review some seminal works or those works that coincide with their point of view and/or those, which are coherent with their findings. Meta-analysis, on the other hand, seems more objective (Florax et al., 2002; Stanley, 2001; Stanley and Jarrell, 1989) because, in most cases, it tries to review all the published and unpublished empirical studies, therefore connecting all findings even the conflicting ones. It does so within a unique statistical and econometric framework. The EGN can be a particularly fruitful area for the application of the meta-analysis technique. This is shown through the increasing number of metaanalytic reviews carried out in the last years. Obviously, the number of metaanalyses is important with regard to the time frame of their publication. They have been carried out in the past 5 years. This may partly explain the urgent need to summarize the huge and cumulative literature on the EGN. On the other hand, as will be shown later, all the undertaken metaanalyses reveal that the plethora of literature on the EGN is subject to major methodological deficiencies, publication, and misspecification biases.

2.2 An Insufficient Theoretical Basis

The EGN is based on the empirical investigation of four basic causal hypothesis and relied heavily on the Granger causality concept. Main methodological developments of the EGN can be derived from the criticism toward inadequate previous econometric model specifications. This is undoubtedly a good way toward gradual improvement. However, this gradual development still leaves substantial margin for further improvements, because: (1) the insufficient econometric specifications, that have been criticized and corrected for, are still finding their way into journals, and (2) often bring about new problems, such as the assumption of homogeneity of the panel cross-sections. The improvements generated through the gradual development of econometric specifications cannot be diminished in their merit. Nevertheless, inadequate model specifications are still a problem haunting the EGN literature even today.

What is perhaps more troubling than the issue of the adequate use of econometrics is the lack of sufficient theoretical underpinnings for the topic itself. As criticized in the methodological critique by Beaudreau (2010), there is no theoretical model for the EGN. This makes the interpretation of the results problematic, and he also provides a good illustration of the problem in asking "what does the presence of Granger causality in this case mean? For example, if energy consumption in $t - 2$ is found to Granger-cause GDP in t, then what are the implications? Furthermore, given the lack of any structure, there is high probability that such a result is spurious and of little to no consequence."

An attempt to provide a link to the neoclassical production function was provided in Ghali and El-Sakka (2004). However, this link is very weak, and basically constitutes just a rewording of how the dynamic interactions between a set of variables, which typically appear in economic production function, can be modeled by VAR. But even this very simple theoretical settings disproves the validity of bivariate models due to potentially substantial omitted variable bias. The mainstream production function theory provides no definite explanation for the role of energy in the setup, and there is no consensus even on the simple issue of whether capital and energy are complements or substitutes. This is evidenced by the mixed results based on the econometric estimates generated at the industry level (Stern, 2004). Yet, given that there is no prior information regarding the influence of additional control variables (typically capital and labor), they cannot be safely omitted. This omission renders bivariate models invalid.

Surprising enough (given the main motivation of the EGN), many authors do not pay significant attention on what the presumed relationship identified in the data may mean. They simply approach the problem as data-driven model fitting, with little to no regard on what the data and the identified relationships should actually represent. However, such an approach provides little value, particularly if the results are complex in their forms(such as disagreements in the short and long run causality directions, multiple intertwined causal or cointegration relationships, and so on) and if the period under investigation is rather long or if the data frequency is low. Note that annual frequency is used in most papers, which in practice also leads to relatively small samples, commonly 30 or 40 observations for a given country.

The foundations of the topic are rooted in rather general macroeconomic estimates and projections. In the introduction of relevant papers, it is indicated that energy consumption is caused by changes in national product. This reasoning was largely based on the fact that the energy is a consumed good itself. As such, it provides a means to satisfy production needs, even though indirectly (in terms of the derived demand).

However, given the way physics defines energy, it is clear that energy can be used to perform work, in a sense similar to the macroeconomic view of labor. In other words, people can perform more work, if additional forms of energy are added, apart from human labor alone. As such, energy is likely capable of improving the productivity of traditional production factors. Given the general availability of country energy balances, this has motivated the enrichment of the traditional macroeconomic production functions with other factors. It may range from general terms, such as total factor productivity or technological improvement, to more specific ones, such as material enhancement, differentiated types of capital, and energy. For example, see KLEMS (capital, labor, energy, materials, services) production function specification in Berndt and Wood (1975).

Some authors also argue that there is a strong need for a distinction between various types of energy. For example, Toman and Jemelkova (2003) argue for a production function specification with a distinction between physical capital, human capital, high-quality energy, and low-quality energy. Particular attention to differentiated characteristics of energy types is present in the estimation of exergy. Exergy is a term from thermodynamics that describes the ability to perform useful work, first proposed in Rant (1956) Exergy was used only in few empirical papers such as Ayres et al. (1996, 2003); Ayres and Warr (2005); Warr et al. (2010); and Warr and Ayres (2010). Further discussion of exergy importance can be found in Stern (2004, 2011). While the use of exergy measurements remains rare, the distinction of different characteristics and use of various energy types has motivated the emergence of specific strands in the EGN, which focus only on the consumption of certain energy types. Notably, this applies to electricity (but also renewables, nuclear heat, fossil fuels, etc.).

Why should electricity enjoy such particular attention? Electricity is high-quality energy, which means it is a type of energy that is easily converted to other types of energy. It is relatively easy to convert electricity to heat; however, it is a lot more difficult to turn heat into electricity. Hence, electricity is higher quality energy than heat. For example, see Wall (1977) for a comparison of exergy quality of different energy types. As such, electricity is easily employed to improve production capabilities, particularly, with engines and machinery using electricity. Some authors argue that availability of energy of sufficient quality is an important aspect of economic development (Stern, 2011). On the other hand, electricity represents only a fraction of total energy use, and the sense of importance can be similarly shared, namely, about the role of petroleum products in relation to the use of combustion engines in transportation.

Some authors also argue for the presence of the so-called energy ladder. This is a description of the typical fuel transition pattern, associated with the increasing economic level in a typically developing country and it can be used to illustrate the differentiated roles of various energy types in economic output generation; see for example, Leach (1992), Barnes and Floor (1996), IEA (2002); and the discussion of a traditional model and energy stacking in Kowsari and Zerriffi (2011).

2.3 Methods

Two econometric methods have been extensively applied in testing for the causal relationship between energy and growth, which are the cointegration analysis and Granger causality tests (Stern, 2011). The assertion that energy and growth are cointegrated has become somewhat a stylized fact (Smyth and Narayan, 2015) and is frequently confirmed in empirical studies. But when it comes to causality, conclusions are mixed and interpretations are many. This mixture in outcomes of the causality tests is attributed to a number of factors, including, among others, institutional differences between countries, model specification, data characteristics, and econometric approach (Sebri, 2015; Smyth and Narayan, 2015). As the choice of appropriate econometric approach is a prerequisite for the correct outcome, it is desirable that the known deficiencies in the methods are taken into consideration.

There are multiple major empirical innovations that have gradually developed in the EGN field. Despite the chronological order of their appearance, the use of the new methods can be often taking place simultaneously, rather than sequentially to old ones. Thus, new methods do not always appear in all new papers, despite the arguably better properties.

2.3.1 Major approaches

The major innovations can be classified as follows:

1. The bivariate causality estimations between energy and economic output (GDP or GNP),
2. a. Cointegration pretesting and VAR/VECM framework,
    b. Enhanced by impulse response analysis, or
    c. Based on unrestricted error-correction models, in particular the ARDL, (Pesaran and Shin, 1998; Pesaran et al., 2001) with better properties to handle small samples,
3. Modified Granger causality tests, notably Toda and Yamamoto (1995) procedure, with better properties and ability to test for Granger causality in non-stationary time series, even with different orders of integration, and
4. Panel data cointegration tests and panel data causality tests, with the ability to increase the statistical efficiency by employing more observations.

The earliest papers relied on standard Granger and Sims causality tests, that is, bivariate tests of two time series. Later on the Granger causality has been based on the results of VAR models. Cointegration testing appeared in the EGN relatively early, and gained popularity during the 1990s, probably as a side effect of the increasing general popularity of the Johansen procedure.[6] The natural extension of cointegration is to model the short run dynamics. The long run dynamics are described by the cointegration relationship(s) in the vector error correction models. This is essentially a VAR model used on differenced data to yield stationary series, enhanced by error correction terms, describing the deviation from the long-run equilibrium/equilibria. This is useful to study what variables are influenced by the error–correction terms, but at the same time, make Granger causality difficult, which is why VAR models are often used to complement VECM.

The ARDL approach was proposed by Pesaran and Shin (1998) and Pesaran et al. (2001). Due to the fact that cointegration relationship is not estimated beforehand but enters the short-run dynamics, ARDL model is sometimes called the "unrestricted ECM." It first appeared in the EGN context probably in Fatai et al. (2004), alongside with Johansen cointegration testing and the modified Granger causality testing. ARDL in the context of electricity nexus first appeared in Narayan and Smyth (2005a,b). The main advantage of the ARDL is its better ability to deal with small samples. Due to data limitations, a small sample is a norm, rather than an exception in the EGN research.

---

[6] Among the first were Nachane et al. (1988), Eden and Jin (1992), Van Hoa (1993), and Masih and Masih (1996).

While certainly less popular than cointegration, but likely an advance in the causality estimation approach, was the use of the modified Granger causality tests. The modified Granger causality tests were introduced in the EGN estimations during the 1990s. Hwang and Gum (1991) was among the first studies that used the modified Granger causality testing, based on Hsiao (1981) augmented lag structure.[7] In addition to Hsiao (1981), more advanced modified Granger causality testing procedures, such as Toda and Yamamoto (1995) or Dolado and Lütkepohl (1996); Saikkonen and Lütkepohl (1996) were used notably for their ability to test for causality among *non-stationary* variables, which may or may not be cointegrated. These methods have been applied in the EGN up to date.[8]

In the next decade, panel cointegration and panel causality have gained fame. The first study focusing on panel data was probably Lee (2005), with the results based on Pedroni (1999, 2000) panel cointegration testing in a trivariate framework (energy consumption, real GDP, and real capital stock). Such trivariate panels were also used in Lee et al. (2008); Costantini and Martini (2010) and Lee and Lee (2010). Mahadevan and Asafu-Adjaye (2007) also used trivariate panel but with energy consumption, economic growth, and prices. Mehrara (2007) was also among the early uses of panel framework to study EGN but only in bivariate settings.

While the traditional unit root tests, Johansen–Juselius cointegration procedures and Engle–Granger error–correction models have been the foremost tools employed to study the EGN, they have shown low power due to small samples properties (Harris and Sollis, 2003) along with the economic structural changes in many countries. That is, more recent studies rely on other proposed techniques, including unit root and cointegration tests with structural breaks (Apergis et al., 2010a; Dogan and Ozturk, 2017; Hamit-Haggar, 2016; Narayan et al., 2010; Sebri and Ben-Salha, 2014), the ARDL model, and bounds testing approach (Pesaran and Shin, 1999; Pesaran et al., 2001), which can be performed irrespective of whether the variables are stationary (Dogan and Ozturk, 2017; Gross, 2012; Ozturk and Acaravci, 2011; Sebri and Ben-Salha, 2014; Zachariadis, 2007), and (Dumitrescu and Hurlin, 2012) whereby the causality test is applicable to stationary series and works well for heterogeneous panel data (Dogan and Seker, 2016; Dogan et al., 2016).

---

[7] We can also observe this relatively simple modification being used later in multiple studies up until recently (Altinay and Karagol, 2004; Aqeel and Butt, 2001; Cheng, 1997, 1998, 1999; Cheng and Lai, 1997; Dagher and Yacoubian, 2012; Yang, 2000c; Yoo, 2006; Yoo and Kim, 2006).

[8] For instance in Fatai et al. (2004); Wolde-Rufael (2004, 2005); Altinay and Karagol (2005); Soytas and Sari (2006); Yoo (2006); Squalli (2007); Wolde-Rufael (2009); Menyah and Wolde-Rufael (2010); Tsani (2010); Adom (2011); Lee and Chiu (2011); Shahiduzzaman and Alam (2012); Jafari et al. (2012); Stern and Enflo (2013); Naser (2014).

2.4 Problems in the EGN research

2.4.1 Limited data availability

The motivation for panel estimation lies mostly in limited data availability. There is only limited number of observations available for each individual country, as annual frequency of data is most typical. Even rather long time periods thus generate statistically small samples with undesirable statistical properties. This is further complicated when the author wishes to employ either additional control variables (capital, labor, prices) or structural breaks, or both.

This limitation, having been present since the very beginning of the EGN research, leads to data samples commonly spanning the periods of 30 or 40 years. While from a purely statistical perspective, a larger sample of observations should yield more information; this will not be the case, if the underlying data generating process changes during the sample period. Such changes are not very easy to incorporate in the models. In the case of structural changes, the most flexible approach is to perform the estimation on subsamples. This is not often feasible due to the low number of observations. Hence, the flexibility gets reduced somehow by the incorporation of structural break dummies. Despite the lower demands on degrees of freedom for the estimation, they still complicate the estimation framework enough, so that the authors often forego their inclusion altogether. This is understandable from the statistical perspective; however, less understandable, given the economic background of what is being modeled. In addition to this, it is well known that Granger causality testing should be undertaken for stationary series to avoid spurious results (Granger and Newbold, 1974). Even though newer modifications of Granger causality tests such as the method of Toda and Yamamoto (1995) can account for non-stationary series, caution should be applied in all cases relying on other methods, particularly in the presence of long time periods and with the absence of structural breaks embedded into the stationarity tests, models, and causal inference tests.

In addition to the discussion on decoupling, fuel substitution is a point of concern in the analysis. The interested reader can go back to the discussion on the energy ladder and the associated exergy differences, leading to differentiated energy use and differentiated energy prices. Also, with the structural economic reforms of the investigated countries, such as in post-communist countries or in developing countries, there is a strong concern on the changes of data coverage and the methodology over the decades and long time spans. In short, the easy solution to incorporate as long time series as possible might not be the best choice. If very long time periods are considered, the presence of structural breaks should be tested both for stationarity and incorporated for causality testing too. If panels are considered, homogeneity of the cross-sections should be verified or panel estimators with random coefficients should be employed.

2.4.2 Bivariate versus multivariate studies and omitted variable bias

Small samples are also causing additional problems in the EGN research. The majority of studies have relied on a simple bivariate model framework. Given that Granger causality is closely tied to the lag structure employed, and there is general tendency for a steep increase in the number of the estimated coefficients both with inclusion of additional variables and with increased lag orders then bivariate framework is understandable from the practical viewpoint and feasibility of obtaining the results. However, given the ambition for policy recommendation, this limitation strongly inhibits the applicability of such results as bivariate framework is associated with a significant potential for omitted variable bias and invalidity of the model.

Despite the concerns associated with bivariate framework, multivariate models are less common, even in recent papers. Again it must be stressed that it is mainly due to limited data availability. While already in 1984, Yu and Hwang (1984) had considered employment besides energy consumption and GNP, Stern (1993) was probably the first true multivariate study (including capital and labor). A similar settings were later employed in Cheng (1998), Cheng (1999), or Stern (2000). Another multivariate study was done by Masih and Masih (1997, 1998) who worked with trivariate settings of energy, GNP, and prices. Similar settings were later employed also in Asafu-Adjaye (2000). It is worth mentioning that trivariate settings were also used in Glasure and Lee (1995), but they focused on the relationship between energy consumption and different measures of employment (i.e. they did not use an economic output variable).

Certain authors decided to choose other macroeconomic variables. For example, Glasure (2002), who in a fashion similar to the so-called "oil-macroeconomy relationship"[9] used multivariate settings with real oil price, government expenditure, and money supply.

The use of multivariate settings is not without its problems too. The main disadvantage is the aforementioned "coefficient hungry" nature of causality tests, leading to a loss of statistical power. An additional difficulty encountered with multivariate models is the insufficient guidance from theory on the selection of control variables. It might be argued that the inclusion of capital and labor variables (or their proxies) stems from the macroeconomic production function theory. At the same time, the oil shocks literature provides numerous explanations for the process of the transmission mechanism of the energy commodity price shock into the economy, or on productivity. This calls for the inclusion of the measure of energy prices in multivariate EGN models. In ideal settings, general inflationary and monetary channels of influence should be incorporated as well. In summary, at least macroeconomic control variables and energy prices should be included in the estimation framework.

---

[9]   A notable area of research focusing on the so-called "oil shocks," prominently represented by Hamilton (1983, 1996, 2003, 2011). In the last 2 decades this research strand has been also suffering from the apparent lack of any fundamental relationship between oil prices and macroeconomic growth, namely, a very similar situation as that in the EGN.

2.4.3 Control for prices

Nevertheless, we have briefly touched the subject of energy quality; and the fact that different energy types are valued differently by consumers (readers might refer to the complex discussion on exergy and energy availability as aforementioned). The use of prices, or better yet, energy prices, is not often found in the studies. Van Hoa (1993) was probably the first paper that controlled for the influence of consumer prices. Glasure and Lee (1995, 1996) used a ratio of wages and energy prices. Masih and Masih (1997) and Masih and Masih (1998) used consumer prices.

The fact that different energy types exhibit different prices for the same calorific amount of energy indicates that different energy types can have different effects on the economy. This issue has motivated authors to focus on differentiated energy types, either individually, leading to specialized, but not numerous EGN branches, or in combination with overall energy consumption or other energy types. Probably the only remarkable "individual EGN" branch is the electricity-growth nexus. Other studies relying only on one particular energy type, such as renewables, nuclear heat, fossil fuels, and so on, are relatively rare to deserve a special category label. However, some studies perform disaggregation of energy consumption by various types. Namely they apply a combination of these specialized branches. However, one might note that, for instance, energy derived from renewable energy sources is a rather popular variable in the EKC literature.

2.4.4 Disaggregation by energy types

Disaggregation of energy to different energy types was already considered in Yu and Choi (1985), who worked with aggregate energy consumption, as well as solid fuels, liquid fuels, natural gas, and others (hydro, nuclear, and electricity). A similar setup was also used in Yang (2000a), who worked with coal, oil, gas, and electricity. Van Hoa (1993) used oil consumption, instead of energy and controlled for prices. Yang (2000b) replaced energy with coal in a bivariate framework. Later, Wolde-Rufael (2004) used only bivariate model setups with coal, coke, electricity, oil, and total energy consumption; however, he also worked on the industrial, not the macroeconomic level of output. Sari and Soytas (2004) used trivariate models, controlling for employment, and disaggregating energy consumption (total, coal, oil, hydraulic power, asphaltite, lignite, waste, wood).

Two main arguments against the disaggregated energy type causal nexus, particularly, combined with the overall GDP, is that any particular energy type only contributes to a small share of the total energy consumption (besides the relatively small share of the energy sector on the overall GDP), and the observed diversity of fuel mixes across individual economic sectors. Selective focus on a particular energy type can lead to incorrect conclusions if not factored for the sectoral differences. Another type of potentially strong bias, stemming from the analysis based on an individual energy type, is rooted in the energy substitution over time and has already been discussed previously with respect to the so-called energy ladder. This is more intense, particularly, if we consider very long time periods (several decades, as is typical due to the limited data frequency and the effort of researchers to increase the number of observations included in their sample). The changes in the consumption of one energy type (for instance, electricity or renewables), can be mistakenly associated with the increased economic growth simply because of the omission of changes in other energy types. In addition, there can also be an unobserved overlap, misattributing the influence—for instance, the renewables are often used for electricity generation in recent years.

2.4.5 Electricity-growth nexus, renewables, nuclear heat, and other specialized branches

The differentiated roles of energy types are more evident if a rather long time horizon is employed. It is clear that the use of electricity has changed quite a lot in last decades, and electricity now has a prominent position in many production processes. For instance, Gross (2012) documents electricity as the energy type with the highest growth rates during the period 1949–2009. Substantial increases in the importance of electricity can also be made apparent in the development of the exergy conversion efficiency trends in the United Stateover the last century (Ayres et al., 2003). The importance of electricity can be expected to rise even more with an increasing popularity in the space cooling, the use of electricity in transportation, etc.

Electricity-economic growth nexus, an important research strand of EGN, appeared relatively early. The early proponent was Sioshansi (1986), who argued that the decoupling of energy and growth did not apply to electricity and economic growth. An explicit assumption for the importance of electricity for growth can be found in Ramcharran (1990), or Huang (1993) and others. However, it should be noted that these papers are based on simple estimates of elasticity coefficients in log–log model in levels, with no concern for nonstationarity. Therefore, it is arguable that the reported results are spurious and also it is clear no inference of causality can be made in such a model. Ferguson et al. (2000) also focused on the correlations between electricity consumption and GDP; however, they also did not investigate causality.[10]

Among the first EGN studies investigating the electricity-growth causal relationship were Murry and Nan (1994); Yang (2000c); Aqeel and Butt (2001); and Ghosh (2002). While it can be argued that even though electricity is important, it still represents only a small fraction of energy consumed. Nevertheless, some studies choose even more stringent limitations in their energy variables, such as those dealing with the nuclear energy-growth nexus. Probably, the first study focusing on nuclear energy was Yoo and Jung (2005), followed 5 years later by a whole set of nuclear energy themed papers by Payne and Taylor (2010);

---

[10] But also note how substantial progress has been made in the quality of journal contributions regarding the topic ever since.

Apergis et al. (2010b); Menyah and Wolde-Rufael (2010); Wolde-Rufael (2010); and Wolde-Rufael (2012). Apart from an inherent selection bias given the limited number of countries that actually use nuclear energy, it is necessary to stress that nuclear heat is used mainly for electricity generation, which makes the limitation to nuclear heat consumption questionable. In addition, only a fraction of electricity is being produced with nuclear heat, casting further doubts on the utility of nuclear heat as the energy variable of choice.

Among the first papers studying the EGN based on renewables were Sari and Soytas (2004); Ewing et al. (2007); Sari et al. (2008), who included them in their multi-energy type investigation. More explicit focus on renewables as a particular energy form can be found in Payne (2009) who distinguished renewable from non-renewable energy sources, Sadorsky (2009) who considered a bivariate framework of renewable energy and income, Menyah and Wolde-Rufael (2010) who included renewable energy alongside with nuclear energy, emissions, and output, Payne (2010c) who focused on biomass and Yildirim et al. (2012), who employed renewable energy and he further controlled for employment and investment.

But renewables-economic growth studies are even weaker in the aforementioned perspective of their miniscule share on total energy consumed. The major motivation for the focus on either nuclear heat or renewables is their low carbon intensity and their presumed importance in the promotion of energy security. See for instance Apergis et al. (2010b) or very similar article by Menyah and Wolde-Rufael (2010) who consider carbon emissions, renewables, nuclear energy and growth, based on the presumed lower carbon intensity of these energy sources. However one should not that lower carbon intensity of renewables or nuclear heat is not guaranteed in all cases, and the calculation of carbon intensities of these energy sources is highly dependent on the method of accounting for life-cycle emissions.

The link between energy security, energy availability, and economic growth is, however, empirically unclear and presently the demand for renewables is driven mainly by government policies (IEA, 2012).

Despite the aforementioned differentiated quality and prices of the different types of energy, the majority of studies in the EGN focus on the thermal aggregate. Thermal aggregate is an overall number taken from the energy balances of a country, summing up the differentiated consumption by using conversion tables, even though there are other methods of aggregation accounting for more than one dimensions of the data, such as exergy (Ahern, 1980) or emergy analysis (Scienceman, 1987). The thermal aggregate measure can therefore cover any energy mix. The problem of aggregation was discussed in detail by Cleveland et al. (2000). They argue that different methods of aggregation can influence the results, not only in terms of significance, but also the direction of the causality. They also discuss the inclusion of energy quality in aggregate analysis. An example of the EGN investigation using quality adjusted Divisia index of the energy consumption can be found first probably in Stern (1993) and Stern (2000).

2.4.6 Sectoral analysis

Another important type of disaggregation does not focus on energy types, but on energy use in the economy. This is labeled as the sectoral approach in this text. Among the first papers in the EGN was probably by Erol and Eden (1990), who used the spectral analysis approach and the energy consumption in residential, industrial, and transportation sectors and total energy consumption. Another sectoral EGN study is Hondroyiannis et al. (2002), who investigated both aggregate and industrial, residential, and transport energy consumption (and also controlled for prices using the consumer price index) Another sectoral contribution was by Liddle (2006) who worked with disaggregated energy types and sectors in bivariate frameworks, and Zachariadis (2007) who also worked with a bivariate setup. A certain mix of fuel-type and sectoral study was presented in Ewing et al. (2007) who studied the relationship between the industrial output and the different energy types in trivariate settings, controlling for employment.

Basically this approach focuses on smaller cross-sections than whole countries. Instead of aggregate total amounts in the macroeconomic sense, it works with individual economy sectors, such as industry or transport. However, due to limited data coverage, often the disaggregation to sectors is limited to energy consumption in the individual sectors, while GDP is taken as an aggregate. It can be argued that the use of energy is rather differentiated in individual sectors. Transport relies dominantly on oil, while industry (particularly manufacturing) mainly needs electricity and heat (e.g., glass production, steel production, etc.), while residential consumption is mainly concerned with heat.[11] What this means for EGN is that there might be some important relationships, which are differentiated enough, so that in their aggregation to higher levels, they remain undiscovered. Gross (2012) argues that such a situation is the example of the so-called Simpson's Paradox (Simpson, 1951). For this reason, he argues that Granger causality in the EGN should be investigated *exclusively*on the sectoral level. A similar conclusion and recommendation can be made from the results in Zachariadis (2007) or Bowden and Payne (2009).

---

[11] Eurostat statistics regarding total residential energy consumption (nrg_100a) and the number of households (lfst_hhnhtych) indicate that in the European Union, household average yearly consumption is about 16.1 MWh per year (it is noteworthy that countries with more favorable weather conditions, such as Spain, Greece, Portugal, show values below 10 MWh per year). Households are among the largest users of energy in the EU, accounting for roughly 30% of final energy consumption. Despite the significant efforts in late years to moderate and reduce energy use, energy consumption per dwelling in the EU has been more or less stable in the past 2 decades. According to the European Environment Agency, nearly 70% of energy consumption goes toward the space heating, followed by water heating (15%), electrical appliances and lightning (11%), and cooking (5%). However, also note that this pattern may be influenced in the future by increased popularity of space cooling, for example, Isaac and van Vuuren (2009) estimate an average 7% yearly growth in electricty used for air conditiong during the period 2020 and 2030 (with projected 40 times as large electricity consumption for air conditioning in the year 2001 compared to 2000).

As was discussed earlier, the EGN research might face several pitfalls. Certainly, there is no value in repeating old mistakes. However, the EGN urgently needs substantial methodological attention, so as to avoid further unflattering evaluations of the EGN research, such as the one found in Karanfil (2009), who argued that additional empirical EGN research papers had *no further research potential*. Literature surveys and meta-analysis indicate that the current situation regarding the existing EGN results is highly unfavorable. New results are needed, which will explicitly take into the account the methodological issues raised throughout the rich history of the field. It is indeed difficult, if not impossible, to always correct for the problems, as the data requirements often overwhelm the data availability. One should note that energy use data are usually made available only in annual frequency, taken from reports on energy balances of the countries, which is hardly sufficient for adequate econometric analysis. As argued in Hajko (2017), the EGN might seek a new success when data coverage of energy sector is vastly improved, for example, with the advent of smart cities and energy demand management systems when the authors will not be burdened by data limitations. However, the current data limitations cannot excuse bad practices. Hopefully, this book will contribute to their elimination.

3 New Challenges: The Water-Energy-Food (WEF) Nexus

Resource use integration, the intensification of production systems, and emerging externalities have long been signs of growth, increasing scarcities, and technological sophistication. Water, energy, and land represent three vital environmental resources that are naturally interlinked. Land with no water is useless for most purposes. We use land to produce food, also utilizing significant amounts of water. While much energy is used in water supply systems, water is also used in power plants directly to produce hydroelectricity and for cooling.

Recently, socioeconomic, demographic, and climatic drivers have increased such interlinkages and have made their consideration in policymaking even more important. The so-called water-energy-food (WEF) nexus constitutes the topic of a relatively recent debate, highlighting resource integration as a global sustainability challenge. The WEF nexus can be best characterized as a sustainability debate first promoted at high-level gatherings, such as the World Economic Forum and the Bonn WEF Nexus conference, both in 2011.

Since then, this debate resulted in hundreds of reports, journal papers, gatherings, and funding programs. Largely, the nexus is well-received as a way of shifting the perspective to analyze neglected issues and increasing interlinkages: hydropower, bioenergy, waste, desalination, irrigation, water use in energy production, resource footprints, energy use for water networks, etc. While many of these issues are not new, the nexus gives more emphasis to the interlinkages and reshapes the current sector-driven debates.

According to Al-Saidi and Elagib (2017), the WEF nexus can be understood as an emergent integrated management paradigm, which includes three aspects, (1) intersectorality of resource use issues, (2) interdependence and interdisciplinarity of management decisions, and (3) interactionality of impacts of resource implications. In this sense, it builds on numerous precursor debates in sustainability science, such as resource valuation, ecological modernization, systems thinking, ecosystems resilience, etc., and goes further than previous management paradigms (ibid).

Continuous growth, together with increasing risks, such as those related to climate change, will lead to a higher-level interdependency in utilizing current resource base. This is also true with the expected increase in renewables use. Renewables require input resources, such as land and water. Such notion of interdependency deserves a higher academic attention even without referring to the WEF nexus debate, which is occasionally criticized as being vague (Bell et al., 2016; Cairns and Krzywoszynska, 2016), neoliberal (Leese and Meisch, 2015), or for becoming an academic or a policy hype (Wichelns, 2017). While the nexus paradigm has been arguably weak on the issue of prioritization, governance, policy, and institutional recommendations, it has incited a great level of experimentation with integrative analysis tools and led to the production of a valuable portfolio of models and case studies (Al-Saidi and Elagib 2017; Wicaksono et al., 2017).

The nexus debate is key to understand the increasingly interdisciplinary and cross-sectoral nature of the impacts of socioeconomic growth on the use of environmental resources. Importantly, it has led to large research initiatives and significantly helped in the mobilization of data to measure resource use challenges. Different approaches are being deployed in quantifying the nexus. The use of indicators to assess resource security, resilience, or vulnerability is one popular approach (e.g., Giupponi and Gain (2016) with regard to WEF nexus and sustainable development goals; Bleischwitz et al. (2014) with regard to vulnerability analysis). Another interesting approach is represented by studies using trade-offs and thresholds (Kurian, 2017) or integrated life cycle-based assessments (Kurian and Ardakanian 2015). The use of linkage analysis and input–output models is also quite suitable for the study of the nexus (Fang and Chen, 2017). Further, various studies have used the nexus approach to provide integrative tools to assess the cross-sectoral impacts of sector specific policies, such as the increase of production of certain crops in Qatar (Daher and Mohtar, 2015) or urban agriculture and water reuse in Munich, Germany (Gondhalekar and Ramsauer, 2017).

Future developments of the WEF nexus idea need to tackle key challenges that hinder both the development of the tools and the implementation of recommendations. Some of the basic challenges are, with regard to the integration of scattered interdisciplinary data and results (Chang et al., 2016), the uncertainties at different scales (Garcia and You, 2016), as well as the availability of data-rich monitoring and modeling systems at different levels backed by wide collaborative efforts between academia and practitioners (Scanlon et al., 2017).

The issue of collaboration and the development of a "community of practice" to address integration of nexus issues is also highlighted by Mohtar and Lawford (2016). Further practical problems of the nexus implementation are related to the adequacy of science-policy integration, inequalities, and path-dependencies in infrastructure and institutional solutions (Romero-Lankao et al., 2017). While the nexus debate is still evolving, prioritization and optimization of nexus issues will be needed. For example, Davis et al. (2016) proposed that nexus systems focus on key issues related to waste, recycling, and the internalization of externalities, while developing new governance arrangements in order to stimulate innovation and development. In this sense, the nexus approach is most useful when it evolves beyond experimentation to a stage where it can offer tested solutions and implementation guidelines.

3.1 Nexus Integration and Assessment

While there has recently been a wealth of evidence about problems and crises arising from the increasing resource integration, sectoral management and policy solutions often fail to capture such issues. Hydropower production for example, is a highly sensitive sector to climatic pressures affecting multiple sectors. The 2010 and 2011 droughts in China led to water shortages and a decrease in hydroelectricity, resulting ultimately in price shocks on the global food markets. Multiple heatwaves in Europe affected power demands due to shortages of cooling water for nuclear reactors. Further, rising energy demands in Europe are shaping land use policies in Latin America and are creating incentives for the production of bioenergy.

There are many similar regional and local examples of increasing interlinkages and failures of one-sided policies. In fact, one can argue that the nexus idea is put forward mainly by water experts, who become aware of the crucial links to the energy and land sectors. Those experts are increasingly frustrated with the multiple failures of water-biased concepts, such as the integrated water resources management (Al-Saidi, 2017; Al-Saidi and Elagib, 2017). The nexus is thus largely water-sector driven and the river basin is still considered as one of the most viable scale for its application (Keskinen and Varis, 2016). In the nexus debate, one should be aware that not all subtopics are equally important or equally investigated. Allan et al. (2015) mentioned three major subnexuses, namely, water–food, water–energy, and energy–climate, with the first one being better conceptualized than the others.

In fact, the two resources, water and energy, play a key role in the WEF nexus and represent the new emphasis in the debate. Water is often the medium through which environmental pressures (e.g. climatic-related impacts, increasing demands for energy) are translated. Under this notion, water is considered by Perrone and Hornberger (2014) as the proximate cause of competition in the nexus. On the other hand, energy growth can be considered as the major driver of pressure on the naturally limited water and land resources. The energy sector is responsible for some of the increased demands for water and land, as they are direct production inputs for energy generation. It also indirectly puts a burden on both resources.

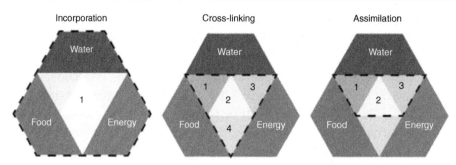

Figure 1.3: Three Understandings of the Water-Energy-Food (WEF) Nexus and Integration.
Adapted from: Al-Saidi, M., Elagib, N.A., 2017. Towards understanding the integrative approach of the water, energy and food nexus. Sci. Total Environ. 574, 1131–1139.

Fossil fuel use is the main driver of climate change, leading to extreme situations, such as droughts, floods, and other human-made climate variability. Therefore, water and energy are particularly interesting to examine in the nexus, and they are very much and increasingly interacting with food issues and external change drivers, such as climate.

The WEF nexus idea of integrating water, energy, and food issues has been promoted for several years now. However, how does the integration under the nexus paradigm look like? According to Al-Saidi and Elagib (2017), one has to differentiate between issue integration on the one hand and people and institutional integration on the other. The latter is related to institutional and participation arrangements, which can extend from joint plans to collective planning or decision making organizations at different levels, depending on the issues and local context. Issue integration can accordingly be understood in three ways: incorporation, cross-linking, and assimilation (Fig. 1.3). Incorporation is the "bird's eye view" of the nexus and encompasses models that achieve a comprehensive integration of issues of the water, energy, and food into one system. Cross-linking is the "inside-out view" and is more concerned with illuminating single or multiple concrete links. Assimilation represents the "prism view" on the nexus from the perspective of one sector, for example, the water sector, in order to integrate food and energy issues into one-water sector strategies. These different understandings of integration under the nexus are being deployed at the same time in the generation of the nexus specific analyses. They are all equally important as they address different needs in the policy value chain, namely, policy, planning and investments (incorporation), coordination and regulation (cross-linking), and operational management (assimilation).

Under each one of the three understandings of the nexus, several methods have been utilized so far, some of which can be briefly mentioned in the following:

1. Incorporation: Vulnerability assessment, integrated models, such as the climate, land, energy, and water (CLEW) model or the integration of the water and energy allocation models (WEAP-LEAP integration); life cycle analyses, metabolism approaches, value chain studies, different institutional analysis studying regimes, and networks.

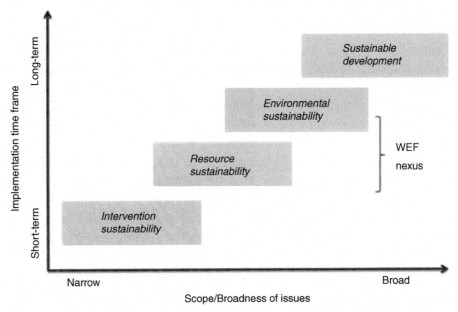

Figure 1.4: WEF Nexus as a Debate About Resource and Environmental Sustainability.

2. Cross-linking: Linkages studies, priority assessments, resilience assessments to certain threats, different models for subnexuses (water–energy, energy–water, energy–food, water–food).
3. Assimilation: environmental footprints, water security assessments, river basin assessments, sectoral policy assessments.

To capture the complexity of interactions and interdependencies of nexus, there is a need for experimentation and method innovation. Nexus-relevant assessment tools can help discover vulnerabilities and risks or deliver practical policy recommendations. The methods illustrated under the different understandings of the WEF nexus depict a wide range of such assessments, which vary a lot in the level of details and captured complexity. A common goal of such assessments is to judge sustainability in a certain region or provide a global or a comparative overview. Nexus-specific assessments are highly useful as overall sustainability assessments of resource use in different basins (Al-Saidi and Ribbe, 2017). Specifically, they address emerging issues often related to resource and environmental sustainability at large (Fig. 1.4).

These two types of sustainability understandings are related to the concrete management and impacts of certain resources use (e.g. water, energy, or land resources), or the broader issue under the environmental sustainability, such as resource securities, societal impacts, and values of the use or misuse of environmental assets. In this sense, nexus assessments often have more of a long-term perspective and entail a broader issue than the evaluation of certain policies or projects. They also aim at providing indicators and guidance for the larger goal of sustainable development.

4 The Hajko Critique

As hinted previously, if energy decoupling holds, then any potential "fundamental EGN relationship" is subject to change, with previous instances of statistically significant causality having at best marginal relevance to actual and future situation. In other words, if decoupling holds, there cannot be any such thing as "fundamental EGN relationship" and its investigation does not have substantial policy importance. This throws much skepticism as to whether any such relationship can to be found. Also it is well documented (Ayres et al., 2003) that the volume of energy consumption has changed dramatically over the past century, and how different is the structure of energy used today compared to only some decades ago In addition, there is no reasonable argument why these changes should cease in a current situation (so that a particular EGN relationship would establish), in particular when we consider the rising prominence of renewables, distributed energy generation, smart grids and smart cities and associated energy demand management systems (Fig. 1.5).

The problem of contradictory results was present in EGN from the very beginning. The debate among economists on what role energy plays in the economy was already growing, when Kraft and Kraft (1978) published their seminal paper. They used Sims (1972) method of causality detection on energy consumption and gross national product.[12] Their main empirical finding was the evidence of the unidirectional causality from GNP to energy consumption in the United States. The investigation of the relationship in the United States remained for a long time in the center of the research attention.

Their results were disputed by Akarca and Long (1980), who pointed out that the original results were heavily influenced by sample selection (note this particular argument on temporal instability is still widely applicable to many recent papers), and argued that the original results were spurious and the proclaimed relationship disappeared when the two last sample years (1973–74) were excluded from the original sample ranging from 1947 to 1974.

This "neutrality" of energy was also supported by Eden and Hwang (1984) who confirmed the absence of causality between energy and GNP in the United States on the annual data sample, ranging from 1947 to 1979, but found unidirectional causality from GNP to energy in a quarterly frequency sample ranging from 1973 to 1981. A similar result supporting the original conclusions from Kraft and Kraft (1978) was also provided by Abosedra and Baghestani (1989), using direct Granger causality procedure. A newer reestimation based on exergy (Warr and Ayres, 2010) found the opposite causality direction, that is, from energy consumption to economic output.

---

[12] However, there were several other papers that focused on elasticity estimates between energy consumption and economic output, for instance, Beenstock and Willcocks (1981); Samouilidis and Mitropoulos (1984); or Desai (1986). These papers, however, did not explicitly focus on causality.

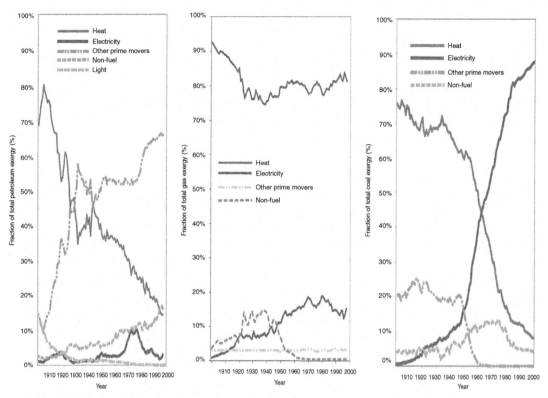

Figure 1.5: Energy Allocation in the Past Century. *Adapted from* Ayres, R.U., Ayres, L.W., Warr, B., 2003. Exergy, power and work in the US economy, 1900–1998. Energy 28(3), 219–273.

The critique of the proclaimed relationship and support for the neutrality of energy and economic output was provided in the articles by Yu and Choi (1985) and Erol and Yu (1987a,b). These articles utilized both Sims' procedure and direct Granger tests [formulated by Sargent (1976)].

The EGN, as an exploratory field, is characteristic with the effort to implement newer econometric techniques to uncover further causal relationships. This is particularly apparent in the upswing of popularity of cointegration pretesting, in particular of Engle and Granger (1987) and Johansen (1988, 1991). We can observe a pattern of cointegration testing in several studies from 1990s (even though in the beginning it was not accompanied with causality tests) and it remains a "standard" approach even today.[13] However, there is a substantial criticism related to the causal inference based on cointegration pretesting. Dolado and Lütkepohl (1996) and Giles and Mirza (1999) recommended to skip cointegration pretesting altogether and instead focus on causality detection using modified causality tests.

---

[13] Among the first EGN studies using cointegration testing along with causality testing was Masih and Masih (1996), followed by Masih and Masih (1997) and Glasure and Lee (1998). However it should be noted that cointegration tests in EGN research had been present earlier in Nachane et al. (1988), Yu and Jin (1992), and Van Hoa (1993).

Given the roots of the EGN, it can be argued that this topic has reflected on various government regulations and policies throughout the years—with renowned examples, such as the gas-guzzler tax, and other energy efficiency targeted measures, typically evoking the needs of energy conservation.

As such it has gradually evolved to a certain kind of policy design research, though typically unsolicited by the policy makers, and more or less ignored in actual energy policies, as evidenced by the lack of research presence in a vast array of energy conservation policy documents (again with a notable example of the main European Energy policy document, Energy Policy for Europe, that ignores the EGN altogether). This is not the case, for instance, for the so-called water–energy nexus (or even the WEF nexus) that enjoys high-profile policy attention despite being much less empirically covered than EGN.

While it has been nearly 40 years since the seminal Kraft and Kraft paper, the topic is still considered controversial, and perhaps rightfully so: the empirical evidence is conflicting, there is an insufficient theoretical framework, and even the availability of the data is poor. Poor data coverage not only limits the estimation scope, but also brings about specific estimation problems. Unfortunately, with an effort to adapt or overcome this situation, additional problems are being introduced into the research papers, such as very long sample periods (several decades) with no regard to structural breaks, inappropriate combination of multiple heteregenous countries in a single dataset, and estimation of causal effects by methods relying on their homogeneity, attributing equal weights to all countries in the estimation despite the huge discrepancies in their actual sizes, an inability to distinguish specific patterns between various different types of energies (evidenced by vastly different prices for equal thermodynamic amount of energy of different types) and so on.

From the very outset, the main attention in the EGN research was focused on detecting the statistically significant Granger causality results. It is well known that correlation does not mean causation, and Granger causality in its essence is based on the ability of improved prediction accuracy using lags of complemental time series. While the statistical/econometric details of such procedures is, undoubtedly, very thoroughly studied, it was never meant to be the main tool to design policy recommendations and any effort to do so should be approached with caution. In addition to that, all Granger causality estimates require solid theoretical support and a justifiable real world explanation. Otherwise, they are reduced to a well known *post hoc ergo propter hoc fallacy* (If event *Y* followed event *X*, event *Y* must have been caused by event *X*).

5 Conclusions

This chapter provides the reader with comprehensive knowledge on what the EGN research is and does, how it started, what progress it has made, and the main problems it is facing today. Emphasis is also laid on the environmental Kuznets curve, which is considered as a special research field close to the EGN field. Readers are also introduced to the new challenges faced in the field, particularly the WEF and the water–energy nexuses. Last and most importantly, the chapter familiarized the readers with the Hajko critique, which introduces a new perspective in the field, hopefully causing a new turn in the field very soon.

References

Abosedra, S., Baghestani, H., 1989. New evidence on the causal relationship between United States energy consumption and gross national product. J. Energy Dev., 285–292.

Adom, P., 2011. Electricity consumption-economic growth nexus: the Ghanaian case. Int. J.Energy Econ. Policy 1 (1), 18–31.

Aghion, P., Hepburn, C., Teytelboym, A., Zenghelis, D., 2014. Path-Dependency, Innovation and the Economics of Climate Change Supporting Paper for New Climate Economy. Grantham Research Institute on Climate Change and the Environment, London School of Economics and Political Science, London.

Aghion, P., Howitt, P., 1998. Endogenous Growth Theory. MIT Press, Cambridge, MA.

Ahern, J.E., 1980. Exergy Method of Energy Systems Analysis. John Wiley & Sons, New York, NY.

Akarca, A.T., Long, T.V., 1980. Relationship between energy and GNP: a reexamination. J. Energy Dev. 5 (2), 326–331.

Allan, T., Keulertz, M., Woertz, E., 2015. The water–food–energy nexus: an introduction to nexus concepts and some conceptual and operational problems. Int. J. Water Resour. Dev. 31 (3), 301–311.

Allen, E.L., Cooper, C., Edmonds, F., Edmonds, J., Reister, D., Weinberg, A., Whittle, C., Zelby, L., 1976. US Energy and Economic Growth, 1975–2010 Technical Report. Institute for Energy Analysis, Oak Ridge, TN.

Al-Mulali, U., Saboori, B., Ozturk, I., 2015. Investigating the environmental Kuznets curve hypothesis in Vietnam. Energy Policy 76, 123–131.

Al-Saidi, M., 2017. Conflicts and security in integrated water resources management. Environ. Sci.Policy 73, 38–44.

Al-Saidi, M., Elagib, N.A., 2017. Towards understanding the integrative approach of the water, energy and food nexus. Sci. Total Environ. 574, 1131–1139.

Al-Saidi, M., Ribbe, L., 2017. Nexus Outlook: Assessing Resource Use Challenges in the Water, Energy and Food Nexus Nexus Research Focus. TH-Köln, University of Applied Sciences, Available from: https://www.water-energy-food.org/fileadmin/user_upload/files/documents/others/Outlook-Nexus_Assessing_Resource_Use_Challenges.pdf.

Altinay, G., Karagol, E., 2004. Structural break, unit root, and the causality between energy consumption and GDP in Turkey. Energy Econ. 26 (6), 985–994.

Altinay, G., Karagol, E., 2005. Electricity consumption and economic growth: evidence from Turkey. Energy Econ. 27 (6), 849–856.

Álvarez, A., Balsalobre, D., Cantos, J.M., Shahbaz, M., 2017. Energy innovations-GHG emissions nexus: fresh empirical evidence from OECD countries. Energy Policy 101, 90–100.

Andreoni, J., Levinson, A., 1998. The Simple Analytics of the Environmental Kuznets Curve. NBER Working Papers, Cambridge, MA, 6739.

Andreoni, J., Levinson, A., 2001. The simple analytics of the environmental Kuznets curve. J. Public Econ. 80, 269–286.

Ang, J.B., 2007. CO2 emissions, energy consumption, and output in France. Energy Policy 35, 4772–4778.

Apergis, N., Loomis, D., Payne, J.E., 2010a. Are shocks to natural gas consumption temporary or permanent? Evidence from a panel of US states. Energy Policy 38, 4734–4736.

Apergis, N., Payne, J.E., 2009. CO2 emissions, energy usage, and output in Central America. Energy Policy 37, 3282–3286.

Apergis, N., Payne, J.E., 2010. The emissions, energy consumption, and growth nexus: evidence from the commonwealth of independent states. Energy Policy 38, 650–655.

Apergis, N., Payne, J.E., Menyah, K., Wolde-Rufael, Y., 2010b. On the causal dynamics between emissions, nuclear energy, renewable energy, and economic growth. Ecol. Econ. 69, 2255–2260.

Aqeel, A., Butt, M.S., 2001. The relationship between energy consumption and economic growth in Pakistan. Asia Pac. Dev. J. 8 (2), 101–110.

Asafu-Adjaye, J., 2000. The relationship between energy consumption, energy prices and economic growth: time series evidence from Asian developing countries. Energy Econ. 22, 615–625.

Ayres, R., Ayres, L., Martinas, K., 1996. Eco-Thermodynamics: Exergy and Life Cycle Analysis. Center for the Management of Environmental Resources Working Paper 96/04/INSEAD.

Ayres, R.U., Ayres, L.W., Warr, B., 2003. Exergy, power and work in the US economy 1900–1998. Energy 28 (3), 219–273.

Ayres, R.U., Warr, B., 2005. Accounting for growth: the role of physical work. Struct. Change Econ. Dyn. 16 (2), 181–209.

Ayres, R.U., Warr, B., 2009. The Economic Growth Engine: How Energy and Work Drive Material Prosperity. Edward Elgar, Cheltenham.

Balsalobre, D., Álvarez, A., Cantos, J.M., 2015. Public budgets for energy RD&D and the effects on energy intensity and pollution levels. Environ. Sci. Pollut. Res. 22 (7), 4881–4892.

Balsalobre, D., Álvarez, A.P., 2016. Economic growth and energy regulation in the environmental Kuznets curve. Environ. Sci. Pollut. Res. 23 (16), 16478–16494.

Barnes, D.F., Floor, W.M., 1996. Rural energy developing countries challenge for economic development. Annu. Rev. Energy Environ. 21, 497–530.

Barro, R.J., Sala-i-Martin, X., 2003. Economic Growth, second ed. MIT Press, Cambridge, MA.

Beaudreau, B.C., 2010. On the methodology of energy-GDP Granger causality tests. Energy 35, 3535–3539.

Beenstock, M., Willcocks, P., 1981. Energy consumption and economic activity in industrialized countries: the dynamic aggregate time series relationship. Energy Econ. 3 (4), 225–232.

Begley, C.G., Ellis, L.M., 2012. Drug development: raise standards for preclinical cancer research. Nature 483 (7391), 531–533.

Bell, A., Matthews, N., Zhang, W., 2016. Opportunities for improved promotion of ecosystem services in agriculture under the water-energy-food nexus. J. Environ. Stud. Sci. 6, 183–191.

Berkhout, P.H., Muskens, J.C., Velthuijsen, J.W., 2000. Defining the rebound effect. Energy Policy 28 (6), 425–432.

Berndt, E.R., Wood, D.O., 1975. Technology, prices, and the derived demand for energy. Rev. Econ. Stat. 57, 259–268.

Bithas, K., Kalimeris, P., 2013. Re-estimating the decoupling effect: is there an actual transition towards a less energy-intensive economy? Energy 51, 78–84.

Bleischwitz, R., Johnson, C.M., Dozler, M.G., 2014. Re-assessing resource dependency and criticality linking future food and water stress with global resource supply vulnerabilities for foresight analysis. Eur. J. Future. Res. 2 (1), 1–12.

Bouoiyour, J., Selmi, R., Ozturk, I., 2014. The nexus between electricity consumption and economic growth: new insights from meta-analysis. Int. J. Energy Econ. Policy 4 (4), 621–635.

Bowden, N., Payne, J.E., 2009. The causal relationship between U.S. energy consumption and real output: a disaggregated analysis. J. Policy Modeling 31, 180–188.

Bruns, S.B., Gross, C., Stern, D.I., 2014. Is there really granger causality between energy use and output? Energy J. 35, 101–134.

Cairns, R., Krzywoszynska, A., 2016. Anatomy of a buzzword: the emergence of 'the water-energy-food nexus' in UK natural resource debates. Environ. Sci. Policy 64, 1462–9011.

Caruso, G., 2005. When will world oil production Peak? In: 10th Annual Asia Oil and Gas Conference, Kuala Lumpur, Malaysia. p. 13.

Chang, Y., Li, G., Yao, Y., Zhang, L., Yu, C., 2016. Quantifying the water-energy-food nexus: current status and trends. Energies 9, 65.

Chen, P.Y., Chen, S.T., Chen, C.C., 2012. Energy consumption and economic growth-new evidence from meta-analysis. Energy Policy 44, 245–255.

Cheng, B.S., 1997. Energy consumption and economic growth in Brazil, Mexico and Venezuela: a time series analysis. Appl. Econ. Lett. 4 (11), 671–674.

Cheng, B.S., 1998. Energy consumption, employment and causality in Japan: a multivariate approach. Indian Econ. Rev. 33, 19–29.

Cheng, B.S., 1999. Causality between energy consumption and economic growth in India: an application of cointegration and error-correction modeling. Indian Econ. Rev. 34, 39–49.

Cheng, B.S., Lai, T.W., 1997. An investigation of co-integration and causality between energy consumption and economic activity in Taiwan. Energy Econ. 19, 435–444.

Choumert, J., Motel, P.C., Dakpo, H.K., 2013. Is the environmental Kuznets curve for deforestation a threatened theory? A meta-analysis of the literature. Ecol. Econ. 90, 19–28.

Cleveland, C.J., Costanza, R., Hall, C.A.S., Kaufmann, R., 1984. Energy and the U.S. economy: a biophysical perspective. Science 225, 890–897.

Cleveland, C.J., Kaufmann, R.K., Stern, D.I., 2000. Aggregation and the role of energy in the economy. Ecol. Econ. 32, 301–317.

Cook, J., Nuccitelli, D., Green, S.A., Richardson, M., Winkler, B., Painting, R., Way, R., Jacobs, P., Skuce, A., 2013. Quantifying the consensus on anthropogenic global warming in the scientific literature. Environ. Res. Lett. 8 (2), 024024.

Cook, J., Oreskes, N., Doran, P.T., Anderegg, W.R., Verheggen, B., Maibach, E.W., Carlton, J.S., Lewandowsky, S., Skuce, A.G., Green, S.A., et al., 2016. Consensus on consensus: a synthesis of consensus estimates on human-caused global warming. Environ. Res. Lett. 11 (4), 048002.

Costantini, V., Martini, C., 2010. The causality between energy consumption and economic growth: a multi-sectoral analysis using non-stationary cointegrated panel data. Energy Econ. 32 (3), 591–603.

Dagher, L., Yacoubian, T., 2012. The causal relationship between energy consumption and economic growth in Lebanon. Energy Policy 50, 795–801.

Daher, B.T., Mohtar, R.H., 2015. Water-energy-food (WEF) Nexus Tool 2.0: guiding integrative resource planning and decision-making. Water Int. 40 (5-6), 748–771.

Davis, S.C., Kauneckis, D., Kruse, N.A., Miller, K.E., Zimmer, M., Dabelko, G.D., 2016. Closing the loop: integrative systems management of waste in food, energy, and water systems. J. Environ. Stud. Sci. 6 (1), 11–24.

Desai, D., 1986. Energy-GDP relationship and capital intensity in LDCs. Energy Econ. 8 (2), 113–117.

Dinda, S., 2004. Environmental Kuznets curve hypothesis: a survey. Ecol. Econ. 49 (4), 431–455.

Dogan, E., Ozturk, I., 2017. The influence of renewable and non-renewable energy consumption and real income on CO_2 emissions in the USA: evidence from structural break tests. Environ. Sci. Pollut. Res. 24 (11), 10846–10854.

Dogan, E., Seker, F., 2016. Determinants of CO_2 emissions in the European Union: the role of renewable and non-renewable energy. Renew. Energy 94, 429–439.

Dogan, E., Sebri, M., Turkekul, B., 2016. Exploring the relationship between agricultural electricity consumption and output: new evidence from Turkish regional data. Energy Policy 95, 370–377.

Dolado, J.J., Lütkepohl, H., 1996. Making Wald tests work for cointegrated VAR systems. Econom. Rev. 15 (4), 369–386.

Dumitrescu, E.I., Hurlin, C., 2012. Testing for Granger non-causality in heterogeneous panels. Econ. Modeling 29 (4), 1450–1460.

Eden, S., Hwang, B.-K., 1984. The relationship between energy and GNP: further results. Energy Econ. 6 (3), 186–190.

Eden, S., Jin, J.C., 1992. Cointegration tests of energy consumption, income, and employment. Resour. Energy 14 (3), 259–266.

Energy Information Administration (EIA), 2016. International Energy Outlook 2016. US Energy Information Administration, Office of Energy Analysis. US Department of Energy, Washington, DC.

Engle, R.E., Granger, C., 1987. Cointegration and error-correction: representation, estimation and testing. Econometrica 55, 251–276.

Erol, U., Eden, S., 1990. Spectral analysis of the relationship between energy consumption, employment, and business cycles. Resour. Energy 11 (4), 395–412.

Erol, U., Yu, E.S., 1987a. On the causal relationship between energy and income for industrialized countries. J. Energy Dev. 13, 113–122.

Erol, U., Yu, E.S.H., 1987b. Time series analysis of the causal relationships between US energy and employment. Resour. Energy 9 (1), 75–89.

Evans, K.L., Jackson, S.F., Greenwood, J.J., Gaston, K.J., 2006. Species traits and the form of individual species–energy relationships. Proceedings of the Royal Society of London B: Biological Sciences, vol. 273 (1595), pp. 1779-1787.

Ewing, B.T., Sari, R., Soyta, U., 2007. Disaggregate energy consumption and industrial output in the United States. Energy Policy 35, 1274–1281.

Fanelli, D., 2011. Negative results are disappearing from most disciplines and countries. Scientometrics 90 (3), 891–904.

Fang, D., Chen, B., 2017. Linkage analysis for the water–energy nexus of city. Appl. Energy 189, 770–779.

Farhani, F., Chaibi, A., Rault, C., 2014. CO2 emissions, output, energy consumption, and trade in Tunisia. Econ. Modelling 38, 426–434.

Fatai, K., Oxley, L., Scrimgeour, F., 2004. Modelling the causal relationship between energy consumption and GDP in New Zealand, Australia, India, Indonesia. Philipp. Thai. Math. Comput. Simul. 64 (3), 431–445.

Ferguson, R., Wilkinson, W., Hill, R., 2000. Electricity use and economic development. Energy Policy 28 (13), 923–934.

Fiorito, G., 2013. Can we use the energy intensity indicator to study "decoupling" in modern economies? J. Clean. Prod. 47, 465–473.

Florax, R., de Groot, H., de Mooij, R., 2002. Meta-Analysis in Policy Oriented Economic Research. CBP Netherlands Bureau of Economic Policy Analysis Report 2002/1.

Garcia, D.J., You, F., 2016. The water-energy-food nexus and process systems engineering: a new focus. Comput. Chem. Eng. 91, 49–67.

Ghali, K.H., El-Sakka, M.I.T., 2004. Energy use and output growth in Canada: a multivariate cointegration analysis. Energy Econ. 26, 225–238.

Ghosh, S., 2002. Electricity consumption and economic growth in India. Energy Policy 30 (2), 125–129.

Giles J.A., Mirza, S., 1999. Some Pretesting Issues on Testing for Granger Noncausality. Econometrics Working Paper EWP9914 ISSN. pp. 1485-6441.

Gillingham, K., Kotchen, M.J., Rapson, D.S., Wagner, G., 2013. Energy policy: the rebound effect is overplayed. Nature 493 (7433), 475–476.

Gillingham, K., Rapson, D., Wagner, G., 2015. The rebound effect and energy efficiency policy. Rev. Environ. Econ. Policy 10 (1), 68–88.

Giupponi, C., Gain, A.K., 2016. Integrated spatial assessment of the water, energy and food dimensions of the Sustainable Development Goals. Reg. Environ. Change 17, 1881–1893.

Glasure, Y.U., 2002. Energy and national income in Korea: further evidence on the role of omitted variables. Energy Econ. 24 (4), 355–365.

Glasure, Y.U., Lee, A.-R., 1995. Relationship between US energy consumption and employment: further evidence. Energy Source. 17 (5), 509–516.

Glasure, Y.U., Lee, A.-R., 1996. Macroeconomic effects of relative prices, money, and federal spending on the relationship between US energy consumption and employment. J. Energy Dev. 22 (1).

Glasure, Y.U., Lee, A.-R., 1998. Cointegration, error-correction, and the relationship between GDP and energy: the case of South Korea and Singapore. Resour. Energy Econ. 20 (1), 17–25.

Gondhalekar, D., Ramsauer, T., 2017. Nexus City: operationalizing the urban water-energy-food nexus for climate change adaptation in Munich, Germany. Urban Climate 19, 28–40.

Granger, C.W., Newbold, P., 1974. Spurious regressions in econometrics. J. Econom. 2 (2), 111–120.

Gross, C., 2012. Explaining the (non-) causality between energy and economic growth in the U.S.: a multivariate sectoral analysis. Energy Econ. 34 (2), 489–499.

Grossman, G., Helpman, E., 1991. Innovation and growth in the global economy. MIT Press, Cambridge, MA.

Grossman, G.M., Krueger, A.B., 1995. Economic growth and the environment. Quarter. J. Econ. 110 (2), 353–377.

Hajko, V., 2017. The failure of energy-economy nexus: a meta-analysis of 104 studies. Energy 125, 771–787.

Halicioglu, F., 2009. An econometric study of CO2 emissions, energy consumption, income and foreign trade in Turkey. Energy Policy 37, 1156–1164.

Hall, C.A.S., Cleveland, C.J., Kaufmann, R., 1986. Energy and Resource Quality: The Ecology of the Economic Process. Wiley Interscience, New York, NY.

Hall, C.A.S., Lindenberger, D., Kümmel, R., Kroeger, T., Eichhorn, W., 2001. The need to reintegrate the natural sciences with economics. BioScience 51, 663–673.

Hamilton, J.D., 1983. Oil and the macroeconomy since World War II. J. Polit. Econ. 91, 228–248.

Hamilton, J.D., 1996. This is what happened to the oil price macroeconomy relationship. J. Monet. Econ. 38, 215–220.

Hamilton, J.D., 2003. What is an oil shock? J. Econom. 113, 363–398.

Hamilton, J.D., 2011. Historical Oil Shocks. Working paper, prepared for the Handbook of Major Events in Economic History. Available from: http://dss.ucsd.edu/ jhamilto/oil_history.pdf.

Hamit-Haggar, M., 2016. Clean energy-growth nexus in Sub-Saharan Africa: evidence from cross-sectionally dependent heterogeneous panel with structural breaks. Renew. Sustain. Energy Rev. 57, 1237–1244.

Harris, R.I.D., Sollis, R., 2003. Applied Time Series Modelling and Forecasting. Wiley, Hoboken, NJ.

Hettige, H., Mani, M., Wheeler, D., 1998. Industrial Pollution in Economic Development: Kuznets Revisited. World Bank Working Paper. p. 1876.

Hirsch, R.L., Bezdek, R., Wendling, R., 2005. Peaking of world oil production. Proceedings of the IV International Workshop on Oil Gas Depletion. pp. 19-20.

Hitch, C.J. (Ed.), 1978. Energy conservation and economic growth. Westview Press, Boulder, CO, (AAAS Selected Symposium 22, OSTI Identifier:6472502).

Hoffert, M.I., Caldeira, K., 2004. Climate change and energy. Encyclopedia of Energy. Elsevier Academic Press, Cambridge, Massachusetts, (Overview 359–380).

Hondroyiannis, G., Lolos, S., Papapetrou, E., 2002. Energy consumption and economic growth: assessing the evidence from Greece. Energy Econ. 24, 319–336.

Hotelling, H., 1931. The economics of exhaustible resources. J. Polit. Econ. 39 (2), 137–175.

Hsiao, C., 1981. Autoregressive modeling and money-income causality detection. J. Monet. Econ. 7, 85–106.

Huang, J.-P., 1993. Electricity consumption and economic growth a case study of China. Energy Policy 21 (6), 717–720.

Hunter, J.E., Schmidt, F.L., Jackson, G.B., 1982. Cumulating research findings across studies. Studying Organizations: Innovations in Methodology. Sage, Beverly Hill, CA.

Hwang, D.B., Gum, B., 1991. The causal relationship between energy and GNP: the case of Taiwan. J. Energy Dev. 16, 219–226.

IEA, 2002. World Energy Outlook 2002. International Energy Agency, Paris.

IEA, 2012. World Energy Outlook 2012. International Energy Agency, Paris.

Isaac, M., van Vuuren, D.P., 2009. Modeling global residential sector energy demand for heating and air conditioning in the context of climate change. Energy Policy 37, 507–521.

Jafari, Y., Othman, J., Nor, A.H.S.M., 2012. Energy consumption, economic growth and environmental pollutants in Indonesia. J. Policy Modeling 34 (6), 879–889.

Jalil, A., Mahmud, S.F., 2009. Environment Kuznets curve for CO2 emissions: a cointegration analysis for China. Energy Policy 37, 5167–5172.

Jevons, W.S., 1865. The Coal Question: An Inquiry Concerning the Progress of the Nation, and the Probable Exhaustion of Our Coal-Mines. Macmillan Publishers, London.

Johansen, S., 1988. A statistical analysis of co-integration vectors. J. Econ. Dyn. Control 12, 231–254.

Johansen, S., 1991. Estimation and hypothesis testing of cointegrating vectors in Gaussian vector autoregressive models. Econometrica 59, 1551–1580.

Kalimeris, P., Richardson, C., Bithas, K., 2014. A meta-analysis investigation of the direction of the energy-GDP causal relationship: implications for the growth-degrowth dialogue. J. Clean. Prod. 67, 1–13.

Karanfil, F., 2009. How many times again will we examine the energy-income nexus using a limited range of traditional econometric tools? Energy Policy 37, 1191–1194.

Kasman, A., Duman, Y.S., 2015. CO2 emissions, economic growth, energy consumption, trade and urbanization in new EU member and candidate countries: a panel data analysis. Econ. Modelling 44, 97–103.

Keskinen, M., Varis, O., 2016. Water-energy-food nexus in large Asian river basins. Water 8 (10), 446.

Khazzoom, J., 1976. Proceedings of the Workshop on Modeling the Interrelationships Between the Energy Sector and the General Economy Conference. Washington, DC, pp. 29–30.

Koirala, B.S., Li, H., Berrens, R.P., 2011. Further investigation of environmental Kuznets curve studies using meta-analysis. Int. J. Ecol. Econ. Stat. 22 (S11), 13–32.

Kowsari, R., Zerriffi, H., 2011. Three dimensional energy profile: a conceptual framework for assessing household energy use. Energy Policy 39 (12), 7505–7517.

Kraft, J., Kraft, A., 1978. On the relationship between energy and GNP. J. Energy Dev. 3, 401–403.

Kurian, M., 2017. The water-energy-food nexus: trade-offs, thresholds and transdisciplinary approaches to sustainable development. Environ. Sci. Policy 68, 97–106.

Kurian, M., Ardakanian, R., 2015. Governing the Nexus. Water Soil and Waste Resources Considering Global Change. Springer International Publishing, Switzerland.

Kuznets, S., 1955. Economic growth and income inequality. Am. Econ. Rev. 45 (1), 1–28.

Lau, L.S., Choong, C.K., Eng, Y.K., 2014. Investigation of the environmental Kuznets curve for carbon emissions in Malaysia: do foreign direct investment and trade matter? Energy Policy 68, 490–497.

Leach, G., 1992. The energy transition. Energy Policy 20, 116–123.

Lee, C.-C., 2005. Energy consumption and GDP in developing countries: a cointegrated panel analysis. Energy Econ. 27 (3), 415–427.

Lee, C.-C., Chang, C.-P., Chen, P.-F., 2008. Energy-income causality in OECD countries revisited: the key role of capital stock. Energy Econ. 30 (5), 2359–2373.

Lee, C.-C., Chiu, Y.-B., 2011. Nuclear energy consumption, oil prices, and economic growth: evidence from highly industrialized countries. Energy Econ. 33 (2), 236–248.

Lee, C.-C., Lee, J.-D., 2010. A panel data analysis of the demand for total energy and electricity in OECD countries. Energy J. 31, 1–23.

Leese, M., Meisch, S., 2015. Securitising sustainability? Questioning the "water, energy and food-security nexus". Water Altern. 8, 695–709.

Liddle, B., 2006. How linked are energy and GDP: reconsidering energy-GDP cointegration and causality for disaggregated OECD country data. Int. J. Energy Environ. Econ. 13 (2), 97–113.

Littell, J.H., Corcoran, J., Pillai, V., 2008. Systematic reviews and meta-analysis. Oxford University Press, New York, NY.

Long, T.V., Schipper, L., 1978. Resource and energy substitution. Energy 3 (1), 63–82.

López-Menéndez, A.J., Pérez, R., Moreno, B., 2014. Environmental costs and renewable energy: re-visiting the environmental Kuznets curve. J. Environ. Manag. 145, 368–373.

Mahadevan, R., Asafu-Adjaye, J., 2007. Energy consumption, economic growth and prices: a reassessment using panel VECM for developed and developing countries. Energy Policy 35, 2481–2490.

Mankiw, N.G., 2006. Macroeconomics, sixth ed. Worth Publishers, New York.

Masih, A.M.M., Masih, R., 1996. Energy consumption, real income and temporal causality: results from a multi-country study based on cointegration and error-correction modelling techniques. Energy Econ. 18, 165–183.

Masih, A.M.M., Masih, R., 1997. On the temporal causal relationship between energy consumption, real income, and prices: some new evidence from Asian-energy dependent NICs based on a multivariate cointegration/vector error-correction approach. J. Policy Modeling 19, 417–440.

Masih, A.M., Masih, R., 1998. A multivariate cointegrated modelling approach in testing temporal causality between energy consumption, real income and prices with an application to two Asian LDCs. Appl. Econ. 30 (10), 1287–1298.

Mehrara, M., 2007. Energy consumption and economic growth: the case of oil exporting countries. Energy Policy 35 (5), 2939–2945.

Menegaki, A.N., 2014. On energy consumption and GDP studies; a meta-analysis of the last two decades. Renew. Sustain. Energy Rev. 29, 31–36.

Menyah, K., Wolde-Rufael, Y., 2010. CO2 emissions, nuclear energy, renewable energy and economic growth in the US. Energy Policy 38 (6), 2911–2915.

Nachane, D.M., Nadkarni, R.M., Karnik, A.V., 1988. Co-integration and causality testing of the energy–GDP relationship: a cross-country study. Appl. Econ. 20 (11), 1511–1531.

Narayan, P.K., Smyth, R., 2005a. Electricity consumption, employment and real income in Australia evidence from multivariate Granger causality tests. Energy Policy 33 (9), 1109–1116.

Narayan, P.K., Smyth, R., 2005b. The residential demand for electricity in Australia: an application of the bounds testing approach to cointegration. Energy Policy 33 (4), 467–474.

Nordhaus, W.D., 2010. Economic aspects of global warming in a post-Copenhagen environment. Proceedings of the National Academy of Sciences. 107 (26), pp. 11721–11726.

Mielnik, O., Goldemberg, J., 2002. Foreign direct investment and decoupling between energy and gross domestic product in developing countries. Energy Policy 30 (2), 87–89.

Mohtar, R.H., Lawford, R., 2016. Present and future of the water-energy-food nexus and the role of the community of practice. J. Environ. Stud. Sci. 6, 192–199.

Murphy, D.J., Hall, C.A.S., 2010. Year in review: EROI or energy return on (energy) invested. Ann. N. Y. Acad. Sci. 1185, 102–118.

Murry, D.A., Nan, G.D., 1994. A definition of the gross domestic product-electrification interrelationship. J. Energy Dev. 19 (2), 275–283.

Narayan, P.K., Narayan, S., Popp, S., 2010. Energy consumption at the state level: the unit root null hypothesis from Australia. Appl. Energy 85, 1082–1089.

Naser, H., 2014. Oil market, nuclear energy consumption and economic growth: evidence from emerging economies. Int. J. Energy Econ. Policy 4 (2), 288.

Omri, A., 2014. An international literature survey on energy-economic growth nexus: evidence from country-specific studies. Renew. Sustain. Energy Rev. 38, 951–959.

Ozturk, I., 2010. A literature survey on energy–growth nexus. Energy Policy 38 (1), 340–349.

Ozturk, I., Acaravci, A., 2011. Electricity consumption and real GDP causality nexus: evidence from ARDL bounds testing approach for 11 MENA countries. Appl. Energy 88, 2885–2892.

Ozturk, I., Acaravci, A., 2013. The long-run and causal analysis of energy, growth, openness and financial development on carbon emissions in Turkey. Energy Econ. 36, 262–267.

Ozturk, I., Al-Mulali, U., 2015. Investigating the validity of the environmental Kuznets curve hypothesis in Cambodia. Ecol. Indic. 57, 324–330.

Panayotou, T., 1993. Empirical Test and Policy Analysis of Environmental Degradation at Different Stages of Economic Development. Technology and Environment Programme, International Labour Office, Geneva, Working Paper, 238.

Pao, H.T., Tsai, C.M., 2011a. Multivariate Granger causality between CO2 emissions, energy consumption, FDI (foreign direct investment) and GDP (gross domestic product): evidence from a panel of BRIC (Brazil, Russian Federation, India, and China) countries. Energy 36, 685–693.

Pao, H.T., Tsai, C.M., 2011b. Modelling and forecasting the CO2 emissions, energy consumption, and economic growth in Brazil. Energy 36, 2450–2458.

Pao, H.T., Yu, C.H., Yang, Y.H., 2011. Modelling the CO2 emissions, energy use, and economic growth in Russia. Energy 36, 5094–5100.

Payne, J.E., 2009. On the dynamics of energy consumption and output in the US. Appl. Energy 86, 575–577.

Payne, J.E., 2010a. On biomass energy consumption and real output in the US. Energy Source. Part B 6, 47–52.

Payne, J.E., 2010b. Survey of the international evidence on the causal relationship between energy consumption and growth. J. Econ. Stud. 37, 53–95.

Payne, J.E., 2010c. A survey of the electricity consumption–growth literature. Appl. Energy 87, 723–731.

Payne, J., Taylor, J., 2010. Nuclear energy consumption and economic growth in the US: an empirical note. Energy Source. Part B 5 (3), 301–307.

Pedroni, P., 1999. Critical values for cointegration tests in heterogenous panels with multiple regressors. Oxf. Bull. Econ. Stat. 61, 653–670.

Pedroni, P., 2000. Full modified OLS for heterogeneous cointegrated panels. Adv. Econom. 15, 93–130.

Perrone, D., Hornberger, G.M., 2014. Water, food, and energy security: scrambling for resources or solutions? Wiley Interdiscip. Rev. 1 (1), 49–68.

Pesaran, M.H., Shin, Y., 1998. An autoregressive distributed-lag modelling approach to cointegration analysis. Econom. Soc. Monogr. 31, 371–413.

Pesaran, M., Shin, Y., 1999. An autoregressive distributed lag modelling approach to cointegrated analysis. In: Strom, S. (Ed.), Econometrics and Economic Theory in the 20th Century: The Ragnar Frisch Centennial Symposium. Cambridge University Press, Cambridge, MA.

Pesaran, M., Shin, Y., Smith, R., 2001. Bounds testing approaches to the analysis of level relationships. J. Appl. Econom. 16, 289–326.

Ramcharran, H., 1990. Electricity consumption and economic growth in Jamaica. Energy Econ. 12 (1), 65–70.

Rant, Z., 1956. Exergie, Ein neues Wort für 'technische Arbeitsfähigkeit'. Forschung auf dem Gebiete des Ingenieurswesens 22, 36–37.

Richmond, A., Kaufmann, R., 2006. Energy prices and turning points: the relationship between income and energy use/carbon emissions. Energy J. 27, 157–180.

Romero-Lankao, P., McPhearson, T., Davidson, D.J., 2017. The food-energy-water nexus and urban complexity. Nat. Climate Change 7, 233–235.

Rotemberg, J.J., Woodford, M., 1996. Imperfect competition and the effects of energy price increases on economic activity. J. Money Credit Bank. 28, 550–577.

Saboori, B., Sulaiman, J., 2013. Environmental degradation, economic growth and energy consumption: evidence of the environmental Kuznets curve in Malaysia. Energy Policy 60, 892–905.

Saboori, B., Sulaiman, J., Mohd, S., 2012. Economic growth and CO2 emissions in Malaysia: a cointegration analysis of the environmental Kuznets curve. Energy Policy 51, 184–191.

Sadorsky, P., 2009. E-energy consumption and income in emerging economies. Energy Policy 37, 4021–4028.

Saikkonen, P., Lütkepohl, H., 1996. Infinite-order cointegrated vector autoregressive processes. Econom. Theory 12 (05), 814–844.

Samouilidis, J.-E., Mitropoulos, C., 1984. Energy and economic growth in industrializing countries: the case of Greece. Energy Econ. 6 (3), 191–201.

Sargent, T.J., 1976. The observational equivalence of natural and unnatural rate theories of macroeconomics. J. Polit. Econ. 84, 631–640.

Sari, R., Ewing, B.T., Soytas, U., 2008. The relationship between disaggregate energy consumption and industrial production in the United States: an ARDL approach. Energy Econ. 30 (5), 2302–2313.

Sari, R., Soytas, U., 2004. Disaggregate energy consumption, employment and income in Turkey. Energy Econ. 26, 335–344.

Sari, R., Soytas, U., 2007. The growth of income and energy consumption in six developing countries. Energy Policy 35, 889–898.

Saunders, H.D., 1992. The Khazzoom-Brookes postulate and neoclassical growth. Energy J. 13, 131–148.

Scanlon, B.R., Ruddell, B.L., Reed, P.M., Hook, R.I., Zheng, C., Tidwell, V.C., Siebert, S., 2017. The food-energy water nexus: transforming science for society. Water Resour. Res. 53 (5), 3550–3556.

Scienceman, D., 1987. Energy and emergy. Environmental Economics: The Analysis of a Major Interface. R. Leimgruber, Geneva, pp. 257–276.

Sebri, M., 2015. Use renewables to be cleaner: meta-analysis of the renewable energy consumption–economic growth nexus. Renew. Sustain. Energy Rev. 42, 657–665.

Sebri, M., Ben-Salha, O., 2014. On the causal dynamics between economic growth, renewable energy consumption, CO2 emissions and trade openness: fresh evidence from BRICS countries. Renew. Sustain. Energy Rev. 39, 14–23.

Selden, T., Song, D., 1994. Environmental quality and development: is there a Kuznets curve for air pollution emissions? J. Environ. Econ. Manag. 27 (2), 147–162.

Shafik, N., Bandyopadhyay, N., 1992. Economic Growth and Environmental Quality: Time-Series and Cross-Country Evidence. World Bank Working Papers, vol. 904, pp. 1-6.

Shahbaz, M., Khraief, N., Uddin, G.S., Ozturk, I., 2014a. Environmental Kuznets curve in an open economy: a bounds testing and causality analysis for Tunisia. Renew. Sustain. Energy Rev. 34, 325–336.

Shahbaz, M.H.H., Lean, H.H.M.S., Shabbir, M.S., 2012. Environmental Kuznets curve hypothesis in Pakistan: cointegration and Granger causality. Sustain. Energy Rev. 16, 2947–2953.

Shahbaz, M., Loganathan, N., Zeshan, M., Zaman, K., 2015a. Does renewable energy consumption add in economic growth? An application of auto-regressive distributed lag model in Pakistan. Renew. Sustain. Energy Rev. 44, 576–585.

Shahbaz, M., Sbia, R., Hamdi, H., Ozturk, I., 2014b. Economic growth, electricity consumption, urbanization and environmental degradation relationship in United Arab Emirates. Ecol. Indic. 45, 622–631.

Shahbaz, M., Solarin, S.A., Sbia, R., Bibi, S., 2015b. Does energy intensity contribute to CO_2 emissions? A trivariate analysis in selected African countries. Ecol. Indic. 50, 215–224.

Shahbaz, M., Solarin, S.A., Mahmood, H., Arouri, M., 2013a. Does financial development reduce CO_2 emissions in Malaysian economy? A time series analysis. Econ. Modelling 35, 145–152.

Shahbaz, M., Tiwari, A.K., Nasir, M., 2013b. The effects of financial development, economic growth, coal consumption and trade openness on CO_2 emissions in South Africa. Energy Policy 61, 1452–1459.

Shahiduzzaman, M., Alam, K., 2012. Cointegration and causal relationships between energy consumption and output: assessing the evidence from Australia. Energy Econ. 34 (6), 2182–2188.

Simpson, E.H., 1951. The interpretation of interaction in contingency tables. J. R. Stat. Soc. 13, 238–241.

Sims, C.A., 1972. Money, income, and causality. Am. Econ. Rev. 62 (4), 540–552.

Sioshansi, F.P., 1986. Energy, electricity, and the US economy: emerging trends. Energy J. 7 (2), 81–90.

Smyth, R., Narayan, P.K., 2015. Applied econometrics and implications for energy economics research. Energy Econ. 50, 351–358.

Solow, R., 1956. A contribution to the theory of economic growth. Q. J. Econ. 70, 65–94.

Soytas, U., Sari, R., 2006. Can China contribute more to the fight against global warming? J. Policy Modeling 28 (8), 837–846.

Squalli, J., 2007. Electricity consumption and economic growth: bounds and causality analyses of OPEC members. Energy Econ. 29 (6), 1192–1205.

Stanley, T.D., 2001. Wheat from chaff: meta-analysis as quantitative literature review. J. Econ. Perspect. 15, 131–150.

Stanley, T.D., Jarrell, S.B., 1989. Meta regression analysis: a quantitative method of literature surveys. J. Econ. Surv. 3 (2), 161–170.

Steinberger, J.K., Roberts, J.T., 2010. From constraint to sufficiency: the decoupling of energy and carbon from human needs, 1975–2005. Ecol. Econ. 70 (2), 425–433.

Stern, D.I., 1993. Energy and economic growth in the USA: a multivariate approach. Energy Econ. 15 (2), 137–150.

Stern, D.I., 2000. A multivariate cointegration analysis of the role of energy in the US macroeconomy. Energy Econ. 22, 267–283.

Stern, D.I., 2004. Economic growth and energy. Encyclopedia of Energy. Elsevier Academic Press, Cambridge, Massachusetts, pp. 35–51.

Stern, D.I., 2010. Energy quality. Ecol. Econ. 69 (7), 1471–1478.

Stern, D.I., 2011. The role of energy in economic growth. Ann. N. Y. Acad. Sci. 1219 (1), 26–51.

Stern, D.I., Enflo, K., 2013. Causality between energy and output in the long-run. Energy Econ. 39, 135–146.

Tan, F., Lean, H.H., Khan, H., 2014. Growth and environmental quality in Singapore: is there any trade-off? Ecol. Indic. 47, 149–155.

Tang, C.F., Tan, B.W., 2015. The impact of energy consumption, income and foreign direct investment on carbon dioxide emissions in Vietnam. Energy 79, 447–454.

Tiwari, A.K., Shahbaz, M., Hye, Q.M.A., 2013. The environmental Kuznets curve and the role of coal consumption in India: cointegration and causality analysis in an open economy. Renew. Sustain. Energy Rev. 18, 519–527.

Toda, H.Y., Yamamoto, T., 1995. Statistical inference in vector autoregressions with possibly integrated processes. J. Econom. 66, 225–250.

Toman, M.T., Jemelkova, B., 2003. Energy and economic development: an assessment of the state of knowledge. Energy J. 24, 93–112.

Torras, M., Boyce, J., 1998. Income, inequality, and pollution: a reassessment of the environmental Kuznets curve. Ecol. Econ. 25, 147–160.

Tsani, S.Z., 2010. Energy consumption and economic growth: a causality analysis for Greece. Energy Econ. 32, 582–590.

Van Hoa, T., 1993. Effects of oil on output growth and inflation in developing countries: the case of Thailand from January 1966 to January 1991. Int. J. Energy Res. 17 (1), 29–33.

Wall, G., 1977. Exergy: A Useful Concept Within Resource Accounting. Chalmers Tekniska Högskola; Göteborgs Universitet, Sweden.

Wang, S.S., Zhou, D.Q.P., Zhou, P., Wang, Q.W., 2011. CO_2 emissions, energy consumption and economic growth in China: a panel data analysis. Energy Policy 39, 4870–4875.

Warr, B.S., Ayres, R.U., 2010. Evidence of causality between the quantity and quality of energy consumption and economic growth. Energy 35, 1688–1693.

Warr, B., Ayres, R., Eisenmenger, N., Krausmann, F., Schandl, H., 2010. Energy use and economic development: a comparative analysis of useful work supply in Austria, Japan, the United Kingdom and the US during 100 years of economic growth. Ecol. Econ. 69 (10), 1904–1917.

Wicaksono, A., Jeong, G., Kang, D., 2017. Water, energy and food nexus: review of global implementation and simulation model development. Water Policy 19 (3), 440–462.

Wichelns, D., 2017. The water-energy-food nexus: is the increasing attention warranted, from either a research or policy perspective? Environ. Sci. Policy 69, 113–123.

Wolde-Rufael, Y., 2004. Disaggregated industrial energy consumption and GDP: the case of Shanghai, 1952–1999. Energy Econ. 26, 69–75.

Wolde-Rufael, Y., 2005. Energy demand and economic growth: the African experience. J. Policy Modeling 27 (8), 891–903.

Wolde-Rufael, Y., 2009. Energy consumption and economic growth: the experience of African countries revisited. Energy Econ. 31 (2), 217–224.

Wolde-Rufael, Y., 2010. Bounds test approach to cointegration and causality between nuclear energy consumption and economic growth in India. Energy Policy 38 (1), 52–58.

Wolde-Rufael, Y., 2012. Nuclear energy consumption and economic growth in Taiwan. Energy Source. Part B 7 (1), 21–27.

Yang, H., 2000a. A note of the causal relationship between energy and GDP in Taiwan. Energy Econ. 22, 309–317.

Yang, H.-Y., 2000b. Coal consumption and economic growth in Taiwan. Energy Source. 22 (2), 109–115.

Yang, H.-Y., 2000c. A note on the causal relationship between energy and GDP in Taiwan. Energy Econ. 22 (3), 309–317.

Yildirim, E., Saraç, Ş., Aslan, A., 2012. Energy consumption and economic growth in the USA: evidence from renewable energy. Renew. Sustain. Energy Rev. 16 (9), 6770–6774.

Yoo, S.H., 2006. The causal relationship between electricity consumption and economic growth in ASEAN countries. Energy Policy 34, 3573–3582.

Yoo, S.-H., Jung, K.-O., 2005. Nuclear energy consumption and economic growth in Korea. Prog. Nucl. Energy 46 (2), 101–109.

Yoo, S.-H., Kim, Y., 2006. Electricity generation and economic growth in Indonesia. Energy 31 (14), 2890–2899.

Yu, E.S.H., Choi, J.Y., 1985. The causal relationship between energy and GNP: an international comparison. J. Energy Dev. 10 (2), 249–272.

Yu, E.S.H., Hwang, B.K., 1984. The relationship between energy and GNP: further results. Energy Econ. 6, 186–190.

Yu, E.S., Jin, J.C., 1992. Cointegration tests of energy consumption, income, and employment. Resour. Energy 14, 259–266.

Zachariadis, T., 2007. Exploring the relationship between energy use and economic growth with bivariate models: new evidence from G-7 countries. Energy Econ. 29, 1233–1253.

Further Readings

Akarca, A.T., Long, T.V., 1979. Energy and employment: a time series analysis of the causal relationship. Resour. Energy 2 (2–3), 151–162.

Allouche, J., Middleton, C., Gyawali, D., 2014. Water and the Nexus, Nexus Nirvana or Nexus Nullity? A Dynamic Approach to Security and Sustainability in the Water-Energy-Food Nexus. STEPS Center. Brighton, UK. Available from: http://steps-centre.org/wp-content/uploads/Water-and-the-Nexus.pdf.

Al-mulali, U., Fereidouni, H.G., Lee, J.Y., 2013. Sab C.N.B.C.examining the bi-directional long run relationship between renewable energy consumption and GDP growth. Renew. Sustain. Energy Rev. 22, 209–222.

Al-mulali, U., Fereidouni, H.G., Lee, J.Y., 2014. Electricity consumption from renewable and non-renewable sources and economic growth: evidence from Latin American countries. Renew. Sustain. Energy Rev. 30, 290–298.

Apergis, N., Ozturk, I., 2015. Testing environmental Kuznets hypothesis in Asian countries. Ecol. Indic. 52, 16–22.

Aslan, A., Ocal, O., 2016. The role of renewable energy consumption in economic growth: evidence from asymmetric causality. Renew. Sustain. Energy Rev. 60, 953–959.

Ben Jebli, M., Ben Youssef, S., 2015b. Output, e and non-e energy consumption and international trade: evidence from a panel of 69 countries. Renew. Energy 83, 799–808.

Bilgili, F., Ozturk, I., 2015. Biomass energy and economic growth nexus in G7 countries: evidence from dynamic panel data. Renew. Sustain. Energy Rev. 49, 132–138.

Club de Rome, Meadows, D.H., 1972. The Limits to gRowth: A Report for the Club of Rome's Project on the Predicament of Mankind. Universe Books, New York, NY.

Dogan, E., 2015. The relationship between economic growth and electricity consumption from renewable and non-renewable sources: a study of Turkey. Renew. Sustain. Energy Rev. 52, 534–546.

He, X., Zhang, Y., 2012. The influence factors of carbon emissions of Chinese industry and the reorganization effects of CKC-an empirical research on the dynamic panel data of different industries based on the STIRPAT model. China Ind. Econ. 1, 26–35.

Heerink, N., Mulatu, A., Bulte, E., 2001. Income inequality and the environment: aggregation bias in environmental Kuznets curves. Ecol. Econ. 38, 359–367.

Hvidt, M., 2013. Economic Diversification in GCC Countries: Past Record and Future Trends. London School of Economics and Political Science, London, pp. 1–55.

Gradus, R., Smulders, S., 1993. The trade-off between environmental care and long-term growth pollution in three prototype growth models. J. Econ. 58, 25–51.

Granger, C.W.J., 1969. Investigating causal relations by econometric models and cross-spectral methods. Econometrica 37, 424–438.

ICF, 2003. The Economic Cost of the Blackout. An Issue Paper on the Northeastern Blackout August 14, 2003.

Johansen, S., Juselius, K., 1990. Maximum likelihood estimation and inference on cointegration with applications to the demand for money. Oxf. Bull. Econ. Stat. 52 (2), 169–210.

Kleemann, L., Abdulai, A., 2013. The impact of trade and economic growth on the environment: revisiting the cross-country evidence. J. Int. Dev. 25, 180–205.

Koçaka, E., Şarkgüneşi, A., 2017. The renewable energy and economic growth nexus in black sea and Balkan Countries. Energy Policy 100, 51–57.

Menegaki, A.N., 2011. Growth and renewable energy in Europe: a random effect model with evidence for neutrality hypothesis. Energy Econ. 33 (2), 257–263.

Menyah, K., Wolde-Rufael, Y., 2010. Energy consumption, pollutant emissions and economic growth in South Africa. Energy Econ. 32 (6), 1374–1382.

Mezher, T., Fath, H., Abbas, Z., Khaled, A., 2011. Techno-economic assessment and environmental impacts of desalination technologies. Desalination 266, 263–273.

Munroe, R., 2011. XKCD Comic 882 Significant. Available from: https://xkcd.com/882/.

Ocal, O., Aslan, A., 2013. Renewable energy consumption–economic growth nexus in Turkey. Renew. Sustain. Energy Rev. 28, 494–499.

Oh, W., Lee, K., 2004a. Causal relationship between energy consumption and GDP revisited: the case of Korea 1970–1999. Energy Econ. 26 (1), 51–59.

Oh, W., Lee, K., 2004b. Energy consumption and economic growth in Korea: testing the causality relation. J. Policy Modeling 26 (8-9), 973–981.

Ozcan, B., 2013. The nexus between carbon emissions, energy consumption and economic growth in Middle East countries: a panel data analysis. Energy Policy 62, 1138–1147.

Sadorsky, P., 2011. Trade and energy consumption in the Middle East. Energy Econ. 33, 739–749.

Sadorsky, P., 2012. Energy consumption, output and trade in South America. Energy Econ. 34, 476–488.

Sebri, M., Abid, M., 2012. Energy use for economic growth: a trivariate analysis from Tunisian agriculture sector. Energy Policy 48, 711–716.

Smulders, S., Bretschger, L., 2000. Explaining Environmental Kuznets Curves: How Pollution Induces Policy and New Technologies. Tilburg University CentER Working Paper No. 2000–95. Available from: https://ssrn.com/abstract=249023 or http://dx.doi.org/10.2139/ssrn.249023.

Soytas, U., Sari, R., 2003. Energy consumption and GDP: causality relationship in G7 countries and emerging markets. Energy Econ. 25, 33–37.

Tiwari, A.K., 2011. A structural VAR analysis of renewable energy consumption, real GDP and CO2 emissions: evidence from India. Econ. Bull. 31 (2), 1793–1806.

Tugcu, C.T., Ozturk, I., Aslan, A., 2012. Renewable and non-renewable energy consumption and economic growth relationship revisited evidence from G7 countries. Energy Econ. 34, 1942–1950.

CHAPTER 2

Disaggregation in the Energy-Growth Nexus: An Indicative Literature Review

Alper Aslan, Ebru Topcu
Nevşehir Hacı Bektaş Veli University, Nevşehir, Turkey

Chapter Outline

1 Introduction: The Need for Disaggregation in the Energy-Growth Nexus (EGN)

The energy economics literature presents a wide range of empirical relationships between a set of macroeconomic variables and energy consumption.[1] However, the causal relationship between energy consumption and economic growth has been very important for policy makers, given the multidimensional relationship that involves an interdisciplinary approach from the fields of economics, environment, and energy. Theoretically, previous literature

---

[1]   See, for example: trade-energy nexus (Cole, 2006; Ghani, 2012; Halicioglu, 2009; among others), finance-energy nexus (Aslan et al., 2014a; Topcu and Altay, 2017; Topcu and Payne, 2017; among others), and urbanization-energy nexus (Fan et al., 2017; Franco et al., 2017; Yang et al., 2017; among others).

has addressed four hypotheses regarding this causal nexus: (1) The Growth Hypothesis, unidirectional causality from energy consumption to economic growth, indicates that an increase in energy use resulted from increased production level leads to a rise in economic activity; (2) The Conservation Hypothesis, unidirectional causality from economic growth to energy consumption, indicates that higher economic growth leads to an increase in energy use; (3) The Feedback Hypothesis, bidirectional causality between energy consumption and economic growth, indicates that an increase in energy use affects economic activity, which, in turn affects energy use; (4) The Neutrality Hypothesis, which supports absence of causality between energy consumption and economic growth, a situation that indicates that energy consumption and economic growth are irrelevant to each others growth.

The energy-growth nexus (EGN) has been certainly well documented for nearly 4 decades, as it involves a huge number of worldwide empirical work, which causes conflicting results in energy-growth literature. As energy consumption is a significant determinant of economic growth, this research field keeps drawing attention. However, EGN studies have most frequently involved aggregate magnitudes, such as in Apergis and Danuletiu (2012), Apergis and Payne (2009a,b,c), Apergis and Tang (2013), Aslan et al. (2014a), Narayan and Popp (2012), Narayan and Smyth (2008), Oh and Lee (2004), Ozturk et al. (2010), Paul and Bhattacharya (2004), Pirlogea and Cicea (2012), Soytas and Sari (2006), Tsani (2010), Yildirim and Aslan (2012), and Yildirim et al. (2014). The number of studies using disaggregate magnitudes are relatively limited, such as in Abbas and Choudhury (2013), Bernard and Kenneth (2016), Hamzacebi (2007), Hu and Lin (2008), Mucuk and Sogozu (2011), Squalli (2007), and Tang and Shahbaz (2013).

As discussed by Hajko (2017), studies looking into EGN at an aggregate level have not been able to disaggregate the impact across various industries and are likely to lead to an aggregation bias. Under the existence of aggregation bias, findings might be inconsistent and maybe not so useful for policy making. As each industry has its own technology and energy behavior, it is more enlightening and meaningful to disaggregate both at sectoral level and energy type (Fig. 2.1), as well as at a fuel level; albeit this is not always feasible from a data availability point of view for all countries. Therefore, the goal of this study is to examine the relationship between energy consumption and economic growth in the United States at the disaggregated level over the period 1962–2015.

This chapter is divided into five parts. Besides the introduction (Section 1) that motivates the chapter, Section 2 provides a selected and representative literature review for energy disaggregation in the EGN, Section 3 provides a selected and representative literature review for sector disaggregation, Section 4 presents case study for the United States, and Section 5 concludes the chapter.

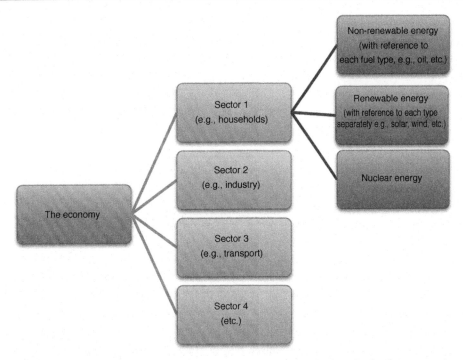

Figure 2.1: Disaggregation by Sectors and Energy Types is the Ideal Disaggregation.

2 Disaggreggation Based on Energy Types and Fuel Types: A Literature Map

The theoretical relationship between energy consumption and economic growth was initially investigated in Kraft and Kraft's (1978) seminal paper. Following this study the literature has expanded quickly with conflicting results related to the direction of the causal relationship between energy consumption and economic growth. This conflict has occurred due to the different econometric methods applied in each study, on various country samples, and various time periods. Recently, the controversy and conflict have drawn the interest of researchers in the field (through extensive literature surveys and metaanalyses or new sophisticated econometric methods), who try to investigate the specific factors that affect the causality between these so important variables, energy, and economic growth.

In energy economics, the relationship between disaggregated energy consumption and economic growth has attracted interest only recently. Despite of the large number of empirical studies on aggregate EGN, much less is known on the relationship between economic growth and disaggregated energy consumption in the empirical literature. Most disaggregated energy studies are based on different energy sources such as nuclear energy, natural gas, renewable energy, and fossil energy (Fig. 2.2). Namely, the studies disaggregate energy consumption

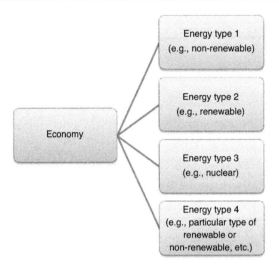

Figure 2.2: Disaggregation on Energy Types is the Most Frequent Type of Disaggregation in the Energy-Growth Nexus (EGN).

into the particular energy types but, to a large degree, they remain aggregate studies, as they are using aggregate energy–type consumption and the aggregate economic output. The remaining part of this section reviews relevant literature with respect to this incomplete disaggregation framework.

2.1 Nuclear Energy

According to the International Atomic Energy Agency (IAEA), nuclear generating capacity is expected to play a major role in the energy mix in the long term because of the population growth and the electricity demand growth in the developing world, as well as the climate change concerns, security of energy supply, and price volatility for other fuels.[2]

There exists a wide empirical literature on nuclear EGN (Naser, 2014; Nazlıoğlu et al., 2011; Ozcan and Ari, 2015; Wolde-Rufael, 2012; Yoo and Jung, 2005; Yoo and Ku, 2009). Some indicative studies are presented in Fig. 2.3. In the context of 16 countries, Apergis and Payne (2010a) investigate the relationship between nuclear energy and economic growth for the period 1980–2005, by employing panel data techniques. According to panel error correction model results, there is a bidirectional relationship between nuclear energy and economic growth in the short term. In terms of the long-run horizon, there is a unidirectional relationship from nuclear energy to economic growth.

---

[2] See: https://www.iaea.org/newscenter/news/iaea-issues-projections-nuclear-power-2020-2050, for extensive information on the topic.

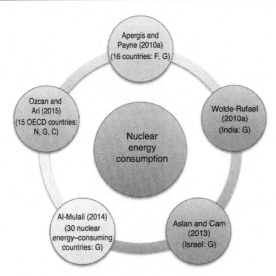

Figure 2.3: Indicative Selection of Studies (Produced From 2010 to 2015) That Deal With Disaggregated Energy Consumption: Nuclear Energy.
C, Conservation; *F*, feedback; *G*, growth hypothesis; *N*, neutrality.

Over the period 1969–2006, Wolde-Rufael (2010a) explores the dynamic connection between nuclear energy consumption and economic growth in India. Results from the Granger causality test show that there is a positive and undirectional relationship from nuclear energy consumption to economic growth. Al-Mulali (2014) reports that nuclear energy consumption has a positive long-term effect on gross domestic product (GDP) in 30 major nuclear energy–consuming countries from 1990 to 2010. He also finds a Granger Causality running from energy consumption to economic growth in the short run. Furthermore, in the case of Israel, Aslan and Cam (2013) investigate the causal connection between nuclear energy consumption and economic growth from 1985 to 2009 employing a bootstrap-corrected causality. According to the results, the direction of causality is from nuclear energy to GDP.

Last, by using the bootstrap causality test, Ozcan and Ari (2015) examine the causal link between nuclear energy consumption and economic growth in 15 OECD countries from 1980 to 2012. They find that there is no causal link between nuclear energy consumption and economic growth in 10 countries. For other five countries, there is a significant causal relationship (of various directions) between nuclear energy and economic growth.

2.2 Natural Gas

In terms of domestic, household consumption, natural gas is definitely one of the cleanest and the most useful energy sources. In the empirical literature, there exists a great amount of studies investigating the relationship between natural gas consumption and economic growth across the globe (Chang et al., 2016; Dogan, 2015; Rafindadi and Ozturk, 2015; Shahbaz et al., 2011).

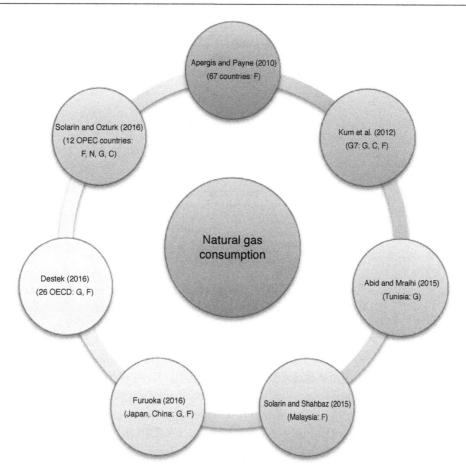

Figure 2.4: Indicative Selection of Studies (Produced From 2010 to 2016) That Deal With Disaggregated Energy Consumption: Natural Gas.
C, Conservation; F, feedback; G, growth hypothesis; N, neutrality.

Over the period 1970–2008, Kum et al. (2012) investigate the causal nexus between natural gas consumption and economic growth in G7 countries by employing bootstrap-corrected causality test. Results of causality test reveal that there exits a unidirectional relationship between natural gas consumption and growth in Italy, while there is a reverse relationship for the United Kingdom. Also, there is a bidirectional relationship between natural gas and growth, which is in line with the feedback hypothesis in France, Germany, and the United States.

Following a chronological order for an indicative selection of studies (Fig. 2.4), within a multivariate framework Apergis and Payne (2010b) investigate the nexus between natural gas consumption and economic growth in 67 countries, from 1992 to 2005. Panel vector error correction model results show a bidirectional causal relationship between natural gas consumption and economic growth both in the short and the long terms. Moreover, for the Gulf Cooperation Council (GCC) countries, Ozturk and Al-Mulali. (2015) evaluate the connection between natural gas consumption and economic growth over the period 1980–2012. Based on their dynamic panel data results (DOLS) and the fully modified least squares (FMOLS) results, natural gas consumption has a positive impact on GDP growth in the long run. Besides, Granger causality results show support for the feedback hypothesis. As far as the work by Abid and Mraihi (2015) is concerned, they examine the causal link between natural gas consumption and economic growth for Tunisia, from 1980 to 2012. The results from Granger causality test provide evidence in favor of the neutrality hypothesis in the short run whereas the long run results indicate a unidirectional causality running from energy consumption to economic growth. For Malaysia, Solarin and Shahbaz (2015) examine the linkages between natural gas consumption and economic growth from 1971 to 2012. The findings reveal that a bidirectional relationship is present, between natural gas consumption and economic growth. In other words, they find evidence for the feedback hypothesis.

More energy-growth research in the particular field of natural gas energy is done by Furuoka (2016). Using the ARDL method he investigates the relationship between natural gas and economic growth in Japan and China from 1980 to 2012. In China, the results indicate a unidirectional relationship between natural gas consumption and economic growth. In the case of China, there exits a bidirectional relationship between natural gas consumption and economic growth. In addition to this study, Destek (2016) investigates the relationship between natural gas and economic growth for 26 OECD member countries over the period 1991–2013. Dynamic panel data results and fully modified least squares results reveal that consumption of natural gas has a positive impact on GDP growth in the long term. The VECM Granger causality test shows evidence for a unidirectional causality running from natural gas to GDP growth in the short run, while there is a bidirectional causal relationship in the long run.

Solarin and Ozturk (2016) investigate the linkages between natural gas consumption and economic growth from 1980 to 2012 in 12 OPEC member countries. When selected countries are evaluated as a whole, results of panel Granger causality test reveal there is evidence for a bidirectional relationship between natural gas consumption and economic growth. When all countries are examined separately, different results are found. In Iraq, Kuwait, Libya, Nigeria, and Saudi Arabia the growth hypothesis is supported, while the conservation hypothesis is supported in Algeria, Iran, United Arab Emirates, and Venezuela. Furthermore, they find support for the presence of the neutrality hypothesis in Angola and Qatar and feedback hypothesis in Ecuador.

2.3 Various Types of Fossil Energy (Oil and Derivatives, Coal, Etc.)

Oil, its derivatives, and coal are among the most commonly used fossil energy sources. Oil, as a major production input, has played a crucial role throughout the industrial revolution up

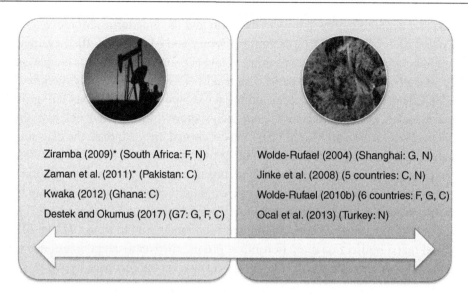

Figure 2.5: Indicative Selection of Studies (Produced From 2004 to 2017) That Deal With Disaggregated Energy Consumption: Fossil Fuel Energy.
An *asterisk* indicates studies with both oil and coal. *C*, Conservation; *F*, feedback; *G*, growth hypothesis; *N*, neutrality.

to the recent economic history. Given this importance, a high number of EGN papers have employed oil and its derivates in this literature. Referring to some of the most representative recent studies, more or less with a chronological order (Fig. 2.5), Wolde-Rufael (2004) assesses the link between disaggregated energy consumption and economic growth over the period 1952–99 for Shanghai. Using the Granger causality test, his empirical findings show a unidirectional causality running from coal consumption to economic growth, while there is no causal relationship between oil consumption and economic growth.

For major OECD and non-OECD countries, Jinke et al. (2008) investigate the causal relationship between coal consumption and economic growth from 1980–2005. Results of the Granger causality test reveal a one-sided causal relationship from GDP to coal consumption in Japan and China. Also, their results support the neutrality hypothesis for India, South Korea, and South Africa. Next, Ziramba (2009) examines the relationship between disaggregate energy consumption and industrial output over the period 1980–2005 for South Africa. This is a study that disaggregates not only at an energy type or fuel level, but also at an economic sector level. The results obtained from Granger causality analysis show a two-sided causal relationship between oil consumption and industrial production. Also, no causal relationship is observed between coal consumption and industrial production.

Another fossil energy study that disaggregates both at sectoral and energy type levels is by Zaman et al. (2011). They investigate the EGN between sectoral oil consumption and economic growth in Pakistan during the period 1972–2008. The findings reveal that oil consumption in primary sectors, such as transportation, power generation, and industry, promotes economic growth. However, oil consumption in secondary sectors, such as households, government, and agriculture, has a negative impact on economic growth. Results of Granger causality test indicate a unidirectional causality from real GDP to transport and industrial sectors. To this group of studies also belongs the study by Kwakwa (2012) who explores the causality between disaggregated energy consumption and growth in Ghana from 1971 to 2007. The results of Granger causality test reveal that a unidirectional causality from total growth to electricity and fossil fuel consumption. While growth is disaggregated, there is a unidirectional short- and long-run causality from agriculture to electricity consumption and a bidirectional causal relationship between manufacturing and electricity consumption.

The empirical literature related to coal EGN has also grown in a comprehensive way both with respect to the countries it involves as well as the methods it applies. In six major coal-consuming countries, Wolde-Rufael (2010b) investigates the causal nexus between coal consumption and real GDP from 1965 to 2005 using VAR method. The results reveal a unidirectional causal relationship from coal consumption to economic growth in India and Japan, while there is a reverse causality between variables in China and South Korea. Moreover, there exists a two-sided causal relationship between coal consumption and economic growth in South Africa and the United States.

Within the asymmetric causality framework, Ocal et al. (2013) explore the relationship between coal consumption and GDP growth for Turkey, over the period 1980–2006. The findings reveal no causal relationship between the coal consumption and GDP growth relationship in Turkey. In other words, the neutrality hypothesis did hold for Turkey during that period. Last, by employing the panel bootstrap causality approach, Destek and Okumus (2017) examine the disaggregated energy consumption and economic growth in G7 countries. The results reveal a unidirectional relationship between oil consumption and growth in Italy, Japan, and the United States, while there is a two-sided relationship in Germany and the United Kingdom, which supports the feedback hypothesis. In addition to that, the direction of causality is from growth to coal consumption in the United States, while there is an adverse relationship in Canada.

2.4 The Importance of Renewables

According to the Kyoto Protocol, countries should use renewable energy sources rather than fossil fuels to reduce their CO_2 emissions. It is well known that fossil fuels are the major contributor of CO_2 emissions and therefore, the most effective way to reduce carbon emissions is to reduce fossil fuel consumption. With at Kyoto Protocol, countries aimed at lower carbon emissions worldwide, regarding the environmental issues. Instead of burning fossil fuels, this Protocol induced signatory countries to adopt technologies that promoted renewable and clean energy resources to retard and eventually reverse the climate change problem. The number of the studies looking into the relationship between renewable energy consumption and economic growth has grown steadily since the 1990s and has expanded even more rapidly nowadays. Therefore, given that this subsection of literature is vast, it is deemed worthwhile to provide a short review of the most representative literature in chronological order, such as Payne (2009), Sadorsky (2009), Menegaki (2011), and Yildirim et al. (2012).

Thus, Bowden and Payne (2010) explore the causal linkages between sectoral renewable and non-renewable energy consumption and real GDP growth in the United States, over the period 1949–2006. The results show that a bidirectional causality exists both between commercial non-renewable energy consumption and real GDP growth and residential non-renewable energy consumption and real GDP growth. Besides, there is a unidirectional causality from residential renewable energy consumption to real GDP growth and industrial non-renewable energy consumption to real GDP growth.

Next, Apergis and Payne (2010c) investigate the renewable energy EGN from 1985 to 2005 for 20 OECD countries. Results from the Granger causality test show a two-sided causal relationship between renewable energy consumption and economic growth in both the short and the long terms. Within the multivariate panel framework, Apergis and Payne (2011) investigate renewable and non-renewable electricity consumption and economic growth in 16 emerging economies from 1990 to 2007. They find a unidirectional causal link from economic growth to renewable electricity consumption in the short term and bidirectional causal relationship in the long term. In the case of non-renewable electricity consumption, they obtain a bidirectional causality in the short and long terms.

Moreover, for the period 1990–2007, Apergis and Payne (2012) explore the connection between renewable and non-renewable energy consumption and economic growth for 80 countries, using the panel error correction model. The findings support the feedback hypothesis between renewable and non-renewable energy consumption and economic growth in both the short and long terms. Also, Tugcu et al. (2012) examine the causal and the long-run links between renewable and non-renewable energy consumption and economic growth in G7 countries for 1980–2009 using the ARDL framework. According to results of augmented production function, there is a bidirectional causal relationship between renewable energy consumption and economic growth in England and Japan. On the other hand, the neutrality hypothesis is valid for France, Italy, Canada, and the United States. Conservation hypothesis is only valid in Germany. The obtained results from classical production function show a bidirectional relationship between renewable energy consumption and economic growth in G7 countries.

Next, Ocal et al. (2013) and Ocal and Aslan (2013) investigate the between renewable EGN over the period 1990–2010 in Turkey. Results of the ARDL approach showed that renewable energy consumption had a negative impact on energy consumption. They further found a unidirectional causal ordering from economic growth to energy consumption based on Toda

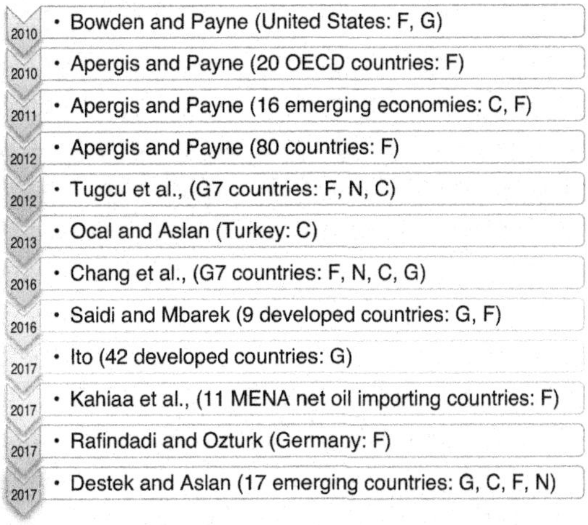

2010	• Bowden and Payne (United States: F, G)
2010	• Apergis and Payne (20 OECD countries: F)
2011	• Apergis and Payne (16 emerging economies: C, F)
2012	• Apergis and Payne (80 countries: F)
2012	• Tugcu et al., (G7 countries: F, N, C)
2013	• Ocal and Aslan (Turkey: C)
2016	• Chang et al., (G7 countries: F, N, C, G)
2016	• Saidi and Mbarek (9 developed countries: G, F)
2017	• Ito (42 developed countries: G)
2017	• Kahiaa et al., (11 MENA net oil importing countries: F)
2017	• Rafindadi and Ozturk (Germany: F)
2017	• Destek and Aslan (17 emerging countries: G, C, F, N)

Figure 2.6: Indicative Selection of Studies (Produced From 2010 to 2017) That Deal With Disaggregated Energy Consumption: Renewable Energy.
C, Conservation; F, feedback; G, growth hypothesis; N, neutrality.

Yamamoto results. More recently, within a heterogeneous panel noncausality framework, Chang et al. (2016) explore the causal nexus between renewable energy consumption and economic growth in G7 countries for the years 1965–2011. The results indicated a bidirectional causal relationship for the pooled panel. At the cross-section level, there was no causal link for Canada, Italy, and the United States; while unidirectional causal relationship from GDP to renewable energy was found for France and United Kingdom. In the case of Germany and Japan, the direction of causal relationship was from renewable energy to growth.

For the period 1990–2013, Saidi and Mbarek (2016) examine the causal relationship between nuclear energy consumption, CO_2 emissions, renewable energy, and real GDP per capita in nine developed countries. The results from a dynamic panel indicated a unidirectional causal link running from renewable energy consumption to real GDP per capita in the short term. In addition to that, there is a bidirectional relationship between renewable energy consumption to real GDP per capita in the long run, which means that the feedback hypothesis applied (Fig. 2.6).

Closing with some of the most recent studies, using the panel data of 42 developed countries, Ito (2017) investigates the relationship between CO_2 emissions, renewable and non-renewable energy consumption, and economic growth for the period 2002–11. According to his results, non-renewable energy consumption affects economic growth negatively, while renewable energy consumption caused a positive impact on economic growth in the long term. Furthermore, within a multivariate panel framework, Kahiaa et al. (2017) investigate the relationship between disaggregate energy consumption (renewable and non-renewable) and economic growth over the period 1980–2012 in 11 MENA, net oil importing countries. Obtaining results from an error correction model showed a bidirectional causal relationship between renewable energy consumption, non-renewable energy consumption, and economic growth.

During the period 1971 Q1 to 2013 Q4, Rafindadi and Ozturk (2017) examine the impact of economic gains in renewable energy use on economic growth in Germany. Results from a VECM Granger causality test confirm the presence of a bidirectional relationship between renewable energy consumption and economic growth. Last, using the bootstrap panel causality approach, Destek and Aslan (2017) explore the impact of renewable and non-renewable consumption on the economic growth for 17 emerging economies over the period 1980–2012. The results showed a unidirectional relationship renewable energy consumption to economic growth in Peru whereas there is a unidirectional relationship from growth to energy consumption in Colombia and Thailand. Moreover, a bidirectional relationship is found for Greece and South Korea. They also find evidence in favor of the neutrality hypothesis for other 12 emerging economies.

2.5 The Flourishing Interest in Electricity

According to the Energy Information Administration (EIA), electricity is the flow of electrical power or charge, which is both a basic part of nature and one of the most widely used forms of energy. As the electricity is produced by converting primary sources of energy such as coal, natural gas, nuclear energy, solar energy, and wind energy into electrical power, it is a secondary energy source.[3] In addition to this, electricity is also a crucial part of daily life in the United States as well as the whole world. The main usage areas of electricity are lighting, heating, cooling, refrigeration, and for operating appliances, computers, electronics, machinery, and public transportation systems, etc.[4]

Electricity, which plays a major role in an economy in either production or consumption side, has been widely used in the energy-growth literature as a proxy for energy (Ghosh, 2002; Ho and Siu, 2007; Jumbe, 2004; Odhiambo, 2009; Pempetzoglou, 2012; Ramcharran, 1990; Shiu and Lam, 2004; Solarin, 2011; Tang and Tan, 2012; Yoo, 2005).

Within a multivariate framework, Narayan and Singh (2007) explore the casual relationship between electricity consumption and economic growth in Fiji. According to their results of Granger causality test, there exits a long-run unidirectional causality from electricity consumption to GDP growth. In another study Narayan and Prasad (2008) investigate the causal relationship between electricity consumption and real GDP growth using a bootstrap causality testing technique for 30 OECD countries. Electricity consumption Granger causes real GDP growth in Australia, Iceland, Italy, the Slovak Republic, the Czech Republic, Korea, Portugal, and the United Kingdom. Later, for a panel of Middle East countries, for the years 1974–2002, Narayan and Smyth (2009) investigate the causality between electricity consumption and GDP growth. At a panel level, they confirm a bidirectional relationship between the variables they used.

---

[3] See: https://www.eia.gov/energyexplained/index.cfm?page=electricity_home.

[4] See: https://www.eia.gov/energyexplained/index.cfm?page=electricity_use.

Enriching their multivariate framework, Shahbaz and Ozturk (2012) investigate the relationship between electricity consumption and economic growth by adding financial development, capital, and labor into their model for Turkey. Empirical results covering the period 1971–2009 indicated that electricity consumption has a positive impact on economic growth. In addition to that, a bidirectional causality is found between electricity consumption and economic growth according to VECM results.

In the case of Turkey, Nazlıoğlu et al. (2014) investigate the causal nexus between electricity consumption and economic growth from 1967 to 2007 by using linear and nonlinear Granger causality tests. Findings from the linear Granger causality test reveal a bidirectional causal relationship in the short and the long terms. On the contrary, results of nonlinear causality test indicate that there is no causal relationship between electricity consumption and economic growth. Also, Aslan et al. (2014b) for the United States, in the years 1973 Q1 to 2012 Q1 investigated the linkage between electricity consumption and economic growth and concluded that there is a long-run relationship between the variables. In terms of causality, a bidirectional causality is found in the long run, while no relationship is detected in the short run. Last, Rodríguez-Caballeroa and Ventosa-Santaulària (2016) investigate casual connection between Electric Power Consumption and GDP growth for Canada, United States, and 17 Latin American countries for the years 1971–2011. The results support the growth hypothesis for eight countries while conversation hypothesis is valid in three countries. Also, results support the presence of neutrality hypothesis for three countries (Fig. 2.7).

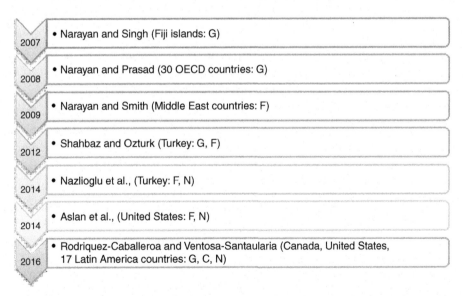

Figure 2.7: Indicative Selection of Studies (Produced From 2007 to 2016) That Deal With Disaggregated Energy Consumption: Electricity.
C, Conservation; *F*, feedback; *G*, growth hypothesis; *N*, neutrality.

3 Disagreggation Based on Economic Sectors: A Literature Map

Another branch of the disaggregated energy consumption-growth studies is based on a sectoral decomposition of the economy. This literature on the connection between disaggregated energy consumption, based on energy sources and economic growth includes a wide range of empirical studies focusing on different countries with different time periods, through the use of different techniques and variables. Although a large number of studies exists on the relationship between aggregated and disaggregated (based on energy sources) energy consumption and economic growth, the number of studies looking into the nexus at the sectoral level is very limited. Providing extensive evidence from sectoral analysis is expected to contribute substantially to the existing empirical literature from the policy-making ability point of view. Therefore, this chapter looks into literature investigation for five economic sectors, that is, industry, commercial, transportation, residential, and the electric power sectors.

3.1 The Transportation Sector

According to the International Energy Outlook 2016 (IEO2016) Reference Case, the transportation sector is an energy-intensive industry. Its energy consumption increased by an annual average rate of 1.4% from 2012 and projects to 155 quadrillion Btu[5] in 2040. Also, an increase in the demand for personal transportation, with an improvement in living standards, has increased total energy consumption. Due to the importance of this fact, the nexus between energy consumption and economic growth in the transportation sector has grown steadily over time.

The literature on the impact of energy consumption on economic growth, in the transportation sector, has yielded mixed results. In the case of G7, Zachariadis (2007) investigates the nexus between sectoral energy consumption and energy growth over the period 1949 to 2004. He finds evidence in favor of the conversation hypothesis, which refers to a unidirectional relationship from GDP to the energy consumption of transport sectors. Costantini and Martini (2010) examine the causal relationship between energy consumption and economic growth for a huge number of developing and developed countries. According to their results, there is no relationship between energy consumption and economic growth in transportation sector for all samples. Moreover, Sultan (2011) examines the causal relationship between economic growth and energy for transport in Mauritius for the time period of 1970–2010, by employing the ARDL cointegration technique. The results indicate a unidirectional causality from economic growth to transport energy in the long run.

---

[5] British Thermal Unit (Btu) equal to about 1055 J.

Furthermore, Abid and Sebri (2012), in Tunisia, examine the causal nexus between energy consumption and economic growth in the transport sector from 1980 to 2007 within a VECM framework. The results indicate that there is no causal link between variables. Zaman et al. (2011) investigate the relationship between sectoral oil consumption and economic growth over the period 1972–2008 in Pakistan, using time series methods. According to results, the oil consumption in the transportation sector has a positive impact on Pakistan's economic growth. Results of the Granger causality test also indicate that there is no causal relationship between these variables. Last, using the ARDL bounds testing approach, Gross (2012) examines the causality between energy consumption and economic growth from 1970 to 2007 in the United States. He found evidence in favor of the feedback hypothesis in the transportation sector.

3.2 The Commercial Sector

The commercial energy use takes place in both profit-seeking and nonprofit enterprises that are confronted with commercial-scale activities. These activities are often the service sector. Energy used for commercial sector buildings, consists of retail stores, office buildings, government buildings, restaurants, hotels, schools, hospitals, leisure, and recreational facilities.[6] The energy use by the commercial sector is mostly dominated by the building-related consumption. It also includes energy consumption for street and other outdoor lighting, and for water and sewage treatments, which however are relatively smaller contributors to the commercial sector's total energy consumption.[7] According to the IEO2016 Reference Case, worldwide delivered energy consumption of the commercial sector increased at an average of 1.6% per year from 2012 to 2040. Commercial sector is the fastest-growing energy demand sector.[8]

Within a multivariate framework, Bowden and Payne (2009) investigate the causal link between sectoral primary energy consumption and economic growth in the United States over the period 1949–2006. The results of Granger causality test confirm that there is a bidirectional relationship between the commercial primary energy consumption and economic growth. In the case of the United States, Gross (2012) in the same country, examined the causal relationship between energy consumption and economic growth from 1970 to 2007 using the ARDL bounds testing approach. The results indicate a unidirectional long-run causal relationship from economic growth to energy consumption in the commercial sector.

Continuing the bibliographic survey in a chronological order, Sami (2012) investigates the causality between the electricity usage from the commercial sector and economic growth

---

[6] See: https://www.eia.gov/outlooks/ieo/buildings.cfm.

[7] The content of this industry is consistent with the information from the US Energy Information Administration (EIA, https://www.eia.gov/tools/faqs/faq.php?id=86&t=1).

[8] See: https://www.eia.gov/outlooks/ieo/buildings.cfm.

over the period 1973–2008 in Philippines, using a bounds testing framework. He finds a unidirectional causality running from electricity consumption to economic growth, which supports the growth hypothesis. In another publication the same year, Zhang and Xu (2012) investigate the causality between energy consumption of services (commercial) sector and the economic growth over the period 1995–2008. They find a unidirectional casual relationship from economic growth to energy consumption in China.

Tang and Shahbaz (2013) investigate the causal nexus between the electric consumption of the commercial sector (services) and real output in Pakistan over the period 1972–2010. The results show a unidirectional causality from electricity consumption to real output service sectors. For the period 1970–2009, Saunoris and Sheridan (2013) investigate the causal link between electricity consumption in the commercial sector and the economic growth in 48 US states. They find support in the favor of the conversation hypothesis for the commercial sector both for the short and the long runs.

3.3 The Industry Sector

The energy-intensive industrial sector uses energy for different purposes, such as steam, heating, cooling, process, and air conditioning. The industrial sector can generally be separated into three groups: the energy-intensive manufacturing group of subsectors, nonenergy-intensive manufacturing group of subsectors, and the nonmanufacturing group of subsectors. According to the IEO2016 Reference Case, the energy consumption of the industrial sector across the globe is predicted to increase at an average of 1.2% per year, from 222 quadrillion Btu in 2012 to 309 quadrillion Btu in 2040.[9]

This survey of some representative studies of the industrial sector with chronological order starts with Jobert and Karanfil (2007), who investigate the relationship between energy consumption and economic growth, over the period 1960–2003, in Turkey. They examine this relationship, at the whole economy level, as well as the industrial sector. The findings indicate the existence of the neutrality hypothesis, either at the aggregate or at the industrial level. It is also found that contemporaneous values of energy consumption and income are correlated. Next, within a multivariate framework, Bowden and Payne (2009) investigate the causal relationship between sectoral primary energy consumption and growth for United States over the period 1949–2006. They find evidence in favor of the growth hypothesis between the industrial primary energy consumption and real GDP. In the case of Greece, Tsani (2010) examines the causal nexus between disaggregated energy consumption and economic growth from 1960 to 2006, using time series techniques. Results reveal that a bidirectional relationship exists between industrial energy consumption and real GDP.

---

[9] See: https://www.eia.gov/outlooks/ieo/pdf/industrial.pdf.

Cheng-Lang et al. (2010) examine the linear and nonlinear causal nexus between industrial sector electricity consumption and the real gross domestic production for Taiwan. In the case of linear Granger causality test, they found evidence in favor of a bidirectional relationship between the electricity consumption of the industrial sector and the real GDP. Afterward, Zaman et al. (2011), explore causality, in Pakistan, between the industrial oil consumption and economic growth during the period 1972–2008, employing time series methods. The results revealed that the industrial sector oil consumption has a positive impact on Pakistan's growth. In addition to that, the obtained findings from Granger causality test show that there is no causal relationship between these variables.

The year 2012 has spawned numerous studies on sectoral EGN. Over the period 1973–2008, Sami (2012) investigates the causal nexus between the industrial sector electricity consumption and economic growth in Philippines, employing the bounds testing method. He finds a unidirectional causality between electricity consumption and economic growth. The direction of causality is from economic growth to industrial sector energy consumption. Zhang and Xu (2012) investigate the causal nexus between energy consumption and economic growth over the period 1995–2008, in the case of China. The obtained results reveal that the economic growth has a positive impact on economic growth, both at national and sectoral levels. In the case of the industrial sector, there is a bidirectional relationship between energy consumption and economic growth in China.

In the case of the Canadian industrial sector, Hamit-Haggar (2012) examine the causal link between greenhouse gas emissions, energy consumption, and economic growth from 1990 to 2007. The empirical results indicate that there is a weak unidirectional relationship from economic growth to energy consumption in the short run. Using the Granger causality test, Abid and Sebri (2012) investigate the casual link between the industry sector energy consumption and the economic growth in Tunisia, from 1980 to 2007. Results reveal a unidirectional causal link from economic growth to energy consumption in the short run, while there is no relationship in the long run. Moreover, Gross (2012) explores the causality between energy consumption and economic growth from 1970 to 2007 in the United States. The results from an ARDL bounds testing approach indicate a unidirectional long-run causal relationship from energy consumption to economic growth in the industrial sector.

Closing the review with studies from 2013, for 48 US states, Saunoris and Sheridan (2013) investigate the causal link between electricity consumption for the industry sector and economic growth during the period 1970–2009. The findings show a unidirectional causality from energy consumption to economic growth in the short run, while there is an adverse relationship in the long run. Last, Tang and Shahbaz (2013) examine the causal nexus between the industrial sector electric consumption and real output for Pakistan, over the period 1972–2010. The results confirmed the validity of the growth hypothesis.

3.4 The Residential Sector

Likewise for the commercial sector, energy use from the residential sector is mostly building related. As stated by Swan and Ugursal (2009), the residential sector is a substantial consumer of energy in every country, given its complex and interrelated characteristics. In the IEO2016 Reference Case, the projected energy consumed by the residential sector is explained to be approximately by 13% of the total world delivered energy consumption in 2040. The delivered energy consumption of the residential sector will increase by an average of 1.4% per year from 2012 to 2040.[10] In the United States, Bowden and Payne (2009) examine the causal nexus between sectoral primary energy consumption and economic growth within a multivariate framework. They used data spanning from 1949 to 2006. The Granger causality results support the validity of the feedback hypothesis between the residential primary energy consumption and real GDP.

Costantini and Martini (2010) investigate the causality between energy consumption and economic growth for OECD and non-OECD countries. They obtained different results for two global samples. Over the period 1960–2006, Tsani (2010) examines the causal link between the residential energy consumption and the economic growth in Greece, using time series techniques. She found evidence for the feedback hypothesis. In the same year another publication by Cheng-Lang et al. (2010) examine the linear and nonlinear causality between the residential sector electricity consumption and real GDP in Taiwan. Findings of the linear Granger causality test revealed that there was no causal relationship between the residential sector electricity consumption and real GDP. In the case of nonlinear causality, they found support in favor of a unidirectional causality from electricity consumption of the residential sector and real GDP.

Sami (2012) assesses the causal nexus between the residential sector electricity consumption and economic growth from 1973 to 2008, in Philippines, using bounds testing method. He found a unidirectional casual relationship from economic growth to electricity consumption in this sector. This means that the conversation hypothesis applied. Using the Granger causality test, Abid and Sebri (2012) investigate the casual link between the residential sector energy consumption and economic growth in Tunisia, from 1980 to 2007. Findings prove that there is a bidirectional relationship between the residential sector energy consumption and the household income, in the short run, whereas there was a unidirectional relationship from household income to energy consumption in the long run. Last, for 48 US states, Saunoris and Sheridan (2013) investigate the causal link between the residential sector electricity consumption and economic growth for the period 1970–2009.The results proved that the conversation hypothesis was valid for the residential sector both in the short and the long runs (Fig. 2.8).

---

[10] See: https://www.eia.gov/outlooks/ieo/buildings.cfm.

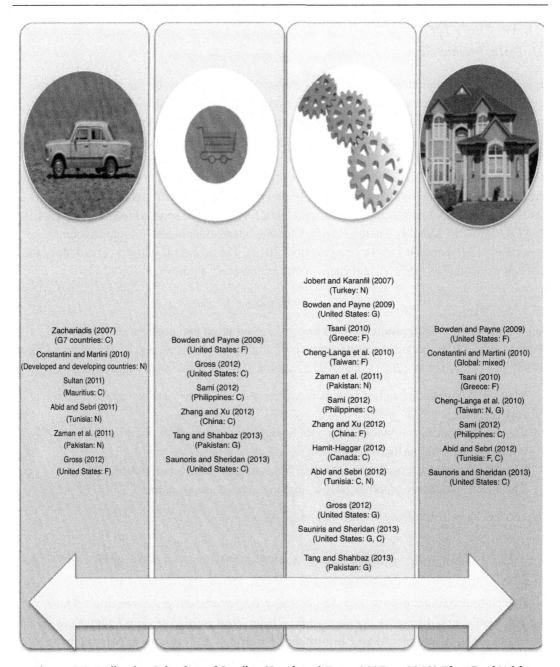

Figure 2.8: Indicative Selection of Studies (Produced From 2007 to 2013) That Deal With Disaggregated Energy Consumption: By Economic Sectors.
C, Conservation; F, feedback; G, growth hypothesis; N, neutrality.

4 Case Study for the United States

4.1 Data, Methodology, and Results

To investigate the relationship between disaggregated EGN, we focus on the US data over the period 1962–2015. The data set includes GDP in constant 2010 US dollars, which represents economic growth; total primary energy consumption; total energy consumption of the residential, commercial, industrial, electric power, and transportation sectors; gross fixed capital formation in constant 2010 US dollars, which stands for the capital variable and labor force (15+) that is measured as the number of workers over the age of 15. GDP and gross fixed capital formation data are collected from the World Bank (World Development Indicators Database) (World Bank, 2017). Energy consumption and labor force data are sourced from the US EIA (2017) (April 2017 Monthly Energy Review) and the International Labour Organization, respectively (International Labour Organization, 2017). The economic growth (y) is defined as a function of capital (k), labor (l), and energy consumption (ec) as shown in Eq. (2.1):

$$y = f(k,l,ec) \qquad (2.1)$$

To investigate the cointegration between energy consumption and economic growth, we employed the ARDL bounds testing approach of cointegration developed by Pesaran and Shin (1999) and Pesaran et al. (2001). The ARDL approach has a wide range of advantages:

1. The ARDL approach can be implemented irrespective of whether the regressors are I(1) and/or I(0).
2. The ARDL approach is statistically a more significant approach to determine the cointegration relationship in small samples.
3. The ARDL approach permits that the variables may have different optimal lags.
4. The ARDL procedure employs only a single reduced form equation.

The model can be formulated as is shown in Eq. (2.2):

$$\Delta \mathrm{In} y_t = \alpha_0 + \sum_{i=1}^{k} \alpha_1 \Delta \mathrm{In} y_{t-i} + \sum_{i=0}^{l} \alpha_2 \Delta \mathrm{In} ec_{t-i} + \alpha_3 \mathrm{In} y_{t-1} + \alpha_4 \mathrm{In} ec_{t-1} + u_t \qquad (2.2)$$

where y is the economic growth, ec is the vector of energy consumption variable.[11] The symbol Δ is the first difference operator while u is the error term. The term k represents the lag length for each variable. Akaike Information Criterion (AIC) is used to determine the appropriate lag length.

In Eq. (2.2), α_1 and α_2 represent the short-term dynamics; α_3 and α_4 represent the long-term dynamics. The null hypothesis in Eq. (2.2) is $\alpha_3 = \alpha_4 = 0$, which implies that no long-run

---

[11] Total primary energy consumption and total energy consumptions by the residential, commercial, industrial, electric power, and transportation sectors.

Table 2.1: ARDL bound test results.

Models	Sectors	F-Statistics	χ^2BG	χ^2ARCH
Model 1	TPEC	14.775***	0.518 (0.59)	0.223 (0.63)
Model 2	CSEC	8.379***	0.624 (0.54)	0.319 (0.57)
Model 3	EPSEC	13.490***	0.487 (0.61)	0.317 (0.57)
Model 4	ISEC	24.541***	0.117 (0.88)	0.072 (0.78)
Model 5	RSEC	8.855***	0.529 (0.59)	0.394 (0.53)
Model 6	TSEC	13.240***	0.213 (0.80)	0.343 (0.56)
Critical Value Bounds				
		I0 Bound (Lower Bond)	I1 Bound (Upper Bond)	
%1		3.65	4.66	
%5		2.79	3.67	
%10		2.37	3.2	

F-statistic refers to cointegration analysis.
χ^2BG represents the diagnostic tests of Breusch–Godfrey Serial Correlation LM test.
χ^2ARCH represents the diagnostic tests of ARCH heteroskedasticity test.
Probability values are in the parentheses.
Triple asterisks (***) indicate significance at 1%.

relationship exists between the variables. If the *F*-test statistic overcomes the upper critical value (I1) derived from Pesaran et al. (2001), the null hypothesis of no cointegration relationship can be rejected.

The ARDL bound test results for each model are presented in Table 2.1. According to Table 2.1, there is a long-run relationship among the variables. Breusch–Godfrey Serial Correlation LM test and ARCH heteroskedasticity test results indicate no serial correlation and heteroskedasticity in all models, respectively.

Table 2.2 reports the long- and short-run coefficients for each model. The estimated long-run energy consumption coefficients are not statistically significant in all models except for Model 3, which represents the energy power sector. In the energy power sector, energy

Table 2.2: ARDL estimation results.

Dependent Variable: y	Total Primary Energy	Commercial Sector	Electric Power Sector	Industrial Sector	Residential Sector	Transportation Sector
Long-run coefficients						
k	0.702 (0.00)***	0.735 (0.00)***	0.582 (0.00)***	0.681 (0.00)***	0.795 (0.00)***	0.632 (0.00)***
l	0.496 (0.10)*	0.533 (0.27)	1.339 (0.03)**	0.152 (0.33)	0.299 (0.49)	0.873 (0.07)*
ec	−0.361 (0.28)	−0.304 (0.40)	−0.433 (0.08)*	0.106 (0.39)	−0.499 (0.34)	−0.487 (0.25)
Short-run coefficients						
k	0.195 (0.00)***	0.255 (0.00)***	0.214 (0.00)***	0.173 (0.00)***	0.275 (0.00)***	0.200 (0.00)
l	−0.340 (0.10)*	−0.577 (0.03)**	0.462 (0.05)**	−0.523 (0.02)**	−0.562 (0.04)**	−0.321 (0.14)
ec	0.338 (0.00)***	0.178 (0.00)***	0.289 (0.00)***	0.261 (0.00)***	0.038 (0.41)	0.329 (0.00)***

Numbers in parentheses are probability values (*P*-values).
The economic growth (*y*) is defined as a function of capital (*k*), labor (*l*), and energy consumption (*ec*).
Single, double, and triple asterisks (*,**,***) represent significance at 10, 5, and 1%, respectively.

consumption has a negative impact on economic growth at a 10% significance level. In terms of the short-run coefficients, energy consumption has a positive impact on economic growth except for the residential sector. In the residential sector, the short-term energy consumption coefficient is not statistically significant. Total primary energy consumption also has a positive impact on the economic growth in the short run.

When we investigated the short- and long-term coefficients of control variables (*k* and *l*), we find for all models that capital has a positive impact on growth in the long run. Capital also affects economic growth positively with the exception of the transportation sector in the short run. The labor variable is statistically significant in the electric power and the transportation sectors in the long run. This finding is valid for Model 1 where energy consumption is proxied by total primary energy consumption. In the short run, labor is statistically significant except for the transportation sector. Labor has a negative impact on the economic growth in the short run in the commercial, industrial, and residential sectors, while there is a positive impact in the electric power sector. Model 1 also shows that labor has a negative impact on economic growth in the short run.

5 Conclusions

This chapter describes the various levels and types of disaggregation in the EGN and provides selected recent, chronologically ordered, literature examples from the main types of disaggregation: energy and sectoral. The formers include: nuclear energy, natural gas, various fossil, renewable, and electricity. The latter encompass: transport, commercial, industry, and residential sectors. The ideal but most difficult type of analysis in the EGN is the disaggregated, both at a sectoral level and at an energy level. Dearth of data can be a hindrance for this analysis, particularly in the underdeveloped countries. In the case advanced countries such as the United States as a typical example, however, disaggregated data is mostly available. Therefore, we investigate the impact of both total and sectoral energy consumption on economic growth in the United States over the period 1962–2015 using the ARDL bound test. The ARDL bound test results reveal a cointegration relation among the variables both at total and sectoral levels. In addition, total primary energy consumption and sectoral energy consumption do not have an impact on economic growth except for electric power sector. Electric power sector energy consumption affects growth negatively. In the short run, energy consumption has a positive impact on economic growth both at total and sectoral levels, with the exception of the residential sector. Overall, results as whole indicate that energy consumption by sectors might have different impacts on economic growth unlike the aggregate impact of total energy consumption. Thus, this study emphasizes the importance of utilization disaggregated data once available by referring to the aggregation bias in the energy economics.

References

Abbas, F., Choudhury, 2013. Electricity consumption-economic growth nexus: an aggregated and disaggregated causality analysis in India and Pakistan. J. Policy Model. 35, 538–553.

Abid, M., Mraihi, 2015. Disaggregate energy consumption versus economic growth in Tunisia: cointegration and structural break analysis. J. Knowledge Econ. 6 (4), 1104–1122.

Abid, M., Sebri, M., 2012. Energy consumption-economic growth nexus: does the level of aggregation matter? Int. J. Energy Econ. Policy 2 (2), 55–62.

Al-Mulali, 2014. Investigating the impact of nuclear energy consumption on GDP growth and CO_2 emission: a panel data analysis. Progr. Nucl. Energy 73 (2014), 172–178.

Apergis, N., Danuletiu, D., 2012. Energy consumption and growth in Romania: evidence from a panel error correction model. Int. J. Energy Econ. Policy 2 (4), 346–348.

Apergis, N., Payne, J.E., 2009a. Energy consumption and economic growth in Central America: evidence from a panel cointegration and error correction model. Energy Econ. 31 (2), 211–216.

Apergis, N., Payne, J.E., 2009b. CO_2 emissions, energy usage, and output in Central America. Energy Policy 37 (8), 3282–3286.

Apergis, N., Payne, J.E., 2009c. Energy consumption and economic growth: evidence from the Commonwealth of Independent States. Energy Econ. 31 (5), 641–647.

Apergis, N., Payne, J.E., 2010a. A panel study of nuclear energy consumption and economic growth. Energy Econ. 32 (3), 545–549.

Apergis, N., Payne, J.E., 2010b. Natural gas consumption and economic growth: a panel investigation of 67 countries. Appl. Energy 87 (8), 2759–2763.

Apergis, N., Payne, J.E., 2010c. Renewable energy consumption and economic growth: evidence from a panel of OECD countries. Energy Policy 38 (1), 656–660.

Apergis, N., Payne, J.E., 2011. Renewable and non-renewable electricity consumption–growth nexus: evidence from emerging market economies. Appl. Energy 88 (12), 5226–5230.

Apergis, N., Payne, J.E., 2012. Renewable and non-renewable energy consumption-growth nexus: evidence from a panel error correction model. Energy Econ. 34, 733–738.

Apergis, N., Tang, C.F., 2013. Is the energy-led growth hypothesis valid? New evidence from a sample of 85 countries. Energy Econ. 38, 24–31.

Aslan, A., Apergis, N., Topcu, M., 2014a. Banking development and energy consumption: evidence from a panel of Middle Eastern countries. Energy 72, 427–433.

Aslan, A., Apergis, N., Yildirim, S., 2014b. Causality between energy consumption and GDP in the U.S.: evidence from wavelet analysis. Front. Energy 8 (1), 1–8.

Aslan, A., Cam, S., 2013. Alternative and nuclear energy consumption–economic growth nexus for Israel: evidence based on bootstrap-corrected causality tests. Progr. Nucl. Energy 62, 50–53.

Bernard, O.A., Kenneth, O.O., 2016. Sectoral consumption of non-renewable energy and economic growth in Nigeria. Int. J. Res. Manag. Econ. Comm. 6 (7), 15–22.

Bowden, N., Payne, J.E., 2009. The causal relationship between U.S. energy consumption and real output: a disaggregated analysis. J. Policy Model. 31 (2), 180–188.

Bowden, N., Payne, J.E., 2010. Sectoral analysis of the causal relationship between renewable and non-renewable energy consumption and real output in the U.S. Energy Sour. B 5, 400–408.

Chang, T., Gupta, R., Inglesi-Lotz, R., Masabala, L.S., Simo-Kengne, B.D., Weideman, J.P., 2016. The causal relationship between natural gas consumption and economic growth: evidence from the G7 countries. Appl. Econ. Lett. 23 (1), 38–46.

Cheng-Lang, Y., Lin, H.P., Chang, C.H., 2010. Linear and nonlinear causality between sectoral electricity consumption and economic growth: evidence from Taiwan. Energy Policy 38 (11), 6570–6573.

Cole, M.A., 2006. Does trade liberalization increase national energy use? Econ. Lett. 92 (1), 108–112.

Costantini, V., Martini, C., 2010. The causality between energy consumption and economic growth: a multisectoral analysis using non-stationary cointegrated panel data. Energy Econ. 32, 591–603.

Destek, M.A., 2016. Natural gas consumption and economic growth: panel evidence from OECD countries. Energy 114 (2016), 1007–1015.

Destek, M.A., Aslan, A., 2017. Renewable and non-renewable energy consumption and economic growth in emerging economies: evidence from bootstrap panel causality. Renew. Energy 111, 757–763.

Destek, M.A., Okumus, I., 2017. Disaggregated energy consumption and economic growth in G-7 countries. Energy Sources 12 (9), 808–814, Part B.

Dogan, E., 2015. Revisiting the relationship between natural gas consumption and economic growth in Turkey. Appl. Econ. Lett. 10 (4), 361–370.

Fan, J.L., Zhang, Y.J., Wang, B., 2017. The impact of urbanization on residential energy consumption in China: an aggregated and disaggregated analysis. Renew. Sust. Energy Rev. 75, 220–233.

Franco, S., Mandla, V.R., Rao, K.R.M., 2017. Urbanization, energy consumption and emissions in the Indian context: a review. Renew. Sust. Energy Rev. 71, 898–907.

Furuoka, F., 2016. Natural gas consumption and economic development in China and Japan: an empirical examination of the Asian context. Renew. Sust. Energy Rev. 56, 100–115.

Ghosh, S., 2002. Electricity consumption and economic growth in India. Energy Policy 30, 125–129.

Gross, C., 2012. Explaining the (non-) causality between energy and economic growth in the U.S.A multivariate sectoral analysis. Energy Econ. 34 (2), 488–499.

Hajko, V., 2017. The failure of energy-economy nexus: a meta-analysis of 104 studies. Energy 125, 771–787.

Halicioglu, F., 2009. An econometric study of CO_2 emissions, energy consumption, income and foreign trade in Turkey. Energy Policy 37 (3), 1156–1164.

Hamit-Haggar, M., 2012. Greenhouse gas emissions, energy consumption and economic growth: a panel cointegration analysis from Canadian industrial sector perspective. Energy Econ. 34 (1), 358–364.

Hamzacebi, C., 2007. Forecasting of Turkey's net electricity energy consumption on sectoral bases. Energy Policy 35, 2009–2016.

Ho, C.-Y., Siu, K.W., 2007. A dynamic equilibrium of electricity consumption and GDP in Hong Kong: an empirical investigation. Energy Policy 35 (4), 2507–2513.

Hu, J.L., Lin, C.H., 2008. Disaggregated energy consumption and GDP in Taiwan: a threshold cointegration analysis. Energy Econ. 30, 2342–2358.

International Labour Organization, 2017. Available from: www.ilo.org.

Ito, K., 2017. CO_2 emissions, renewable and non-renewable energy consumption, and economic growth: evidence from panel data for developing countries. Int. Econ. 151, 1–6.

Jinke, L., Hualing, S., Dianming, G., 2008. Causality relationship between coal consumption and GDP: difference of major OECD and non-OECD countries. Appl. Energy 85, 421–429.

Jobert, T., Karanfil, F., 2007. Sectoral energy consumption by source and economic growth in Turkey. Energy Policy 35, 5447–5456.

Jumbe, C.B.L., 2004. Cointegration and causality between electricity consumption and GDP: empirical evidence from Malawi. Energy Econ. 26, 61–68.

Kahiaa, M., Aïssaa, M.S.B., Lanouarb, C., 2017. Renewable and non-renewable energy use -economic growth nexus: the case of MENA Net Oil Importing Countries. Renew. Sust. Energy Rev. 71 (2017), 127–140.

Kraft, J., Kraft, A., 1978. On the relationship between energy and GNP. J. Energy Dev. 3, 401–403.

Kum, H., Ocal, O., Aslan, A., 2012. The relationship among natural gas energy consumption, capital and economic growth: bootstrap-corrected causality tests from G-7 countries. Renew. Sust. Energy Rev. 16 (5), 2361–2365.

Kwakwa, P.A., 2012. Disaggregated energy consumption and economic growth in Ghana. Int. J. Energy Econ. Policy 2 (1), 34–40.

Menegaki, A.N., 2011. Growth and renewable energy in Europe: a random effect model with evidence for neutrality hypothesis. Energy Econ. 33, 257–263.

Mucuk, M., Sogozu, İ.H., 2011. Sectoral energy consumption and economic growth nexus in Turkey. Energy Educ. Sci. Technol. B 3 (4), 441–448.

Narayan, P.K., Popp, S., 2012. The energy consumption-real GDP nexus revisited: empirical evidence from 93 countries. Econ. Model. 29 (2), 303–308.

Narayan, P.K., Prasad, A., 2008. Electricity consumption–real GDP causality nexus: evidence from a bootstrapped causality test for 30 OECD countries. Energy Policy 36, 910–918.

Narayan, P.K., Singh, B., 2007. The electricity consumption and GDP nexus for the Fiji Islands. Energy Econ. 29 (6), 1141–1150.

Narayan, P.K., Smyth, R., 2008. Energy consumption and real GDP in G7 Countries: new evidence from panel cointegration with structural breaks. Energy Econ. 30 (5), 2331–2341.

Narayan, P.K., Smyth, R., 2009. Multivariate Granger causality between electricity consumption exports and GDP: evidence from a panel of Middle Eastern Countries. Energy Policy 37 (1), 229–236.

Naser, H., 2014. Oil market, nuclear energy consumption and economic growth: evidence from emerging economies. Int. J. Energy Econ. Policy 4 (2), 288–296.

Nazlıoğlu, S., Kayhan, S., Adıgüzel, U., 2014. Electricity consumption and economic growth in Turkey: cointegration, linear and nonlinear Granger causality. Energy Sour. B 9 (4), 315–324.

Nazlıoğlu, S., Lebe, F., Kayhan, S., 2011. Nuclear energy consumption and economic growth in OECD countries: cross-sectionally dependent heterogeneous panel causality analysis. Energy Policy 39, 6615–6621.

Ocal, O., Aslan, A., 2013. Renewable energy consumption–economic growth nexus in Turkey. Renew. Sustain. Energy Rev. 28, 494–499.

Ocal, O., Ozturk, I., Aslan, A., 2013. Coal consumption and economic growth in Turkey. Int. J. Energy Econ. Policy 3 (2), 193–198.

Odhiambo, N.M., 2009. Electricity consumption and economic growth in South Africa: a trivariate causality test. Energy Econ. 31 (5), 635–640.

Oh, W., Lee, K., 2004. Energy consumption and economic growth in Korea: testing the causality relation. J. Policy Model. 26 (8–9), 973–981.

Ozcan, B., Ari, A., 2015. Nuclear energy consumption-economic growth nexus in OECD: a bootstrap causality test. Proc. Econ. Finance 30, 586–597.

Ozturk, I., Al-Mulali, 2015. Natural gas consumption and economic growth nexus: Panel data analysis for GCC countries. Renew. Sust. Energy Rev. 51 (2015), 998–1003.

Ozturk, I., Aslan, A., Kalyoncu, H., 2010. Energy consumption and economic growth relationship: evidence from panel data for low and middle income countries. Energy Policy 38 (8), 4422–4428.

Paul, S., Bhattacharya, R.N., 2004. Causality between energy consumption and economic growth in India: a note on conflicting results. Energy Econ. 26 (6), 977–983.

Payne, J.E., 2009. On the dynamics of energy consumption and output in the US. Appl. Energy 86 (4), 575–577.

Pempetzoglou, M., 2012. Electricity consumption and economic growth: a linear and nonlinear causality investigation for Turkey. Int. J. Energy Econ. Policy 4 (2), 263–273.

Pesaran, M.H., Shin, Y., 1999. An autoregressive distributed lag modelling approach to cointegration analysis. In: Steinar, S. (Ed.), Econometrics and Economic Theory in the 20th Century: The Ragnar Frish Centennial Symposium. Cambridge University Press, Cambridge, pp. 371–413.

Pesaran, M.H., Shin, Y., Smith, R.J., 2001. Bounds testing approaches to the analysis of level relationships. J. Appl. Econ. 16, 289–326.

Pirlogea, C., Cicea, C., 2012. Econometric perspective of the energy consumption and economic growth relation in European Union. Renew. Sust. Energy Rev. 16 (8), 5718–5726.

Rafindadi, A.A., Ozturk, I., 2015. Natural gas consumption and economic growth nexus: is the 10th Malaysian plan attainable within the limits of its resource? Renew. Sust. Energy Rev. 49, 1221–1232.

Rafindadi, A.A., Ozturk, I., 2017. Impacts of renewable energy consumption on the German economic growth: evidence from combined cointegration test. Renew. Sust. Energy Rev. 75 (2017), 1130–1141.

Ramcharran, H., 1990. Electricity consumption and economic growth in Jamaica. Energy Econ. 12, 65–70.

Rodríguez-Caballeroa, C.V., Ventosa-Santaulària, D., 2016. Energy-growth long-term relationship under structural breaks. Evidence from Canada, 17 Latin American economies and the USA. Energy Econ. 61 (2016), 121–134.

Saunoris, J.W., Sheridan, B.J., 2013. The dynamics of sectoral electricity demand for a panel of US states: new evidence on the consumption–growth nexus. Energy Policy 61, 327–336.

Sadorsky, P., 2009. Renewable energy consumption and income in emerging economies. Energy Policy 37, 4021–4028.

Saidi, K., Mbarek, M.B., 2016. Nuclear energy, renewable energy, CO_2 emissions, and economic growth for nine developed countries: evidence from panel Granger causality tests. Progr. Nucl. Energy 88, 364–374.

Sami, J., 2012. Sectoral Electricity Consumption and Economic Growth in Philippines: New Evidence From Cointegration and Causality Results. College of Business, Hospitality and Tourism Studies. Working Paper Series, No. 03/12.

Shahbaz, M., Chandran, V.G. R., Pervaiz, A., 2011. Natural gas consumption and economic growth: cointegration, causality and forecast error variance decomposition tests for Pakistan. MPRA Paper No. 35103.

Shahbaz, M., Ozturk, I., 2012. Electricity Consumption and Economic Growth Causality Revisited: Evidence from Turkey. Munich Personal RePEc Archive, No. 36637, March 2012.

Shiu, A., Lam, P., 2004. Electricity consumption and economic growth in China. Energy Policy 32, 47–54.

Squalli, J., 2007. Electricity consumption and economic growth: bounds and causality analyses of OPEC members. Energy Econ. 29 (6), 1192–1205.

Solarin, S.A., 2011. Electricity consumption and economic growth: trivariate investigation in Botswana with capital formation. Int. J. Energy Econ. Policy 1 (2), 32–46.

Solarin, S.A., Ozturk, I., 2016. The relationship between natural gas consumption and economic growth in OPEC members. Renew. Sust. Energy Rev. 58 (2016), 1348–1356.

Solarin, S.A., Shahbaz, M., 2015. Natural gas consumption and economic growth: The role of foreign direct investment, capital formation and trade openness in Malaysia. Renew. Sust. Energy Rev. 42 (2015), 835–845.

Soytas, U., Sari, R., 2006. Energy consumption and income in G-7 countries. J. Policy Model. 28 (7), 739–750.

Sultan, R., 2011. Dynamic linkages between transport energy and economic growth in Mauritius: implications for energy and climate policy. J. Energy Technol. Policy 2 (1), 24–37.

Swan, L.G., Ugursal, V.I., 2009. Modeling of end-use energy consumption in the residential sector: a review of modeling techniques. Renew. Sustain. Energy Rev. 13 (8), 1819–1835.

Tang, Shahbaz, 2013. Sectoral analysis of the causal relationship between electricity consumption and real output in Pakistan. Energy Policy 60 (2013), 885–891.

Tang, C.F., Tan, E.C., 2012. Electricity consumption and economic growth in Portugal: evidence from a multivariate framework analysis. Energy J. 33 (4), 23–48.

Topcu, M., Altay, B., 2017. New insight into the finance-energy nexus: disaggregated evidence from Turkish sectors. Int. J. Financial Stud. 5 (1), 1–16.

Topcu, M., Payne, J.E., 2017. The financial development–energy consumption nexus revisited. Energy Sour. B 12 (9), 822–830.

Tsani, S.Z., 2010. Energy consumption and economic growth: a causality analysis for Greece. Energy Econ. 32, 582–590.

Tugcu, C.T., Ozturk, I., Aslan, A., 2012. Renewable and non-renewable energy consumption and economic growth relationship revisited: evidence from G7 countries. Energy Econ. 34 (6), 1942–1950.

US Energy Information Administration, 2017. April 2017 Monthly Energy Review. Available from: https://www.eia.gov/totalenergy/data/monthly/.

World Bank, World Development Indicators, 2017. Available from: http://databank.worldbank.org/data/reports.aspx?source=world-development-indicators.

Wolde-Rufael, Y., 2004. Disaggregated industrial energy consumption and GDP: the case of Shanghai 1952-1999. Energy Econ. 26 (2004), 69–75.

Wolde-Rufael, Y., 2010a. Bounds test approach to cointegration and causality between nuclear energy consumption and economic growth in India. Energy Policy 38, 52–58.

Wolde-Rufael, Y., 2010b. Coal consumption and economic growth revisited. Appl. Energy 87 (1), 160–167.

Wolde-Rufael, Y., 2012. Nuclear energy consumption and economic growth in Taiwan. Energy Sourc. B 7 (1), 121–127.

Yang, Y., Liu, J., Zhang, Y., 2017. An analysis of the implications of China's urbanization policy for economic growth and energy consuption. J. Clean. Prod. 161, 1251–1262.

Yildirim, E., Aslan, A., 2012. Energy consumption and economic growth nexus for 17 highly developed OECD countries: further evidence based on bootstrap-corrected causality tests. Energy Policy 51, 985–993.

Yildirim, E., Sukruoglu, D., Aslan, A., 2014. Energy consumption and economic growth in the next 11 countries: the bootstrapped autoregressive metric causality approach. Energy Econ. 44, 14–21.

Yoo, S., 2005. Electricity consumption and economic growth: evidence from Korea. Energy Policy 33, 1627–1632.

Yoo, S.H., Jung, K.O., 2005. Nuclear energy consumption and economic growth in Korea. Progr. Nucl Energy 46 (2), 101–109.

Yoo, S.H., Ku, S.J., 2009. Causal relationship between nuclear energy consumption and economic growth: a multi-country analysis. Energy Policy 37, 1905–1913.

Zachariadis, T., 2007. Exploring the relationship between energy consumption and economic growth with bivariate models: new evidence from G-7 countries. Energy Econ. 29, 1233–1253.

Zaman, B.U., Farooq, M., Ullah, S., 2011. Sectoral oil consumption and economic growth in Pakistan: an ECM approach. Am. J. Sci. Ind. Res. 2 (2), 149–156.

Zhang, C., Xu, J., 2012. Retesting the causality between energy consumption and GDP in China: evidence from sectoral and regional analyses using dynamic panel data. Energy Econ. 34 (6), 1782–1789.

Ziramba, E., 2009. Disaggregate energy consumption and industrial production in South Africa. Energy Policy 37, 2214–2220.

Further Readings

Akhmat, G., Zaman, K., 2013. Nuclear energy consumption, commercial energy consumption and economic growth in South Asia: bootstrap panel causality test. Renew. Sust. Energy Rev. 25 (2013), 552–559.

Apergis, N., Payne, J.E., Menyah, K., Wolde-Rufael, Y., 2010. On the causal dynamics between emissions, nuclear energy, renewable energy, and economic growth. Ecol. Econ. 69 (2010), 2255–2260.

Ghani, G.M., 2012. Does trade liberalization effect energy consumption? Energy Policy 43, 285–290.

Lee, C.-C., Chiu, Y.B., 2011. Nuclear energy consumption, oil prices, and economic growth: evidence from highly industrialized countries. Energy Econ. 33 (2011), 236–248.

Linden, M., Ray, D., 2017. Aggregation bias-correcting approach to the health–income relationship: life expectancy and GDP per capita in 148 countries, 1970-2010. Econ. Model. 61, 126–136.

Payne, J.E., 2010. A survey of the electricity consumption-growth literature. Appl. Energy 87 (3), 723–731.

Reynolds, D.B., Kolodziej, M., 2008. Former Soviet Union oil production and GDP decline: Granger-causality and the multi-cycle Hubert curve. Energy Econ. 30, 271–289.

Yildirim, E., Sarac, S., Aslan, A., 2012. Energy consumption and economic growth in the USA: evidence from renewable energy. Renew. Sust. Energy Rev. 16 (9), 6770–6774.

Yoo, S.H., 2006. Causal relationship between coal consumption and economic growth in Korea. Appl. Energy 83, 1181–1189.

On the Dynamics of Renewable Energy Consumption (Aggregated and Disaggregated) and Economic Growth: An Approach by Energy Sources

Antonio C. Marques, José A. Fuinhas
University of Beira Interior and NECE-UBI, Covilhã, Portugal

Chapter Outline

1 Introduction

The nexus between energy consumption and economic growth deserves to be taken not only as one of the main topics covered in this book, but also requires to take into account the context of energy transition from conventional energy toward renewable energy, within the aim of increasing electrification of economies. This energy transition should ensure that the energy will keep playing a critical role in the process of economic growth. Put in other words, this transition of sources should be made in such a way that economic growth is not compromised.

The diversification of sources cannot be a theoretical goal, without both analyzing and understanding the conditions for a successful diversification. Indeed, one must understand that the transition may be a failure, if attention is not paid to economic growth. In this respect, it should be noted that in this chapter, the ongoing interesting debate between growth and sustainable development will be left aside. It is assumed that there is a collective goal for the enhancement of economic growth, on the assumption that economic growth and sustainable development, albeit being different, do not diverge. These conditions for successful diversification could be of various types, such as backup capability, the availability of natural endogenous resources, the costs of technology development, or even the costs of change in this global paradigm.

Around the world, more and more, energy decisions are no longer between consuming and supplying more energy, but rather the critical decisions are concerned with what type of sources needs to be developed and what the consequences of the deployment of each source are. This is the first reason that motivated this chapter. This chapter also intends to be a contribution to knowledge, through the assessment of these conditions toward a successful diversification mix, within the context of studying the electricity-economic growth nexus, disaggregated by source. The novelty and main contributions of this chapter are as follows: (1) it highlights the appropriateness of studying the nexus by disaggregating the energy consumption by source; (2) it focuses on the German case, which is characterized by a nuclear phaseout, through the assessment of the energy transition from conventional sources toward renewable sources; (3) it provides empirical evidence for the critical factors which make the transition successful; and (4) it discusses some additional measures on the demand side management (DSM) to materialize an energy transition without compromising economic growth.

After this introduction, the chapter continues by reviewing the literature on electricity consumption and economic growth. Afterward Section 3 focuses on the debate on this nexus, by each generation source, with particular emphasis on the interaction of renewable energy sources (RES) and conventional energy sources. Their interaction ability, both by replacing each other and complementing themselves with each other, is essential for the energy transition. Moreover, in this section a discussion will be carried out around the main challenges faced by the transmission system operators (TSO) and the energy policy makers in general, in accommodating the intermittent generation, which is a characteristic of RES. Section 4 is entirely devoted to a case study with transition of energy sources. This is the German case, which hosts the phase-out of nuclear power. Therein, the critical factors, which should be checked for a successful energy transition, are analyzed. These factors are discussed in detail in Section 5, where it becomes clear that the energy consumption–economic growth nexus (EGN) ought to evolve, by taking into account the role of each source and their interactions. Section 6 focuses on demand side management, a very up-to-date topic nowadays. This section explains possible measures and solutions that facilitate the energy transition in countries that do not possess the perfect conditions, such as the backups shared via the integrated electricity market. Finally, Section 7 concludes and presents some implications and guidelines for policy makers.

2 Energy Versus Electricity From Renewables and the Economic Growth Nexus

With the current trend toward RES penetration in the electricity mix, the effects that are resulting for the economic growth, have gained serious attention in literature (Table 3.1). Overall, the findings about the effects of the electricity sources on economic growth have not shown any consensus whatsoever (Omri, 2014).

In general, literature has focused on the analysis of causalities between RES consumption and economic growth (e.g., Apergis and Payne, 2014; Ben Aïssa et al., 2014; Chang et al., 2015; Furuoka, 2016). However, this analysis is not able to capture the sign of such aforementioned transition effects on the corresponding relationships. Consequently, some research has been conducted so as to obtain impacts, that is, analyzing if the relationships have been negative or positive. On the one hand, some authors argue that the RES consumption has a positive impact on economic growth (e.g., Al-mulali et al., 2014; Amri, 2017; Bhattacharya et al., 2016; Inglesi-Lotz, 2015; Koçak and Şarkgüneşi, 2017; Rafindadi and Ozturk, 2016; Shahbaz et al., 2015). On the other hand, some authors argue that the RES have hampered the economic activity (Marques and Fuinhas, 2012; Marques et al., 2016; Ocal and Aslan, 2013). Notwithstanding the above findings, the neutrality hypothesis is also supported in the literature (Ben Aïssa et al., 2014; Payne, 2009). Largely, the doubts about the full effect from RES on economic activity have inspired policy makers to substitute policy-driven guidance by market-driven instruments. The first kind of policies, namely feed-in-tariffs, have been transferred to the economy, namely in the form of electricity tariffs. The market instruments and the new trend to licensing projects at the market price could be seen as a strong incentive to achieve the cost-effectiveness of the technologies, namely solar photovoltaic (PV) and wind.

3 Disaggregating the EGN

3.1 Renewables Versus Nonrenewables

Following the evolution of the literature on the nexus, studies were gradually developed by focusing on renewable and nonrenewable aggregated sources. However, on the one hand, generation technologies are not identical everywhere (countries, sectors, etc.). On the other

Table 3.1: Summary of the literature on renewable and nonrenewable energy consumption and economic growth nexus.

Authors	Periods	Countries	Methodology	Main Results
Rafindadi and Ozturk (2016)	1970QI–2013QIV	Germany	ARDL bounds test VECM (vector error correction mechanism) Granger causality	RE has a positive impact on the GDP (gross domestic product) (LONG-RUN) RE ↔ GDP (long-run) RE ≠ GDP (short-run)
Koçak and Şarkgüneşi (2017)	1990–2012	9 Black Sea and Balkan countries	DOLS FMOLS Heterogeneous panel causality	RE has a positive impact on the GDP RE ↔ GDP
Kahia et al. (2016)	1980–2012	13 MENA Net Oil Exporting Countries (NOECs)	FMOLS PECM	RE and NRE has a positive impact on the GDP Short-Run GDP ↔ NRE RE → GDP RE ≠ NRE Long-Run GDP ↔ NRE RE ↔ GDP RE ↔ NRE
Kahia et al. (2016)	1980–2012	5 MENA Net Oil Exporting Countries (NOECs)	FMOLS PECM	RE and NRE has a positive impact on the GDP Short-, and long-Run GDP ↔ NRE RE ↔ GDP RE ↔ NRE
Furuoka (2016)	1992–2011	3 transition economies in the Baltic region, namely, Estonia, Latvia and Lithuania	Homogeneous and heterogenous panel methods (granger causality)	GDP → NRE GDP → RE NRE → RE
Dogan (2016)	1988–2012	Turkey	ARDL Approach VECM Granger causality test.	NRE has a positive impact on GDP both short-and long run Short-run causality GDP ↔ NRE GDP ← RE RE ≠ NRE Long-run causality GDP ↔ NRE GDP ↔ RE RE ↔ NRE
Bildirici and Gökmenoğlu (2016)	1961–2013	G7	MS-VAR Granger causality	

Table 3.1: Summary of the literature on renewable and nonrenewable energy consumption and economic growth nexus. (*cont.*)

Authors	Periods	Countries	Methodology	Main Results
Asafu-Adjaye et al. (2016)	1990–2012	53 countries	Polled Mean Group	
Amri (2017)	1990–2012	72 countries subdivided into: whole, developed and high income	Method of moment's generalized (GMM)	RE ↔ GDP (positive) (for whole, developed and high-income)
Ito (2017)	2002–2011	42 developing countries	GMM PMG	RE has a positive effect on the economic growth in the long run NRE has a negative effect on economic growth in the long run substitute relationship between RE and NRE
Cerdeira Bento and Moutinho (2016)	1960–2011	Italy	ARDL approach Toda-Yamamoto causality test	RES → NRES
Marques et al. (2016)	2010m1–2014m11	France	ARDL approach	RES $\overset{-}{\to}$ Y NUC $\overset{+}{\to}$ Y NUC $\overset{=}{\to}$ Fossil RES $\overset{+}{\leftrightarrow}$ NUC (short-run)
Marques and Fuinhas (2016)	2006m1–2014m6	Portugal	ARDL approach	Long-run Hydro $\overset{+}{\to}$ IPI The wind and solar PV $\overset{-}{\to}$ IPI IPI, Hydro, and Thermal $\overset{+}{\to}$ SR $\overset{+}{\to}$ Wind $\overset{+}{\to}$ OR
Saidi and Ben Mbarek (2016)	1990–2013	9 developed countries	FMOLS and DOLS VECM Granger causality	RES $\overset{-}{\to}$ Nuclear (only in FMOLS estimation at 10%) Nuclear $\overset{-}{\to}$ RES
Marques and Fuinhas (2015)	2007m1–2012m10	Portugal	VAR Granger Causality	IPI ↔ SR (Special Regime) IPI → OR (Ordinary Regime) IPI → IMP (imports) IPI ↔ SRR (Special Regime Renewable) IPI → SRNR (Special Regime nonrenewable) SR ↔ OR

(Continued)

Table 3.1: Summary of the literature on renewable and nonrenewable energy consumption and economic growth nexus. (*cont.*)

Authors	Periods	Countries	Methodology	Main Results
Dogan (2015)	1990–2012	Turkey	ARDL approach VECM Granger causality	Short-run NRES ↔ RES NRES ≠ Y RES ≠ Y Long-run Y → RES Y ↔ NRES
Ben Jebli and Ben Youssef (2015)	1980–2009	Tunisia	ARDL bounds test VECM Granger causality	NRES → RES (short-run)
Marques et al. (2014)	2004m8–2013m10	Greece	Johansen cointegration VECM	RES → NRES Hydro ↔ NRES Hydro ↔ RES IPI → RES NRES → IPI
Al-mulali et al. (2014)	1980–2010	8 Latin American countries	DOLS Granger causality	RES and NRES $\overset{+}{\rightarrow}$ Y RES ↔ Y NRES → Y RES $\overset{+}{\rightarrow}$ NRES
Salim et al. (2014)	1980–2011	29 OCDE countries	Westerlund cointegration test Pooled Mean Group Panel Granger Causality	Y ↔ NRES (both short-and long-run) Y → RES NRES $\overset{-}{\leftrightarrow}$ RES
Apergis and Payne (2012)	1990–2007	80 countries	Multivariate Panel Error Correction model	Short- and long-run RES ↔ Y NRES ↔ Y RES ↔ NRES
Apergis et al. (2010)	1984–2007	19 developed and developing countries	Panel Error Correction Model	RES → NUC RES ↔ Y NUC → Y

GDP, Gross domestic product; *VECM*, Vector error correction mechanism.
Source: Authors' compilation.

hand, the use of endogenous resources by economies which are traditionally dependent on primary energy imports could change the nature of the nexus, due to the potential direct impact on the domestic economic activity.

In such a way, the analysis of the interaction between the electricity generation sources, namely RES, fossil fuels, and nuclear energy, has been very attractive to the research field connected with studies dealing with the electricity-growth nexus. As summarized in Table 3.1, literature has found dissimilar effects of RES and fossil fuels on economic activity. The identification of the effects of the RES on both fossil fuels and nuclear energy, and vice versa, has proven to be very helpful on the decision-making process.

Cerdeira Bento and Moutinho (2016) investigated the nexus for Turkey from 1960 to 2011, using a Toda–Yamamoto causality test. Both the RES and non-RES electricity production seem to be causing the economic activity. However, the RES production has also led to the electricity production from fossil fuels. Indeed, some literature have proved a unidirectional causality from RES to non-RES (Marques et al., 2014). Conversely, the unidirectional causality from non-RES to RES has also been proven, however, with less frequency (Ben Jebli and Ben Youssef, 2015). Furthermore, the bidirectional causality between RES and non-RES has been the most supported result in the literature (Apergis and Payne, 2012; Dogan, 2015; Salim et al., 2014).

The assessment of short-run and long-run relationships, that is, the dynamic relationships, has also been employed in the EGN, and on the interactions between electricity sources (Sebri, 2015). The empirical literature for Portugal shows that there is: (1) a unidirectional causality running from economic growth to electricity consumption in the short run and (2) a bidirectional causality between both magnitudes in the long run (Shahbaz et al., 2011). However, considering the same country, and a similar time-span, Tang and Tan (2012) do not have apportioned the short-run and long-run effects, but have also confirmed the bidirectional causality between electricity consumption and economic growth. Moreover, Dogan (2015) has used an autoregressive distributed lag (ARDL) bounds test and a Vector Error Correction Mechanism (VECM) Granger causality test for Turkey from 1988 to 2012 and found dissimilar results from the short run to the long run. In fact, with reference to the short run, Dogan (2015) found: (1) a bidirectional causality between economic growth and electricity consumption from non-RES; (2) a unidirectional causality from RES to economic activity; and (3) no relationship between RES and non-RES. However, in the long run a bidirectional causality between all three magnitudes was found, namely RES, non-RES and economic activity.

3.2 The Characteristics and the Role Played by Each Electricity Source

Over the last years, the empirical literature has also resorted to different approaches, disaggregating the energy consumption according to its sources. Indeed, the results of this procedure have produced a more reliable framework for policy makers to elaborate proper energy policies. The integration of new RES raises new challenges both for the electricity production systems (Verzijlbergh et al., 2016) and for the TSO.

Some countries, such as Portugal and Spain, which have been at the forefront of the deployment of RES, have designed a framework to broadly classify RES and non-RES as special regime and ordinary regime, respectively. Based on that framework, Marques and Fuinhas (2015) found a bidirectional causality between special regime, comprised mainly RES and large hydro, and the industrial production index (IPI). However, when they quantified the impact of RES on IPI, they showed that RES has had a negative impact on the economic activity (Marques and Fuinhas, 2016). Moreover, there was a bidirectional causality between special and ordinary regime (Marques and Fuinhas, 2015). Nonetheless, when analyzed by electricity source, only the wind power caused positively the ordinary regime (Marques and Fuinhas, 2016). Another relevant finding is that, within the intermittent renewables, the effects from solar PV and wind energy could be dissimilar. Indeed, the solar PV could be a stimulating factor on economic activity; meanwhile, there is evidence of a restrictive effect from wind power. Such outcome could come from the different characteristics of the two technologies. The first one, namely the solar PV, enjoys a smaller scale of operation with plenty of small players working domestically and locally. On the contrary, just a small number of large international players actually provides wind power. In such a way, and considering the tradition of operating wind farms supported by feed-in-tariffs, the multiplier effects from that source onto the domestic economies are far from being straightforward.

To sum up, the literature has dealt a lot with the causality between economic activity and energy/electricity consumption. More recently, the literature has expanded the EGN by studying these issues in a disaggregated way, namely by energy sources. Finally, the authors expand the framework for policy makers, while at the same time, such studies are developed in the EGN and the interactions between electricity sources. It is worthwhile highlighting that the expected substitution effect of non-RES by RES has not been proven categorically. Indeed, meanwhile the RES technologies do not enhance both the cost-effectiveness and first and foremost their storage ability, then they will need to be backed up by fossil fuels power plants. The critical point, however, is that the backup load causes a negative effect on economic activity, due to the high economic inefficiencies in resources allocation, which are associated with the idle capacity. The effective establishment of the substitution effect of the conventional sources by the new RES is critical for the preservation of high economic activity. This is the main motivation for researchers to keep working, until they fully understand the interactions of energy sources. Policies should be designed to promote that substitution of sources or diversification of the mix, without compromising the economic growth goal.

3.3 The Management of the Electricity System: The Challenges of Supporting Renewables' Intermittency

The diversification of sources, particularly due to the accommodation of new renewable electricity generation from wind and solar PV, leads not only policy makers, but also the TSO to face enormous challenges. Besides the practical question associated with the daily real-time management of the contribution of each source, they also have to deal with new innovative forms of organization and business models of Energy Services Companies (ESCOs). Because of that, some new questions arise, namely: How to integrate or make these new services compatible with the grid as a whole? Let's look, for instance, at the development of local energy communities. These communities are not exempted from the network connection, despite the several questions that have appeared about the characteristics of such connection. For instance, what kind of services should be provided and what is the quality of the network? Moreover, given the autonomy of these communities, does the regulator have to accommodate the cost of providing such service to that community, into the common electricity rates?

Sooner or later, the electricity system management will be confronted with two fundamental principles: reliability versus individual freedom. Which one should prevail? The former? Are the consumers and the manufacturers willing to let go the almost complete reliability of the system in favor of the integration of RES and, in such a way, preserve the environment? Within this framework, where is the principle of freedom in consumption? According to that, consumers should be free to consume any quantity of electricity, at any time they wish.

It is important to note that, in most markets, the principle of freedom is observed. However, in these markets, the pricing mechanism somehow regulates this freedom. In fact, when there is scarcity of goods, the freedom of consumption is self-limited, because of the high price. On the contrary, when the supply of goods is in surplus, the consumer can freely decide on the level of consumption, in accordance either with her/his willingness to pay, or with the restrictions from one's available budget. In the electricity market, however, this mechanism is far from being accomplished. The specific characteristics of the electricity market are an example of this. In fact, although the product is homogeneous, the production process is not. This is a difference of considerable importance when compared to other industries, where the production of goods or services actually relies on quite identical technology with very similar production costs. This means that, unlike other industries, the cost of production in the electricity industry is volatile. Therefore it is really hard or almost impossible to carry out the composite price of instant generation. Despite the restrictions on the freedom of choice, the economic instrument of price should be used in the electricity industry. Consequently, later in this chapter, we will come back to the problem of peak-load pricing and tariff alternative instruments.

Going back to the issue of local energy communities, which are one of the most increasing forms of organization, policy makers need to make decisions for them. For example, let's look at infrastructures such as hospitals in which failures are strictly forbidden. Should they be located within these local communities or should they be primarily supplied by the national grid for security reasons? Another sensitive issue refers to the relocation of production, with the distance narrowing between the consumption and the generation locations. Therefore while losses are minimized, this can also provoke lead to a reduction of the scale effect, which could entail higher average costs for the system as a whole. In fact, any network industry, and in particular energy industry, needs scale economies. However, these strategies of production decentralization, by using locally endogenous resources, may result in additional costs to be borne by all. The expansion of the solar PV self-consumption units can also provoke increased load at the distribution stage, in addition to being able to worsen issues related to income distribution (Frondel et al., 2015).

4 Disaggregating the EGN in Germany; Replacing Nuclear Power

This empirical application focuses on the German electricity generation system, by analyzing the dynamics among the diverse electricity sources and the economic growth. Furthermore, this application analyzes the electricity mix, by observing the impact of the several sources within a system. These sources are traditionally dependent on nuclear power. As widely known, Germany has defined ambitious targets of replacing nuclear power by renewables. Trying to replace a mature source of baseload by a mix of sources with notoriously different characteristics is actually an enormous challenge. For this reason, the analysis of the interactions of sources in the German system is a very interesting task.

Two different methodologies were used. The first one employs an ARDL model, with monthly data. The second one uses daily data through a vector autoregressive (VAR), to check the robustness of findings achieved from the monthly data framework.

4.1 The Monthly Data Framework

We have started by using monthly data ranging from July 2008 to November 2016. The data were sourced both from the European Network of Transmission System Operators (ENTSO-E) and from Eurostat. The definition, the source, and the descriptive statistics of the series are shown in Table 3.2. Thereafter, the operators "L" and "D" denote the natural logarithm and the first differences of logarithms, respectively. All the variables have been selected in accordance with the literature on the EGN. However, it is worth stressing that the IPI is used as a proxy of the economic activity due to the inexistence of an appropriate monthly series for GDP. This proxy of economic activity has been employed several times in literature, when the electricity-growth nexus is studied with a monthly frequency. Please note that the use of this (monthly) framework is plenty of opportunity. Indeed, the management of the system is carried out daily and mostly monthly based. Factors such as the forecast of consumptions, temperatures, availability of natural resources, and capacity factors are essentially carried out on a monthly basis. When the mix and interaction of sources are analyzed under an annual framework, some of the real time critical interactions of the sources and management decisions are lost, because of the average values the econometric methods considered. The summary statistics are displayed in Table 3.2.

A visual inspection of the monthly series was carried out, to assess the stationarity of the series. All the series appear as nonstationary at their levels. The mean and the variance of the variables are not constant throughout the entire time-span. The Augmented Dickey–Fuller (ADF) test (Dickey and Fuller, 1981), the Perron (PP) test (Phillips and Perron, 1988), and the Kwiatkowski Schmidt Shin (KPSS) test (Kwiatkowski et al., 1992) were performed to assess the integration order of the variables. The ADF test was carried out under the null hypothesis of a unit root and following the Schwartz information criterion. The PP test has

Table 3.2: Summary statistics (monthly data).

	Definitions	Sources	Observations	Mean	Maximum	Minimum	Standard Deviation	Jarque–Bera Normality Test
IPI	Industrial Production Index calendar adjusted	EUROSTAT	101	105.4337	119.90	82.10	9.3633	9.1154
FOSSIL	Electricity production from fossil fuels	ENTSO-E	101	28029.90	35871	19382	3592.61	2.4994
NUCLEAR	Electricity production from nuclear	ENTSO-E	101	8632.554	13536	4094	2070.023	4.1441
HYDRO	Electricity production from hydro power	ENTSO-E	101	1928.505	2737	1316	309.5228	3.0038
WIND	Electricity production from wind power	ENTSO-E	101	4298.129	11495	1650	2018.468	60.7713
SOLAR	Electricity production from solar PV	ENTSO-E	101	1987.079	5090	199	1588.841	11.4526
PUMP	Electricity consumption for pumping	ENTSO-E	101	638.6337	838	461	79.0947	0.6761
RXM	Ratio of coverage of electricity imports by exports	ENTSO-E	101	1.6487	2.8158	1.2032	0.3254	27.9319

Source: Authors' compilation.

the same null hypothesis as the ADF test, although one can use the Bartlett Kernel spectral estimation method and the Newey–West-bandwidth. The KPSS test has the null hypothesis of stationarity and also uses the Bartlett Kernel spectral estimation method and the Newey–West-bandwidth.

The unit root tests are inconclusive about the order of integration of variables (Table 3.3). However, it should be stressed that none of the variables are integrated of order two, that is I(2), which allows the use of the ARDL approach. The visual inspection of the series LIPI,

Table 3.3: Order of integration tests (monthly data).

	ADF			PP			KPSS	
	C	CT	None	C	CT	None	C	CT
LIPI	−2.128	−1.21	1.8325	−4.5039***	−5.9647***	0.1109	0.7736***	0.1178
DLIPI	−12.3642***	−12.3105***	−12.4223***	−23.5786***	−22.6888***	−22.3665***	0.0425	0.0279
LFOSSIL	−1.0932	−1.5359	−0.2285	−4.7355***	−4.9425***	0.0713	0.2219	0.0449
DLFOSSIL	−10.5322***	−10.4609***	−10.6005***	−10.258***	−10.3086***	−10.3086***	0.0248	0.0187
LNUCLEAR	−3.1258**	−4.5291***	−0.3128	−3.0195*	−4.5484***	−0.5847	0.9676***	0.1151
DLNUCLEAR	−10.8392***	−10.7923***	−10.8825***	−11.2799***	−11.6093***	−11.3139***	0.0972	0.0904
LHYDRO	−4.6122***	−5.4473***	−0.1883	−4.6853***	−5.0419***	−0.3103	0.7266**	0.0589
DLHYDRO	−11.3197***	−11.2636***	−11.3789***	−14.4125***	−14.3261***	−14.555***	0.1996	0.1644**
LWIND	0.0052	−5.3725***	1.9151	−4.0451***	−5.3665***	0.8673	1.0461***	0.065
DLWIND	−5.6365***	−5.5975***	−5.6547***	−12.7209***	−12.6554***	−12.7695***	0.0755	0.0711
LSOLAR	−3.0703**	−0.7102	3.2623	−2.3546	−2.5289	0.142	0.9991***	0.1406*
DLSOLAR	−10.3769***	−11.3075***	−2.6411***	−4.7982***	−4.7717***	−4.8339***	0.2625	0.1757**
LPUMP	−5.6799***	−5.7379***	0.2206	−4.9359***	−4.9148***	0.2371	0.2656	0.0441
DLPUMP	−4.5295***	−4.6474***	−4.5519***	−21.2257***	−20.7008***	−21.3944***	0.5**	0.5***
LRXM	−3.6684***	−5.8483***	−1.7387*	−3.6684***	−4.6261***	−1.7152*	0.9468***	0.0975
DLRXM	−10.3419***	−10.2869***	−10.3782***	−10.3452***	−10.2899***	−10.382***	0.0424	0.0414

ADF, Augmented Dickey-Fuller test; PP, Phillips Perron test; KPSS, Kwiatkowski Phillips Schmidt Shin; C, constant; CT, constant and trend; None, without constant and trend. ***, **, and * indicate that the statistic is statistically significant at 1%, 5%, and 10%, respectively.

LNUCLEAR, and LWIND suggested the presence of structural breaks. Accordingly, the analysis of the correlograms and partial correlograms both point to the likely occurrence of structural breaks. However, the results from the traditional unit root tests are not reliable in the presence of structural breaks, as noted by Baum (2004). To get over this limitation, Zivot and Andrews (1992) and Phillips and Perron (1988) proposed three models to test the integration order in the presence of structural breaks. Their methods allow investigating whether a break exists and, if so, at what point of time it occurs. This information should be considered by the use of an appropriate dummy variable to estimate parsimonious models.

The integration order tests with structural breaks were performed in the presence of an intercept to the LIPI and LNUCLEAR series and with a linear trend to the LWIND variable. The first differences of these three series have also been tested. This methodology has also been used by Hamdi et al. (2014) to study the electricity-growth nexus in Bahrain.

As displayed in Table 3.4, the three series are stationary in their levels, that is, I(0), or they tend to be stationary in the presence of one break-point. Such information about the structural breaks will be used in the subsequent estimation. Accordingly, shift dummies were used to capture the breaks occurred in the series.

4.2 The Method in the Monthly Data Framework

Faced with a mixture of stationarity and nonstationarity variables in their levels, the ARDL model appeared to be appropriate. Indeed, this model has the advantage of being able to handle series of integration I(0), I(1) or borderline, that is I(0)/I(1), and it creates efficient parameters estimates. Furthermore, this method has other advantages, namely: (1) the ability to analyze both the short-run and long-run relationships individually; (2) the ability to support endogeneity; and (3) being robust in the presence of time-dummies. In fact, this research aims to study the relationships between electricity generation sources and economic growth, and we expect the presence of endogeneity between the series. Besides, as shown by

Table 3.4: Unit root tests with structural breaks (monthly data).

	Zivot and Andews		Perron	
	t-Statistic	Break	*t*-Statistic	Break
LIPI	−7.7218***	2011m12	−5.9913***	2010m04
DLIPI	−11.5075***	2010m03	−8.1121***	2009m02
LNUCLEAR	−6.0151***	2011m04	−6.2498***	2011m05
DLNUCLEAR	−11.0113***	2011m06	−8.0064***	2011m05
LWIND	−6.5939***	2013m08	−6.8467***	2010m03
DLWIND	−1.5943***	2010m05	−5.7327***	2015m01

The null hypothesis is that the series has a unit root with a structural break in intercept or in the trend. ***, denotes statistical significance at 1% level. The LIPI, DLIPI, LNUCLEAR and DLNUCLEAR have been tested with a break in intercept, and LWIND and DLWIND have been tested with a break in trend.

Jouini (2015), the robustness of the ARDL model provides evidence for possible directional causal relationships. Therefore the results of the ARDL model can be compared with the results of the VAR methodology.

Following the main objective of this application, we have analyzed the nature of the electricity source interactions and their short-run and long-run effects. Six models were carried out, namely, IPI, FOSSIL, NUCLEAR, HYDRO, WIND and SOLAR. The ARDL models that were applied actually follow the general form of the unrestricted error correction model (UECM) and may be specified as follows:

$$DY_t = f(DY_t; Y_{t-1}; X_t) \tag{3.1}$$

where the "D" operator represents the first differences of logarithms, Y_t = [LIPI,LFOSSIL,LN UCLEAR,LHYDRO,LWIND,LSOLAR], and X_t is a vector of time-dummies.

The quality and goodness-of-fit of the models were evaluated through a battery of diagnostic tests, namely: (1) the Jarque–Bera normality test (Jarque and Bera, 1980); (2) the Breusch–Godfrey serial correlation LM test (Breusch, 1978); (3) the auto regressive conditional heteroscedasticity (ARCH) test for heteroscedasticity (Bollerslev, 1986); (4) the Ramsey regression equation specification error test (RESET) test for functional misspecification (Ramsey, 1969); (5) the stability of the coefficients was assessed by providing the cumulative sum of recursive residuals (CUSUM) and CUSUM of squares tests (Nyblom, 1989); and (6) the investigation of the existence of cointegration; the ARDL bounds test was performed, as suggested by Pesaran et al. (2001).

4.3 Results From the Monthly Data Framework

In first differences, the coefficients of the variables refer to the short run, and they can be directly interpreted as semi-elasticities. For the long-run, the estimated coefficients were used to compute the elasticities, by dividing them by the coefficient of the error correction mechanism (ECM) and then multiplying by -1. Semi-elasticities and elasticities are displayed in Table 3.5. Once again, they were subject to a battery of tests to assess the goodness-of-fit of the estimations, as stated above, and they are shown both in Table 3.5 and Fig. 3.1. The normality of the residuals is not rejected, thus confirming the normality behavior of the residuals. The Breusch–Godfrey LM and the ARCH test rejected the presence of serial correlation and heteroscedasticity to the third order, respectively. The RESET test proves the appropriate functional form of each of the models.

It is worth noting that fossil fuel electricity sources were included in the monthly estimations in an aggregated form, considering the unavailability of disaggregated data for the entire period. Please note that the daily data framework will allow overcoming this predicament, given that the data by each fossil sources are available for the daily (more recent) period under analysis.

Table 3.5: Elasticities, semi-elasticities and adjustment speeds of models (monthly data).

	IPI	FOSSIL	NUCLEAR	HYDRO	WIND	SOLAR
	Short-run (semi-elasticities)					
DLIPI		0.8359***			1.0404***	2.8192***
DLFOSSIL	0.3787***				−0.6953***	−1.824***
DLNUCLEAR	0.0829**	−0.2026***		0.1705**	−0.9241***	−0.5394**
DLHYDRO					0.6132***	
DLWIND	0.0849***	−0.1515***	−0.0605*	0.1204***		−0.5694**
DSOLAR	0.0539***	−0.0883***	−0.0377*	0.0557**	−0.2351***	
DLPUMP			0.2407***	0.2572***		
DLRXM		−0.3444***	−0.3091***	0.3602***	−1.1433***	
	Speed of adjustment					
ECM	−0.9485***	−0.9127***	−0.7973***	−0.3758***	−0.9401***	−0.4094***
	Long-run (computed elasticities)					
LIPI		1.1017***			1.0184***	2.1959**
LFOSSIL	0.4812***		0.2582**	−0.5513**		
LNUCLEAR	0.1721***	−0.2734***			−1.0815***	−0.9436*
LHYDRO		−0.1608***				
LWIND	0.1373***	−0.1977***	−0.1239**	0.2755***		−0.7249**
LSOLAR	0.1528***	−0.2166***	−0.0867***		−0.3799***	
LPUMP			0.2423**		0.3728*	−0.9656*
LRXM	0.1884***	−0.4194***	−0.2021**		−0.9740***	
C	−3.9456***		5.3864***	4.1011***	13.1788***	6.9998**
TREND						0.0067**
	Time dummies					
SD_2008M07	0.1436***					
SD_2009M01		0.1891***				
SD_2011M05			−0.1921***			
SD_2013M08					0.2345***	
D_2011M05			−0.4723***			
D_2012M04			−0.3347***			
D_2013M02					−0.5937***	
D_2015M07			−0.3136***			
D_2016M04			−0.3347***			
D_2016M05			−0.2602***			
ID11						−0.7143***
	Diagnostic tests					
ARCH	1.0373 (1)	0.4439	1.5123	0.1146	0.0113	0.174
	1.7613 (2)	0.2354	0.7979	0.2819	0.5536	0.0334
	0.9695 (3)	0.3477	0.9643	0.1688	0.4652	0.0575
LM	0.0051 (1)	2.4707	0.3681	0.3184	0.0663	0.0548
	0.1481 (2)	1.4131	0.1838	1.1604	0.4146	2.3037
	1.9348	1.2480	0.1715	0.8919	0.4049	1.5919
JB	1.7897	0.6125	0.6928	4.3017	0.592	0.7705
RESET	0.2242	0.6221	0.4194	0.6682	1.5360	0.6221

*, **, *** denote statistical significance at 10%, 5% and 1% level, respectively. ARCH: test for heteroscedasticity (Bollerslev, 1986); JB: Jarque–Bera normality test (Jarque and Bera, 1980); LM: Breusch–Godfrey serial correlation LM test (Breusch, 1978); RESET: Ramsey regression equation specification error test (Ramsey, 1969); ln (.) lag order.

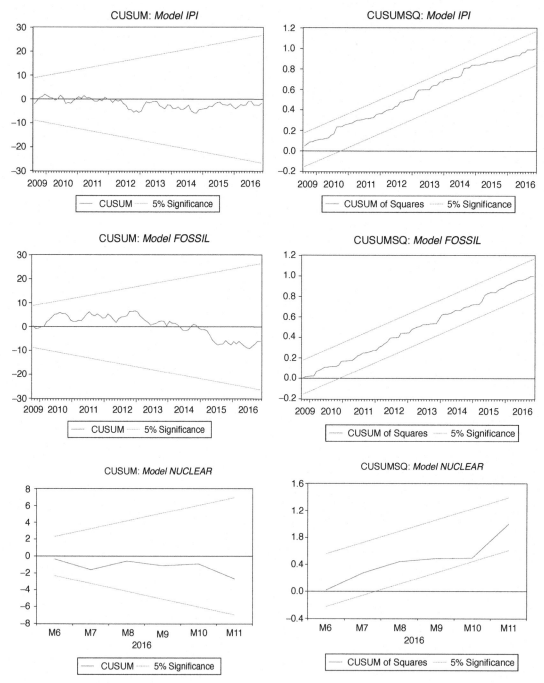

Figure 3.1: Test of Cumulative Sum of Recursive Residuals (CUSUM) and Squared of the Cumulative Sum of Recursive Residuals (CUSUMSQ) (Monthly Data).

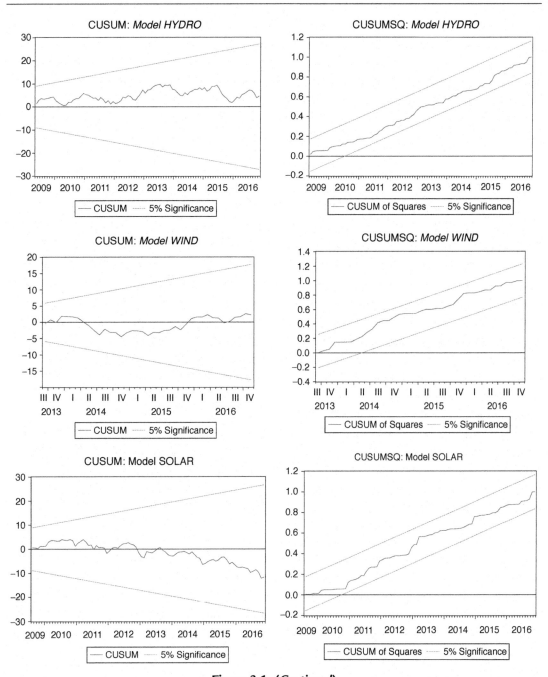

Figure 3.1: (*Continued*)

Overall, the results revealed great consistency, both with the up-to-date literature and with the reality characterizing the German electricity system. All the estimations displayed a negative and highly statistically significant speed of adjustment (ECM). More specifically, they revealed a high speed of adjustment for all the models, with the exception of HYDRO and SOLAR models.

The visual inspection of the series revealed that disturbances were coinciding with the financial crisis, and a shift dummy has been created taking the value of 0 from July 2008 to May 2010, and 1 for the remaining time-span. In fact, this break-point has been confirmed following the Zivot and Andrews' test. The shift dummy on FOSSIL models has been employed to control for the introduction of flexible natural gas plants in the electricity mix. Indeed, the introduction of these flexible generation plants has been helpful to manage the scarcity of the RES electricity production. However, the natural gas plants, being the less pollutant fossil fuel, increase the quota of fossil fuels in the electricity generation systems.

The Fukushima accident brought Germans an additional motivation to replace the nuclear electricity generation source. Consequently, the German electricity system has decommissioned several nuclear power plants, thus initiating the "nuclear power phase-out". The shift dummy used in the NUCLEAR models to capture this negative effect, as well as the Zivot and Andrews and Perron tests, capture a break point close to this date. Besides, the break enumerated as D_2011m05 takes the value of 1 in May 2011 and 0 otherwise. This dummy captures the effect of immediate shutting down or decommissioning of nuclear power plants. The break shown by the dummy encoded as D_2012m04 captures the effect of the German policy to close the rest of nuclear power plants. In July 2015, the RES contribution in German electricity mix reached 78% of their electricity production. The dummies encoded as D_2015m07, D_2016m04 and D_2016m05 also capture the effect of nuclear plants shutdown because their maintenance services also coincide with periods of high exploitation of RES.

In August 2013, the German offshore wind energy foundation and large investments on this RES technology was announced. In the same way, the Zivot and Andrews unit root tests highlight that a structural break starts on this date. Indeed, the increase of wind power electricity production can be visually observed, due to the introduction of the wind offshore technology. The dummy d_2013m02 and the seasonal dummy encoded as id11 capture the effects of wind power and the solar PV electricity production minimums.

Fig. 3.1 shows both the CUSUM and the CUSUMSQ test for the five models. Please note that the period shown in Fig. 3.2 is not the same in all models, given that dummy variables were applied in NUCLEAR and WIND models. Consequently, the CUSUM and CUSUMSQ are computed only for the periods after the inclusion of the dummies. Accordingly, the CUSUM and CUSUMSQ tests confirmed that the models were stable for the entire time-span under analysis.

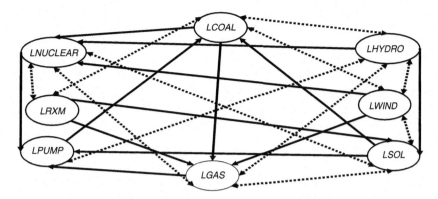

Figure 3.2: Granger Causalities.
Dashed lines denote 10% of significance.

To perform the ARDL bounds test, the F-statistic was used in the Wald test, with the null hypothesis that the long-run coefficients in models are equal to zero, that is, no cointegration, and the alternative coefficients are different from zero (cointegration). The results of the ARDL bounds test in Table 3.6 reveal that in all models the coefficients of the parameter are statistically different from zero. Therefore this means that all variables are cointegrated, that is, they have a long-run relationship.

In brief, the results of monthly data reveal that electricity sources, other than hydropower, have a positive and highly statistical significant effect on economic growth. Moreover, the IPI has also caused the wind power and solar PV exploitation in both the short and long run, supporting the idea that the deployment of renewables requires the adequate economic resources (wealth) to promote them. Besides, the IPI has also increased the fossil fuels combustion to produce electricity. This finding emphasizes that the main RES sources, namely wind power and solar PV, have been replacing for the fossil fuels electricity generation. In fact, these substitution effects have been proved through the models of wind

Table 3.6: The ARDL bounds test.

	F-statistic	k	Critical values	
			Bottom	Top
IPI	19.2369***	5	3.41	4.68
FOSSIL	15.7941***	6	3.15	4.43
NUCLEAR	23.8098***	4	3.74	5.06
HYDRO	9.6494***	2	5.15	6.36
WIND	16.9947***	5	3.41	4.68
SOLAR	9.3774***	4	4.4	5.72

*** Denotes statically significant at 1%. *K* is the number of independent variables. Critical values are derived from Pesaran et al. (2001).

power and solar PV, and in the NUCLEAR and FOSSIL models. With reference to the hydropower, given their low speed of adjustment, the German electricity system appears to use this source as a baseload source.

4.4 Robustness Check: Daily Data Framework

As previously highlighted, we have also used a daily data framework to check the robustness of the findings achieved under the monthly data context. Moreover, we have additionally investigated the relationships among fossil fuels in a disaggregated form. The discussion of the main findings under both frameworks, monthly and daily, will be done in the next section.

The daily data cover the time-span from January 1, 2015 to February 28, 2017, using all available data from the German electricity generation system. The data were sourced from the ENTSO-E transparency. Table 3.7 contains the variables' definition, sources and summary statistics of the natural logarithms series. It should be noted that electricity production from oil sources has been discontinued since the early 2016. Consequently, it has been excluded from this study, because it has been active only for half of the period under analysis.

Working upon daily data leads to several peculiarities, which have been well identified in literature, namely the noise in the series as well as the extreme observations. These outliers

Table 3.7: Summary statistics (daily data).

	Definitions	Sources	Observations	Mean	Maximum	Minimum	Standard Deviation
LCOAL	Electricity production from coal	ENTSO-E transparency	790	14.4683	15.0874	0	1.2507
LGAS	Electricity production from gas	idem	790	11.5489	13.4426	0	1.2652
LNUCLEAR	Electricity production from nuclear	idem	790	13.6186	13.9095	0	0.8743
LHYDRO	Electricity production from hydro power	idem	790	12.2487	12.8098	0	0.8089
LWIND	Electricity production from wind power	idem	790	13.3519	15.0025	0	1.0443
LSOL	Electricity production from solar PV	idem	790	12.4220	13.7309	0	1.2492
LPUMP	Electricity consumption for pumping	idem	790	11.0994	11.8469	0	1.0265
LRXM	Ratio of coverage of electricity imports by exports	idem	790	0.3098	1.2240	0.0025	0.1802
LCONS	Electricity consumption	idem	790	15.1917	15.683	0	2.0483

Source: ENTSO-E transparency, Authors' compilation.

Table 3.8: Summary statistics after outliers correction (daily data).

	Observations	Mean	Maximum	Minimum	Standard Deviation
LCOAL	790	14.5949	15.0874	13.7905	0.3464
LGAS	790	11.6224	13.4426	9.2335	0.7764
LNUC	790	13.6755	13.9095	13.1698	0.1792
LHYDRO	790	12.2997	12.8098	11.6247	0.2344
LWIND	790	13.3829	15.0025	11.1425	0.7966
LSOL	790	12.471	13.7309	9.6735	0.9023
LPUMP	790	11.1928	11.8469	10.5178	0.2574
LRXM	790	0.3076	0.7715	0.0025	0.173
LCONS	790	15.463	15.683	15.0898	0.1333

Source: Authors' compilation.

could have provoked erroneous and biased results. To overcome this problem, the interquartile outlier correction method was used. Table 3.8 shows the summary statistics of the series after the performance of the outliers' correction procedure.

To assess the integration order of the variables, both before and after the use of the interquartile outlier correction method, the most frequent integration order tests of the ADF, PP and KPSS were conducted and displayed in Table 3.9.

Even with a lack of consensus from the KPSS test, which is not robust in the presence of outliers, the tests support that all variables are I(0) in their levels, both with intercept only and with intercept and trend.

4.4.1 Methods in the daily data framework

Bearing in mind that the management of the system is carried out in real time, one would expect that all variables interact with each other leading to the effect of an endogenous adjustment. In other words, the use of some sources during the day is strongly dependent on the availability of other sources and, consequently, the dynamics of such interaction naturally bring endogeneity into the system. As such, the use of the VAR technique is required to handle these data features. This technique treats the series as potentially endogenous and evaluates the relationship without the previous prerequisite to distinguish the endogenous from exogenous series, as required by the simultaneous equations model.

Considering that the variables are I(0) in their levels, this research uses a VAR model with the variables in levels as follows:

$$Y_t = \sum_{p=1}^{k} \gamma * Y_{t-1} + \varphi * X_t + \varepsilon_t, \tag{3.2}$$

where Y_t is the vector of the endogenous variables, and X_t is the vector of the exogenous variables, γ is the coefficient matrix of endogenous variables, φ is the coefficient

Table 3.9: Integration order tests (daily data).

	ADF			PP			KPSS	
	C	CT	None	C	CT	None	C	CT
Before outlier correction								
LCOAL	−13.7866***	−13.7876***	−0.465	−17.1082***	−17.1049***	−1.3316	0.1998	0.1642**
LGAS	−17.0985***	−17.9772***	−0.3851	−18.8416***	−19.1871***	−0.5994	1.4401***	0.1572**
LNUCLEAR	−32.9116***	−32.9865***	−0.2323	−32.0691***	−32.2116***	−0.1661	0.2603	0.2199***
LHYDRO	−19.0415***	−19.0582***	−0.1764	−30.0477***	−30.0547***	−0.1609	0.6623**	0.4335***
LWIND	−20.7714***	−20.8342***	−0.1777	−22.6282***	−22.4787***	−0.5344	0.2462	0.1744**
LSOL	−4.0929***	−4.1756***	−0.0241	−24.7201***	−24.7473***	−0.6876	0.271	0.2138**
LPUMP	−21.5882***	−21.5908***	−0.4574	−21.8591***	−21.8545***	−0.5205	0.1486	0.1371*
LRXM	−4.2212***	−4.6209***	−1.8265*	−15.3335***	−16.2677***	−4.4346***	0.8409***	0.1049
LCONS	−3.2886**	−3.3368*	0.1991	−8.554***	−8.5304***	−0.0192	0.4545*	0.1682**
After outlier correction								
LCOAL	−4.2014***	−4.2373***	−0.1414	−10.6053***	−10.6095***	−0.1998	0.2061	0.1855**
LGAS	−3.6857***	−4.4395***	0.1589	−9.6525***	−11.7421***	0.1039	1.5185***	0.1834**
LNUCLEAR	−4.1281***	−4.5091***	−0.3364	−5.7433***	−6.4715***	0.0755	0.4322*	0.1029
LHYDRO	−4.3533***	−4.3390***	−0.2146	−10.5663***	−10.5884***	0.2435	0.3975*	0.3139***
LWIND	−14.0093***	−14.0676***	−0.1567	−14.3112***	−14.2626***	−0.1194	0.2365	0.1679**
LSOL	−3.4739***	−5.1746***	0.172	−7.7436***	−8.0237***	0.0529	0.2805	0.2079**
LPUMP	−4.7919***	−4.8349***	−0.1365	−28.3386***	−28.3775***	−0.0861	0.6076***	0.4054***
LRXM	−4.1245***	−4.5015***	−1.7622*	−14.4918***	−15.4313***	−4.5639***	0.8239***	0.1116
LCONS	−3.9124***	−3.9871***	0.5839	−17.9169***	−18.0786***	0.8384	0.5487**	0.1548**

ADF, Augmented Dickey-Fuller test; PP, Phillips Perron test; KPSS, Kwiatkowski Phillips Schmidt Shin; C, constant; CT, constant and trend; None, without constant and trend. ***, ** and * indicate that the statistic is significant at 1%, 5% and 10%, respectively.

matrix of exogenous variables, k is the optimal number of lags, and ε_t represents the residuals. The vector of endogenous variables is Y_t = [LCOAL,LGAS,LNUCLEAR, LHYDRO,LWIND,LSOLAR,LPUMP,LRXM], and the vector of exogenous variables is X_t = [LCONS,INTERCEPT,TREND].

The procedures that were used are as follows: (1) the selection of the optimal number of lags; (2) the implementation of Granger causality tests; (3) the implementation of the impulse response function analysis; and (4) the implementation of variance decomposition analysis. The Granger causality test allows identifying the causal relationships between the series, which occur when a particular series in the past, or in the present, helps predicting future values of another variable (Granger, 1969). The impulse response functions make it possible to analyze the behavior of the series ceteris paribus, namely based on the existing impulse in another series. In fact, it actually displays the effect that a shock in the error term, in a given period, has on the values of current and future endogenous series. The forecast error variance decomposition allows the assessment of how a series responds to shocks in the other series.

For the implementation of the VAR estimation, firstly it is important to select the optimal lag structure throughout the sequential modified LR test, the final prediction error, and the Schwartz information criterion. All tests indicate that one lag is the optimal choice. Please note that this short optimal number of lags reveals a parsimonious model and could be a sign of absence of the omission variable bias. Furthermore, so as to validate the estimated VAR model, several diagnostic tests were applied, namely the Jarque–Bera test, the autocorrelation test through the LM test, and heteroscedasticity test through the performance of the White test (without cross terms). However, the VAR specification fails all relevant diagnostic tests, namely that there is strong evidence for the presence of heteroscedasticity, nonnormality of residuals, and autocorrelation, something which is not new when working upon these high frequency data. These violations are of minor importance when a high number of observations are present, as with the case under research.

4.4.2 Results from daily data

Overall, the results of exogeneity blocks reveal that all variables except the LCONS, the constant, and the trend could be considered as endogenous (see Table 3.8 and Fig. 3.2). This supports the option for using the VAR as consistent. The Granger causality, the variance decomposition, and the impulse response functions are displayed in Tables 3.10 and 3.11 and Figs. 3.2 and 3.3.

The Granger causality analysis has detected 11 bidirectional causalities and 13 unidirectional causalities. These results further support the existence of endogeneity in the German electricity mix.

Considering that the focus of this analysis is placed on the relationship between the sources of electricity generation, the variance decomposition was performed for LCOAL,

Table 3.10: Granger causality/block exogeneity (daily data).

	LCOAL	LGAS	LNUCLEAR	LHYDRO	LWIND	LSOL	LPUMP	LRXM
LCOAL does not cause	-	7.1512***	37.8916***	70.9445***	8.0731***	1.0725	0.5029	3.3183*
LGAS does not cause	2.5582	-	11.8757***	22.8736***	0.0734	25.8412***	9.265***	2.6343
LNUCLEAR does not cause	1.6007	3.4479*	-	1.3483	1.8646	12.0744***	4.4425**	17.1099***
LHYDRO does not cause	8.1368***	12.761***	7.2001***	-	8.2374***	21.7637***	3.6128*	1.9473
LWIND does not cause	41.6447***	26.3279***	73.1541***	27.5606***	-	24.6445***	2.3328	0.0128
LSOL does not cause	4.9639**	33.8927***	42.2302***	0.4184	14.5537***	-	20.9315***	0.0002
LPUMP does not cause	13.2133***	0.0838	1.0779	4.1075**	2.0644	0.0907	-	1.3914
LRXM does not cause	19.3321***	19.1717***	31.5205***	2.1836	5.69E−05	7.6152***	0.3279	-
ALL does not cause	82.1799***	118.8194***	177.8144***	233.7692***	50.9151***	94.0554***	94.8034***	51.9457***

"All" denotes the causality test set for all independent variables. ***, **, and * denote statistical significance at 1%, 5%, and 10%, respectively.

Table 3.11: Variance decomposition.

Periods	S.E.	*LCOAL*	*LGAS*	*LNUCLEAR*	*LHYDRO*	*LWIND*	*LSOLAR*	*LPUMP*	*LRXM*
				Decomposition of *LCOAL*					
1	0.1938	100	0	0	0	0	0	0	0
2	0.2308	95.6658	0.3718	0.4529	0.6812	0.7206	0.0139	1.1106	0.9834
5	0.2657	89.7194	0.4637	0.8613	2.5577	1.2247	0.0466	2.1227	3.004
10	0.2771	86.7127	0.4494	0.8362	4.4011	1.1738	0.2139	2.2776	3.9354
15	0.2804	85.3237	0.4775	0.8432	5.3503	1.1488	0.4659	2.2752	4.1154
20	0.282	84.5043	0.4975	0.8843	5.8707	1.1391	0.6767	2.2628	4.1647
25	0.283	83.9954	0.5095	0.9292	6.1723	1.1333	0.8225	2.2526	4.1852
31	0.2837	83.6244	0.5177	0.9741	6.3802	1.1287	0.9324	2.2444	4.1982
				Decomposition of *LGAS*					
1	0.4265	17.7473	82.2528	0	0	0	0	0	0
2	0.4878	15.1955	80.3271	0.8422	0.8893	0.6701	0.9737	0.0076	1.0944
5	0.5399	13.1997	74.3525	1.4469	4.1214	0.7378	3.6064	0.0234	2.5119
10	0.5673	12.5983	68.4925	1.3424	8.1458	0.8625	5.6807	0.0899	2.7879
15	0.5817	12.4614	65.3422	1.3144	10.375	0.9329	6.5749	0.1424	2.8572
20	0.59	12.4163	63.5862	1.3444	11.581	0.9479	7.0396	0.1696	2.9149
25	0.5951	12.3894	62.5626	1.3958	12.256	0.9471	7.3074	0.1826	2.9597
31	0.5986	12.3649	61.8524	1.4597	12.701	0.9422	7.4939	0.1895	2.9964
				Decomposition of *LNUC*					
1	0.0646	13.1033	2.8584	84.0383	0	0	0	0	0
2	0.0814	9.3644	1.8133	83.1269	0.5498	2.5225	1.0409	0.0809	1.5011
5	0.1105	5.4927	0.9872	78.7069	1.8566	5.2834	3.3951	0.3039	3.9741
10	0.1346	3.9178	0.6794	75.3316	3.5312	5.5119	5.5646	0.3829	5.0805
15	0.1471	3.4666	0.606	72.9158	5.0174	5.2257	6.9425	0.3706	5.4555
20	0.1544	3.3295	0.5934	70.9967	6.2609	4.9603	7.8669	0.3477	5.6446
25	0.1589	3.3072	0.5999	69.5188	7.2366	4.7634	8.4902	0.3298	5.7541
31	0.1623	3.3311	0.6136	68.2407	8.0883	4.6041	8.9756	0.3164	5.8303
				Decomposition of *LHYDRO*					
1	0.1262	13.7482	12.6613	0.2409	73.3495	0	0	0	0
2	0.1511	9.6242	9.8259	0.2919	77.7605	2.0156	0.0195	0.3411	0.1214
5	0.1847	8.8637	6.6725	0.2304	78.2015	3.949	0.6414	1.1992	0.2423
10	0.2099	10.6529	5.4593	0.2369	74.3337	3.9467	2.8358	1.5743	0.9603
15	0.2219	11.2062	5.1488	0.2173	72.3282	3.7919	4.3191	1.626	1.3624
20	0.2282	11.3466	5.0154	0.2453	71.3342	3.6825	5.1833	1.6223	1.5703
25	0.2318	11.3762	4.9419	0.3115	70.7696	3.6072	5.6924	1.6096	1.6916
31	0.2343	11.3726	4.8904	0.4039	70.3651	3.5481	6.0449	1.5958	1.7792
				Decomposition of *LWIND*					
1	0.6208	9.0485	0.2131	8.0459	0.0820	82.6063	0.0042	0	0
2	0.7119	12.6544	0.5316	6.8921	0.2427	78.7573	0.7349	0.1869	3.70E−05
5	0.7718	14.9511	0.5812	6.0129	1.1877	74.4062	2.3562	0.3104	0.1942
10	0.7866	14.6747	0.6028	5.8119	2.4065	72.5762	3.3146	0.3059	0.3075
15	0.7918	14.5245	0.6391	5.7616	3.0783	71.7607	3.6098	0.3198	0.3063
20	0.7945	14.4889	0.6577	5.7286	3.4359	71.3106	3.7326	0.3303	0.3153
25	0.796	14.4794	0.6669	5.7078	3.6285	71.0603	3.7956	0.3358	0.3258
31	0.7969	14.4744	0.6722	5.6954	3.7492	70.8982	3.8369	0.3386	0.3352

(Continued)

Table 3.11: Variance decomposition. (*cont.*)

Periods	S.E.	LCOAL	LGAS	LNUCLEAR	LHYDRO	LWIND	LSOLAR	LPUMP	LRXM
				Decomposition of *LSOLAR*					
1	0.4412	0.8676	0.4865	4.5290	0.6217	0	93.4952	0	0
2	0.5465	0.6318	0.5651	3.1984	1.6827	0.8569	92.688	0.007	0.3699
5	0.6803	0.4505	1.4677	2.1015	7.5192	2.3498	85.4827	0.0103	0.6185
10	0.7716	0.6309	2.0348	1.8493	15.7552	2.8892	76.2054	0.0839	0.5512
15	0.8189	1.2507	2.1693	1.8829	19.9909	2.8544	70.9173	0.1653	0.7692
20	0.8467	1.7606	2.2065	1.9997	22.1039	2.7579	67.9515	0.2128	1.0072
25	0.8634	2.0768	2.2174	2.1399	23.2164	2.6813	66.2512	0.2361	1.1809
31	0.8753	2.2868	2.2197	2.2958	23.9187	2.6206	65.0955	0.2484	1.3145
				Decomposition of *LPUMP*					
1	0.2327	1.0101	0.2519	0.7796	0.8032	8.3652	1.5795	87.1816	0.0289
2	0.2392	0.9858	0.5949	0.7869	1.1679	9.1025	2.1122	85.2073	0.0425
5	0.2459	0.9417	1.0774	1.0142	2.5395	9.4226	4.2075	80.7433	0.0537
10	0.2514	1.0386	1.2709	1.091	4.4915	9.4093	5.3482	77.2849	0.0655
15	0.2542	1.2061	1.3249	1.0859	5.5983	9.326	5.7353	75.6238	0.0997
20	0.2557	1.3284	1.3472	1.0738	6.1914	9.2579	5.9202	74.7453	0.1356
25	0.2566	1.4009	1.3578	1.0694	6.5152	9.2122	6.0234	74.2584	0.1628
31	0.2572	1.4471	1.3638	1.0729	6.7225	9.1783	6.0939	73.9369	0.1844
				Decomposition of *LRXM*					
1	0.1216	2.7377	0.7321	0.2269	0.9971	24.8464	1.7147	0	68.7451
2	0.141	2.0488	0.5499	0.6715	1.2766	24.5694	1.6643	0.1233	69.0962
5	0.1532	2.2234	0.4951	1.9142	2.4218	22.618	1.4886	0.1503	68.6887
10	0.1581	2.8019	0.5107	3.4124	3.8414	21.2769	1.7992	0.1445	66.213
15	0.1607	2.9822	0.5258	4.3232	4.7218	20.6097	2.2719	0.1419	64.4234
20	0.1624	3.0532	0.5388	4.8569	5.3081	20.1934	2.6396	0.1402	63.2698
25	0.1635	3.0934	0.549	5.1694	5.7072	19.9249	2.8939	0.1393	62.5229
31	0.1643	3.1247	0.5577	5.3821	6.0221	19.7227	3.091	0.1388	61.9609

Source: Authors' compilation.

LGAS, LNUCLEAR, LHYDRO, LWIND, LSOLAR, LPUMP, and LRXM, thus revealing the impacts from shocks on the other variables. Table 3.10 shows the results of variance decomposition. In fact, the variance decomposition results are in line with the results obtained from the exogeneity tests. Therefore all variables reveal a dynamic behavior, which is a requirement for endogeneity.

With regard to the LCOAL, after a two-day lag, shocks to LCOAL explain about 95% of the forecast error variance. This impact is reduced to 83.6% at the end of day 31. When comparing the shocks to both LPUMP and LRXM and the shocks to both LNUCLEAR, LWIND, and LSOLAR, the shocks to LPUMP and LRXM explain substantially a larger percentage of the forecast error variance than the shocks to both LNUCLEAR, LWIND, and LSOLAR, that is, about 2.25, 4.2, and 1%, respectively at the end of day 31. Shocks to the LHYDRO gain strength consistently, and jump from around 0.7%–6.4% in the explanation of the forecast error variance.

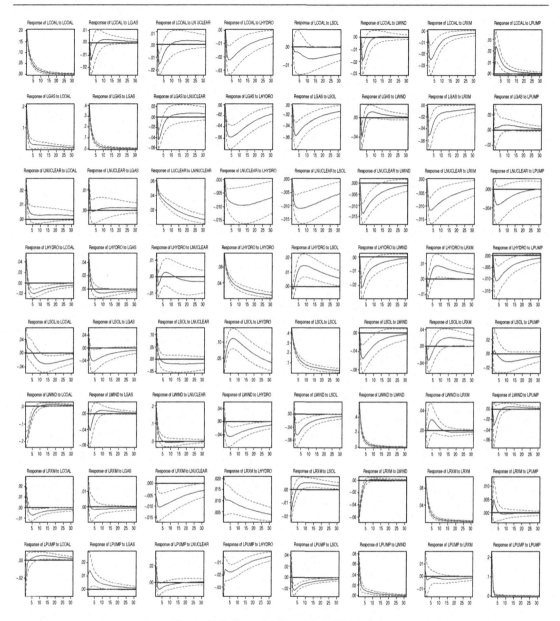

Figure 3.3: Impulse Response Function.
Dashed lines denote innovations at ±2 standard errors.

When considering the LGAS, the shock to LCOAL and LHYDRO accounts for part of the forecast error variance, namely around 12% for each. On the one hand, shocks to LHYDRO gain power considerably. The two-day lag explains 1% of the forecast error variance, whereas the 31st day explains 12%. On the other hand, shocks to LCOAL decrease from 17.7%, in the two-day lag, to 12% in the 31st day. Furthermore, the height of LSOLAR should be emphasized. The shocks on them explain 7.5% of the forecast error variance. Therefore after a two-day lag, shocks to LGAS explain around 80% of the forecast error variance. However, this impact is reduced to about 61.8% at the end of the day 31.

Regarding the LNUCLEAR, after the 31st-day lag, shocks to LNUCELAR explain about 68% of the forecast error variance. In fact, their explanation is also reduced in comparison with the two-day lag. When comparing the shocks to LHYDRO and LSOLAR with the shocks to LWIND and LRXM, the former explains substantially a larger percentage of the forecast error variance than the shocks to both LWIND and LRXM, that is, about 8, 9, 4.6, and 5.8%, respectively on day 31. However, shocks to LCOAL decay consistently, and decrease from 13.1% on the 1-day lag to 3.3% in the 31-day lag.

When considering the LHYDRO, the shocks to both LCOAL, LSOLAR, LGAS, and LWIND explain part of the forecast error variance, around 11, 6, 4.8, and 3.5%, respectively, at the end of the 31-day lag. Usually, this source has a backup role in the electricity mix. However, in the German electricity mix, this source plays more often a base load role more often. With regard to LWIND, after a two-day lag, shocks to LWIND explain around 78% of the forecast error variance. This impact is reduced to about 70% at the end of the 10-day lag. In fact, the high ECM value of the WIND ARDL model also proves that the wind power has a high adjustment speed, while the other sources do not have a larger impact on them, although wind power has a larger impact on the other sources. Accordingly, the electricity production system plans the electricity generation in accordance with the availability of the electricity production from wind power. Furthermore, the shocks to LCOAL and LNUCLEAR explain a larger percentage of the forecast error variance, about 14.4% and 5.7%, respectively, at the end of 31st day. Indeed, this result also proves the existence of a substitution effect between coal and nuclear sources.

When focusing on LSOLAR, after a two-day lag, shocks to LSOALR explain 92% of the forecast error variance. This impact is largely reduced to 65% at the end of the 31st day. However, shocks to LHYDRO gain strength consistently, and jump from about 1.7% to 24% in the explanation of the forecast error variance. When comparing the shocks to LCOAL, LGAS, LNUCLEAR and LWIND, each one explains a reduced percentage of the forecast error variance, about 2% each in the thirty-one-day. With reference to both LPUMP and LRXM, the shocks to LPUMP and LRXM explain around 85% and 69% of the forecast error variance of LPUMP and LRXM, respectively. At the end of the 31st day, these impacts are reduced by around 10%. Regarding the comparison of the shocks to both LHYDRO, LWIND and LSOLAR with the shocks to LCOAL, LGAS and LNUCLEAR on the forecast error variance decomposition of LPUMP and LRXM; Being the cross-border markets and pumping systems used for managing the scarcity and excess of electricity production from RES, the shocks to both LHYDRO, LWIND and LSOLAR explain a substantially a larger percentage of the forecast error variance than the shocks to LCOAL, LGAS and LNUCLEAR.

The impulse response-functions to an innovation are shown in Fig. 3.3. In general, the responses to an innovation in the electricity mix are dissimilar. This means that some sources react and respond more quickly to a shock, thereby returning faster to the equilibrium. However, it should be highlighted that all the impulse response functions tend to converge to zero, which is an additional proof of robustness and appropriateness of the VAR technique used.

Overall, the comparison between the VAR results with the results of the ARDL models support proves that the two methodologies produce actually similar and statistically significant effects. Moreover, it is worth noting that when fossil fuels are analyzed separately, one can perceive the baseload role of nuclear and coal power plants, and the backup role of hydropower in the German electricity production system. In fact, the nuclear phase-out has stimulated the RES deployment. Renewables have required flexible generation plants, like natural gas, to back up the volatile electricity production from wind power and solar PV.

5 The Pertinence of Studying Energy Consumption—Economic Growth Nexus by Source, in Germany: Conditions of a Successful Energy Transition

The analyses of the electricity–growth nexus and the dynamics of the interactions of the electricity generation sources, in a country where the energy transition is being implemented at a high speed, is an extraordinary experience. First, there is strong evidence for the growth hypothesis for all the generation sources, except for hydropower. The electricity consumption causes economic activity. This backs up the growth hypothesis detailed by source, both in the short and in the long-run. Moreover, there is strong support for the feedback hypothesis in the following sources: fossil, wind, and solar PV. This is of particular interest and deserves particular and careful discussion. The effect of the new intermittent renewables on the economic activity is not uniform in literature, as previously stated. There is already evidence for no significant impact or even a negative impact from RES to growth. The negative impact is mainly observed for the wind power. In Germany, both solar PV and wind power increase economic growth, which could be a signal of maturity of these two technologies in Germany, or even of efficiency in their accommodation into the electricity system. This fact represents some support for the decision to replace the conventional nuclear source by RES. This means that the rationale for such replacement is observed in historical data. Moreover, and partially in line with that literature, there is confirmation of the contribution from solar PV, being at a larger amount than wind power, to the economic growth. With great internal consistency, the economic activity stimulates more solar PV, as it actually should, due to the multiplier effect of this source on the regional dynamics, as previously discussed.

Overall, one gets the idea that the country could be less dependent on nuclear power than it might seem. It appears that Germany was preparing the way toward the energy transition from nuclear energy toward renewables in the last years, and this was not only a consequence of the external factors or signals, such as the nuclear accident of Fukushima. Specifically, one is able to confirm the presence of the substitution effect between different sources. Within the renewables, there is evidence for the substitution effect between solar PV and wind power. However, this substitution effect was not observed for the fossil fuels consumption which is connected with renewables' production. This is in accordance with what expected, that is, some fossil sources back up renewables. Conversely, it is remarkable to see that in the nuclear model, there seems to be evidence for a process of preparing the nuclear phase-out and replacement process. This is evident both through the complementarity effect with fossils, namely natural gas as proved in the daily data framework, and by the effect of replacing the role of the nuclear baseload by resorting to the Nord Pool electricity market.

Germany enjoys unique conditions for making a successful energy transition, without compromising economic growth. In fact, there seems to be rationality in the German system. The access to the most developed electricity market of Nord Pool is a crucial piece. As shown in the IPI model, the coverage ratio of the electricity imports by exports feeds the economic activity. Moreover, the feedback hypothesis is not confirmed on the nuclear source, which seems to be somehow compatible with the maintenance of the living standards of German people, given that demand for additional power with the increased economic activity has not been fulfilled by the nuclear energy, both in short and in the long run. Additionally, there is strong evidence for a particular characteristic related with the backup role. Whereas in most countries the backup has to be guaranteed locally, it seems that Germany is also benefiting from shared backups through the Nord Pool market. This backup is famous for the solar PV, which allows this source to be deployed. This creates additional conditions for an efficient transition of sources.

6 The Case for a Demand Side Management (DSM) Intervention

The accommodation of renewables into the national electricity systems often requires the contribution of the demand side, to better match the time consumption with the time of generation (in a better way). This varies in accordance with the availability of natural resources. This match cannot be dissociated from the period of generation. This means that the traditional notion of peak demand needs to be reformulated. The new concept of net demand, by considering the supply of renewables, should become the main reference for all the stakeholders, especially policy makers. This section is dedicated to the debate about possible demand side measures aimed at making the diversification of the electricity mix economically sustainable.

There are three possible ways to deal with the balance in the electric system: (1) managing the supply of electricity, namely by adding generation capacity to satisfy any level of instantaneous demand; (2) using management instruments to shape the consumption, namely by smoothing the load curve during the day; and (3) a combination of both (1) and (2). In a context of energy transition, the greater challenge is to adjust the distribution of loads without compromising economic growth, in particular when the growth hypothesis is confirmed.

The DSM is a well-established solution to lead both the giant growth of electricity as well as the intermittent alternative electricity sources. In fact, the incorporation of renewables on a large-scale requires a new flexibility to balance the demand side with the supply availability. The DSM considers a wide set of means which can influence the duration and the intensity of the final consumption of electricity. Therefore the DSM includes two main areas: (1) the energy efficiency; and (2) the demand response. Energy efficiency consists of using less energy to perform the same tasks. This implies a permanent reduction in demand. Regarding the demand response, it consists of any method to reduce, flatten, or transfer the energy demand. The second area is also known as the Demand Side Response (DSR).

Largely, the DSM, and particularly the DSR, operates at the end users' level. It influences the transformation of the consumption pattern, namely by reallocating the electricity load from periods with a large stress on the system, toward periods in which there is no need to turn on the backup facilities. Therefore it is more important to enlarge the consumption in renewables' high availability periods. However, such strategy would be successful if there is a scale in the response. This means that a substantial number of consumers have the incentives to share the economic benefits to change their consumption habits without a severe loss of welfare and living style. To be effective, regulation is needed so as to involve the largest possible number of consumers.

When designing the regulation measures, the classic forms of smoothing the load curve ought to become priorities. These are the strategies of "peak clipping", "valley filling", and "load shifting". Moreover, the change of loads, namely the forms of strategic load conservation and strategic load growth, should be fully understood. The consumers, who are increasingly "prosumers", should understand the benefits of participating in both the DSM and DSR programs, namely by picking up a share of the released economic benefit, for instance, by paying a higher price when they are consuming above a certain level of installed capacity, and/or having an incentive when they move consumption toward periods of idle capacity.

Another crucial issue in the energy transition refers to the predictability. The predictability in the generation is mainly dependent, on the one hand, on the improvements in the technology of generation, for instance by enlarging the efficiency of the wind turbines or solar PV panels. On the other hand, it is dependent on the capacity of resources forecast, such as wind speed forecast for a long period. Regarding the impact of consumption on the predictability, this is actually crucial. Predictability is a critical factor in any production activity, and it guarantees lower prices by itself. Greater predictability means increased rates of use for the generation facilities. Greater smoothing in the load diagram can be achieved, on the one hand, by allowing the usage of more facilities in off peak time; otherwise they should be shut down. On the other hand, by cutting off peaks and valleys, while dropping new backup facilities which have very low utilization rates.

Price differentiation, depending on the period in which the electricity is consumed, is a form of Peak-Load Pricing, and it is an instrument of demand rationing. In addition to the rates based on the time of consumption, also known as tariffs Time-Of-Use (TOU), other solutions such as Critical Peak Pricing rates (CPP) and real-time Pricing (RTP), should also be available. These last two are considered to be dynamic rates, because the price evolves from nonstandard, static form, as happens with the TOU. Authors such as Park et al. (2015) summarize the advantages and disadvantages of each of these rates. Additionally, another kind of instrument of price-based demand response program is the Peak Time Rebates (PTR). By applying rebates when the consumption is reduced in certain periods with maximum load, the consumers are encouraged to change behaviors and into more efficient equipment (Alasseri et al., 2017). With the exception of the TOU tariffs, the other options do require a communication mechanism between supplier and consumer, leading us to conclude about the high pertinence of using smart metering on a wide basis.

With reference to the notion of net demand previously discussed, the accommodation of intermittent sources without storage capacity of the primary power souparticular emphasis on the trerce, leads to the redefinition of what should be considered as peak period and nonpeak period. Usually, a maximum peak period should not be there, if one considers the net load of the system. Therefore first the demand must be discounted by renewables supply, and after that the period can be classified as a peak one. The net demand of the system is then obtained by the difference between the total load and the instantaneous supply. Consequently, the DSR programs must be focused on instantaneously removing the load when the supply from renewables' decreases, as well as enlarging the load in periods of natural resources abundance, as it happens with wind and solar energy. Adapted to the notion of net demand, the PTR could be a powerful instrument for the load demand shift, such as illustrated in Fig. 3.4.

The dotted arrows represent the load that can be moved from the valley periods toward the peak periods whenever it can be met through the renewable electricity generation. In short,

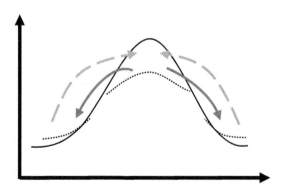

Figure 3.4: Net Demand and Load Shift.

the DSM and DSR instruments could be of outstanding usefulness to diminish the asymmetry of information in the sector, the cause of market power abuse. They could bring additional predictability to the electricity system, thus benefiting the social welfare.

7 Conclusions and Policy Implications

This chapter focuses on the analysis of one of the most exciting research topics in energy economics nowadays, which is the recomposition of the electricity mix, by incorporating the intermittent renewable sources, and the consequences of this diversification in the economic growth. First, the chapter shortly revises the literature about the EGN, with particular emphasis on the trend of the evolution from energy toward electricity and from aggregated consumption toward disaggregated analysis, by considering separately each generation source. It examines the interactions between sources and the relationship between the latter and economic growth, aiming to conclude on the conditions necessary to make the energy transition suitable. To do that, the chapter empirically assesses the case of Germany, one of the countries with greater commitment in this energy transition.

In Germany, there is strong empirical evidence supporting the growth hypothesis for all the generation sources, other than hydropower. Contrary to other countries, usually countries with less wealth, the intermittent renewables sources have not been hampering economic growth. This is a first indication that the substitution of resources has not contributed to the reduction of economic growth. Regarding the new renewables, looking at the situation in this country confirms what the literature has recently shown, namely about the great relevance of the solar PV source on the strengthening of economic activity.

Overall, one proves the need for verification of the appropriate conditions to carry out the energy transition, by particularly ensuring the shared backups through a market with considerable depth. The role of baseload nuclear power could be played both by the natural gas source and by the Nord Pool market. The membership of a developed electricity market seems crucial for the successful energy transition, and as such the energy policies should be focused on encouraging the deployment of electricity markets in places where they are not yet developed. The German system meets privileged conditions to make the phaseout of nuclear power, which doesn't seem to be the case with other European countries, such as the Spanish and the Portuguese case, due to the narrowness of the market. In short, because of the great ability to add value to their products, the Germany productive structure will be likely to transfer some of the additional costs from renewables to the price of products, without compromising its competitiveness. Moreover, they can back up the alternative sources using the electricity market, thus benefiting from a kind of shared backups. Finally, some measures are suggested, particularly in the DSM, given that a critical success factor in the diversification of the mix is the instantaneous equilibrium between supply and demand of electricity.

Acknowledgments

The financial support of the NECE-UBI, Research Unit in Business Science and Economics, sponsored by the Portuguese Foundation for the Development of Science and Technology, project UID/GES/04630/2013, is gratefully acknowledged.

References

Al-mulali, U., Fereidouni, H.G., Lee, J.Y.M., 2014. Electricity consumption from renewable and non-renewable sources and economic growth: evidence from Latin American countries. Renew. Sustainable Energy Rev. 30, 290–298. doi: 10.1016/j.rser.2013.10.006.

Alasseri, R., Tripathi, A., Joji Rao, T., Sreekanth, K.J., 2017. A review on implementation strategies for demand side management (DSM) in Kuwait through incentive-based demand response programs. Renew. Sustainable Energy Rev. 77, 617–635. doi: 10.1016/j.rser.2017.04.023.

Amri, F., 2017. Intercourse across economic growth, trade and renewable energy consumption in developing and developed countries. Renew. Sustainable Energy Rev. 69, 527–534. doi: 10.1016/j.rser.2016.11.230.

Apergis, N., Payne, J.E., 2012. Renewable and non-renewable energy consumption-growth nexus: evidence from a panel error correction model. Energy Econ. 34, 733–738. doi: 10.1016/j.eneco.2011.04.007.

Apergis, N., Payne, J.E., 2014. Renewable energy, output, CO_2 emissions, and fossil fuel prices in Central America: evidence from a nonlinear panel smooth transition vector error correction model. Energy Econ. 42, 226–232. doi: 10.1016/j.eneco.2014.01.003.

Apergis, N., Payne, J.E., Menyah, K., Wolde-Rufael, Y., 2010. On the causal dynamics between emissions, nuclear energy, renewable energy, and economic growth. Ecol. Econ. 69, 2255–2260. doi: 10.1016/j.ecolecon.2010.06.014.

Asafu-Adjaye, J., Byrne, D., Alvarez, M., 2016. Economic growth, fossil fuel and non-fossil consumption: a Pooled Mean Group analysis using proxies for capital. Energy Econ. 60, 345–356. doi: 10.1016/j.eneco.2016.10.016.

Baum, C.F., 2004. A review of Stata 8.1 and its time series capabilities. Int. J. Forecasting 20, 151–161. doi: 10.1016/j.ijforecast.2003.11.007.

Ben Aïssa, M.S., Ben Jebli, M., Ben Youssef, S., 2014. Output, renewable energy consumption and trade in Africa. Energy Policy 66, 11–18. doi: 10.1016/j.enpol.2013.11.023.

Ben Jebli, M., Ben Youssef, S., 2015. The environmental Kuznets curve, economic growth, renewable and non-renewable energy, and trade in Tunisia. Renew. Sustainable Energy Rev. 47, 173–185. doi: 10.1016/j.rser.2015.02.049.

Bhattacharya, M., Paramati, S.R., Ozturk, I., Bhattacharya, S., 2016. The effect of renewable energy consumption on economic growth: evidence from top 38 countries. Appl. Energy 162, 733–741. doi: 10.1016/j.apenergy.2015.10.104.

Bildirici, M.E., Gökmenoğlu, S.M., 2016. Environmental pollution, hydropower energy consumption and economic growth: evidence from G7 countries. Renew. Sustainable Energy Rev.doi: 10.1016/j.rser.2016.10.052.

Bollerslev, T., 1986. Generalized autoregressive conditional heteroskedasticity. J. Econ. 31, 307–327.

Breusch, T.S., 1978. Testing for autocorrelation in dynamic linear models. Aust. Econ. Pap. 17, 334–355.

Cerdeira Bento, J.P., Moutinho, V., 2016. CO_2 emissions, non-renewable and renewable electricity production, economic growth, and international trade in Italy. Renew. Sustainable Energy Rev. 55, 142–155. doi: 10.1016/j.rser.2015.10.151.

Chang, T., Gupta, R., Inglesi-Lotz, R., Simo-Kengne, B., Smithers, D., Trembling, A., 2015. Renewable energy and growth: evidence from heterogeneous panel of G7 countries using Granger causality. Renew. Sustainable Energy Rev. 52, 1405–1412. doi: 10.1016/j.rser.2015.08.022.

Dickey, B.Y.D.A., Fuller, W.A., 1981. Likelihood ratio statistics for autoregressive time series with a unit root. Econometrica 49, 1057–1072.

Dogan, E., 2015. The relationship between economic growth and electricity consumption from renewable and non-renewable sources: a study of Turkey. Renew. Sustainable Energy Rev. 52, 534–546. doi: 10.1016/j.rser.2015.07.130.

Dogan, E., 2016. Analyzing the linkage between renewable and non-renewable energy consumption and economic growth by considering structural break in time-series data. Renewable Energy 99, 1126–1136. doi: 10.1016/j.renene.2016.07.078.

Frondel, M., Sommer, S., Vance, C., 2015. The burden of Germany's energy transition: an empirical analysis of distributional effects. Econ. Anal. Policy 45, 89–99. doi: 10.1016/j.eap.2015.01.004.

Furuoka, F., 2016. Renewable electricity consumption and economic development: new findings from the Baltic countries. Renew. Sustainable Energy Rev. 71, 450–463. doi: 10.1016/j.rser.2016.12.074.

Granger, C.W.J., 1969. Investigating causal relations by econometric models and cross-spectral methods. Econometrica 37, 424–438. doi: 10.2307/1912791.

Hamdi, H., Sbia, R., Shahbaz, M., 2014. The nexus between electricity consumption and economic growth in Bahrain. Econ. Model. 38, 227–237. doi: 10.1016/j.econmod.2013.12.012.

Inglesi-Lotz, R., 2015. The impact of renewable energy consumption to economic growth: a panel data application. Energy Econ. 53, 58–63. doi: 10.1016/j.eneco.2015.01.003.

Ito, K., 2017. CO_2 emissions, renewable and non-renewable energy consumption, and economic growth: evidence from panel data for developing countries. Int. Econ., 1–6. doi: 10.1016/j.inteco.2017.02.001.

Jarque, C.M., Bera, A.K., 1980. Efficient tests for normality, homoscedasticity and serial independence of regression residuals. Econ. Lett. 6, 55–259.

Jouini, J., 2015. Economic growth and remittances in Tunisia: bi-directional causal links. J. Policy Model. 37, 355–373. doi: 10.1016/j.jpolmod.2015.01.015.

Kahia, M., Ben Aïssa, M.S., Charfeddine, L., 2016. Impact of renewable and non-renewable energy consumption on economic growth: new evidence from the MENA Net Oil Exporting Countries (NOECs). Energy 116, 102–115. doi: 10.1016/j.energy.2016.07.126.

Koçak, E., Şarkgüneşi, A., 2017. The renewable energy and economic growth nexus in black sea and Balkan countries. Energy Policy 100, 51–57. doi: 10.1016/j.enpol.2016.10.007.

Kwiatkowski, D., Phillips, P.C.B., Schmidt, P., Shinb, Y., 1992. Testing the null hypothesis of stationary against the alternative of a unit root. J. Econ. 54, 159–178.

Marques, A., Fuinhas, J., 2012. Is renewable energy effective in promoting growth? Energy Policy 46, 434–442. doi: 10.1016/j.enpol.2012.04.006.

Marques, A.C., Fuinhas, J.A., 2015. The role of Portuguese electricity generation regimes and industrial production. Renew. Sustainable Energy Rev. 43, 321–330. doi: 10.1016/j.rser.2014.11.053.

Marques, A.C., Fuinhas, J.A., 2016. How electricity generation regimes are interacting in Portugal. Does it matter for sustainability and economic activity? J. Renewable Sustainable Energy 8, 25902. doi: 10.1063/1.4944959.

Marques, A.C., Fuinhas, J.A., Menegaki, A., 2014. Interactions between electricity generation sources and economic activity in Greece: a VECM approach. Appl. Energy 132, 34–46. doi: 10.1016/j.apenergy.2014.06.073.

Marques, A.C., Fuinhas, J.A., Nunes, A.R., 2016. Electricity generation mix and economic growth: what role is being played by nuclear sources and carbon dioxide emissions in France? Energy Policy 92, 7–19. doi: 10.1016/j.enpol.2016.01.027.

Nyblom, J., 1989. Testing for the constancy of parameters over time. J. Am. Stat. Assoc. 84, 223–230. doi: 10.1080/01621459.1989.10478759.

Ocal, O., Aslan, A., 2013. Renewable energy consumption-economic growth nexus in Turkey. Renew. Sustainable Energy Rev. 28, 494–499. doi: 10.1016/j.rser.2013.08.036.

Omri, A., 2014. An international literature survey on energy-economic growth nexus: evidence from country-specific studies. Renew. Sustainable Energy Rev. 38, 951–959. doi: 10.1016/j.rser.2014.07.084.

Park, S.C., Jin, Y.G., Song, H.Y., Yoon, Y.T., 2015. Designing a critical peak pricing scheme for the profit maximization objective considering price responsiveness of customers. Energy 83, 521–531. doi: 10.1016/j.energy.2015.02.057.

Payne, J.E., 2009. On the dynamics of energy consumption and output in the US. Appl. Energy 86, 575–577. doi: 10.1016/j.apenergy.2008.07.003.

Pesaran, M.H., Shin, Y., Smith, R.J., 2001. Bounds testing approaches to the analysis of level relationships. J. Appl. Econ. 16, 289–326. doi: 10.1002/jae.616.

Phillips, P.C.B., Perron, P., 1988. Testing for a unit root in time series regression. Biometrika 75, 335–346.

Rafindadi, A.A., Ozturk, I., 2016. Impacts of renewable energy consumption on the German economic growth: evidence from combined cointegration test. Renew. Sustainable Energy Rev., 0–1. doi: 10.1016/j. rser.2016.11.093.

Ramsey, J.B., 1969. Tests for specification errors in classical linear least-squares regression analysis. J. R. Stat. Soc. Ser. B 31, 350–371.

Saidi, K., Ben Mbarek, M., 2016. Nuclear energy, renewable energy, CO_2 emissions, and economic growth for nine developed countries: evidence from panel Granger causality tests. Prog. Nucl. Energy 88, 364–374. doi: 10.1016/j.pnucene.2016.01.018.

Salim, R.a., Hassan, K., Shafiei, S., 2014. Renewable and non-renewable energy consumption and economic activities: further evidence from OECD countries. Energy Econ. 44, 350–360. doi: 10.1016/j. eneco.2014.05.001.

Sebri, M., 2015. Use renewables to be cleaner: meta-analysis of the renewable energy consumption–economic growth nexus. Renew. Sustainable Energy Rev. 42, 657–665. doi: 10.1016/j.rser.2014.10.042.

Shahbaz, M., Loganathan, N., Zeshan, M., Zaman, K., 2015. Does renewable energy consumption add in economic growth? An application of auto-regressive distributed lag model in Pakistan. Renew. Sustainable Energy Rev. 44, 576–585. doi: 10.1016/j.rser.2015.01.017.

Shahbaz, M., Tang, C.F., Shahbaz Shabbir, M., 2011. Electricity consumption and economic growth nexus in Portugal using cointegration and causality approaches. Energy Policy 39, 3529–3536. doi: 10.1016/j. enpol.2011.03.052.

Tang, C.F., Tan, E.C., 2012. Electricity consumption and economic growth in Portugal: evidence from a multivariate framework analysis. Energy J. 33, 23–48.

Verzijlbergh, R.A., De Vries, L.J., Dijkema, G.P.J., Herder, P.M., 2016. Institutional challenges caused by the integration of renewable energy sources in the European electricity sector. Renew. Sustainable Energy Rev., 0–1. doi: 10.1016/j.rser.2016.11.039.

Zivot, E., Andrews, D.W.K., 1992. Further evidence on the great crash, the oil-price shock, and the unit-root hypothesis. J. Bus. Econ. Stat. 10, 251. doi: 10.2307/1391541.

Further Reading

Marques, A.C., Fuinhas, J.A., Afonso, T.L., 2015. The dynamics of the Italian electricity generation system: an empirical assessment. WSEAS Trans. Bus. Econ. 12, 229–238.

CHAPTER 4

The Role of Potential Factors/Actors and Regime Switching Modeling

Roula Inglesi-Lotz

University of Pretoria, Pretoria, South Africa

Chapter Outline

1 Introduction

Energy is considered by many as the backbone of the economy, without which, the vast majority of economic activities would not be functional. Appropriate price structures, availability, and sustainability of energy supplies are crucial for the survival and living standards of the global population. The energy, as an industrial sector of the economy as well, creates job opportunities, and a little value is added during the extraction, generation, and distribution of energy goods and services. Peter Voser, in the World Economic Forum report titled "Energy for Economic Growth. Energy Vision Update 2012" (WEF and IHS CERA, 2012) stated:

> Energy is the oxygen of the Economy and the life blood of growth, particularly in the mass industrialization phase that emerging economic giants are facing today.

The Economics and Econometrics of the Energy-Growth Nexus
http://dx.doi.org/10.1016/B978-0-12-812746-9.00004-3

The recent energy literature makes effort to explore and shed light on the dynamics and mechanisms of the relationship between economic growth and energy consumption. Two main opposing perspectives are expressed: (1) energy is a key and invaluable factor of economic production, as none of the other production factors can function without it; and (2) energy is a neutral factor to economic growth as its cost is still low as a share to the total economic output (Ghali and El-Sakka, 2004).

Regardless of the interest of the literature into establishing a specific "rule" for this relationship, consensus can still not be reached. In general, the energy–economic growth nexus (EGN) can potentially be illustrated by the following four hypotheses: Growth, Conservation, Feedback, and Neutrality. According to the growth hypothesis, energy causes economic growth. If, as a result of policies energy use decreases, then GDP growth will also be restricted. The conservation hypothesis suggests that economic growth is a contributing factor to the energy consumption trend, which is positive for energy saving policies that will not affect the levels of economic growth. Next, under the feedback hypothesis, there is a bidirectional relationship between energy and economic consumption, and hence they are interdependent. Finally, the neutrality hypothesis states that the two phenomena are independent to each other.

As Inglesi-Lotz and Pouris (2016) pointed out, various modeling exercise findings varied due to different time periods, econometric, and other quantitative approaches and under a multivariate context due to the group of variables included in the modeling too. Methodologically, Stern (1993, p. 144) identified potential reasons for the disagreement of results in the literature: "errors of measurements in variables, omitted variables, and inappropriate functional form." Also, Rahman and Mamun (2016, p. 809) pointed out that "country heterogeneity with respect to stages of economic agreements, country heterogeneity with respect to stages of economic growth, energy use patterns, trade patterns, and trade volume may also be reasons for not reaching conclusive results with regard to the nature of the relationship between energy and economic growth in the literature." The four hypotheses, hence, are considered easy to investigate; however, the existence of other factors can influence the findings and research setup must be foresightful.

Not only endogenous but also exogenous factors are found to impact on the existence and direction of causality between energy and economic growth, which could be potentially enlightening for the investigation of the relationship (Karanfil, 2009). This chapter aims to critically review the literature and advocates the importance and magnitude of several of these determinants as discussed in various studies. Among others, the list will include production factors, international trade, and financial development as well as less discussed ones such as militarization and tourism. In addition, regime switching and time varying modeling are considered and presented which represent a different cadre within which an economy works.

2 Determinants of the Relationship Between Energy and Economic Growth

2.1 Production Factors

Economic theory traditionally states that capital, labor, and land are the main factors of any productive activity, while goods such as fuels are intermediate inputs—also essential for the production both at micro-level and macro-level (Stern, 2004). In the past, the focus of economic-growth theory literature was placed on main factors: capital and natural resources. The reason for that is primarily that the financial gains of not only the primary factors, but also the intermediate inputs, were received by the owners of the main factors.

Nowadays, with the advancement of technology, hardly any of the primary factors can be used in the production without the use of energy; additionally, even the transport and the distribution of products are impossible without energy. The role of energy in the growth process is appreciated in the current literature (even at a lesser extent) as energy resources are becoming limited, considered no reproducible, compared to the primary factors of production, and their use should be managed efficiently and effectively.

Energy is vital in all production processes as they include transformation at some stage. At a macro-level particularly, all economic processes—even the service activities—require the use of energy even indirectly in maintenance and production of labor or capital goods; or even in the process of knowledge accumulation and technological progress. All in all, energy must be incorporated into machinery, workers, as well as natural resources and materials to be made productive. Most importantly, it should be stressed here that energy cannot substitute the rest of the production factors.

In neoclassical economics, energy is considered endogenous to the economy in any time period: physical and economic constraints restrict its availability and the efficiencies with which these functions operate (Stern, 2004). Alternative growth models suggest that energy is the main factor of production, which has as a consequence the treatment of energy as an exogenous factor in growth models, while capital and labor are treated as flows rather than stocks.

A general production function is used next to represent factors that are conducive or restrictive to the relationship between energy consumption and economic production (adopted by Stern (2004)):

$$(Q_1,....,Q_m)' = f\left(A,X_1,....,X_n,E_1,....E_p\right) \tag{4.1}$$

where Q denotes outputs from several economic sectors; X denotes the inputs/production factors such as capital and labor; E the diverse energy inputs; and A the level of technology in the production. There are a variety of factors that may affect the nexus between energy E and Q from a production perspective (Stern, 2004):

- Substitution of energy and other factors;
- Technological advancements;

- Energy input composition changes;
- Composition shifts of output;
- Composition shifts of other factors (labor-intensive to capital-intensive production techniques).

In this approach, a main question where studies have not reached unanimous agreement is, whether capital and energy are complements or substitutes. At the industry level, they can be both: the degree of complementarity depends on the industry characteristics and development, the time period, and the level of aggregation taken into account. Econometric models should take into account the physical interdependence between capital and natural goods and fuels. On the other side, the substitutability opportunities are less probable and their assessment depends highly on the scale used: the substitution at an individual sector overestimates the savings achieved at the economy-wide level. The trends of the index are not constant and not always predictable. The characteristics of the economic sectors play a significant role too (Stern, 2004).

Focusing on the substitution in production techniques and hence, the different mix of inputs required and the energy demand, the decisions are primarily based on the changes in relative prices of inputs. The *autonomous energy efficiency index* is the one measuring all the changes in the role of energy on economic production (energy input to economic output ratio) that are related primarily to technology changes or other changes, but nonrelated to energy costs.

Only when technological change is considered endogenous, then energy price changes encourage technological improvements. The reason for that is, that energy prices act as an incentive for energy efficiency and energy-saving technologies, to save costs. On the other side, when energy prices are decreasing, their proportion in the total operational cost of businesses is minimal and businesses tend to use energy more intensively.

However, this relationship between prices and other cost incentives from the one side and energy efficiency technological improvements on the other side is not linear. A new energy-saving technology (which can be a program, a tax, or an actual tangible technology) aims at lowering the energy bill of the consumers and hence, eventually, a reduction in emissions. However, such a "lowering of the bill" may be perceived as a reduction of the real price of energy services and hence, a tendency of the consumers to eventually increase their demand for energy which partially offsets the energy-saving potential of the initial technology. Also, by this reduction in energy prices, the real incomes of consumers increase, and the consumers spend the increases in consuming other goods and services. However, Howarth (1997) first explained that these effects do not cancel out the reductions in energy completely from the energy efficiency technology but rather, the impact of the effect is smaller than the initially brought reduction in energy consumption.

Finally, although capital and technology are the main production factors that are considered interlinked with energy, labor is also critical for the future of the energy sector within the economy.

Comparing the economic contribution of the power generation sector to total GDP with that of others', one would expect an equivalent high share in the quantity of labor in the sector; fact that is not confirmed by the reality. Although, however, the number of workers is low, the energy sector requires primarily highly skilled labor, resulting in high salaries in the industry.

Jobs in the energy sector are classified into three main categories: (1) direct (employed by businesses in the energy sector for generation, distribution, or any other function); (2) indirect (employed by businesses that provide the activities of the energy sector with goods and services); and (3) induced jobs (individuals employed in the first two groups, receiving income from salaries, and boost the demand for goods and services in the whole economy). The sector is characterized by supply chains and high remunerations. The indirect and induced jobs have particular importance for boosting economic growth and certainly, create a substantial impact to the economy. Finally, a significant amount of job opportunities are created during the construction of power plants and other facilities. The drawback is that this type of jobs are temporary and not sustainable (WEF and IHS CERA, 2012).

2.2 International Trade

In the economic growth literature, the influence of international trade has been in the spotlight especially recently, when globalization and its consequences are prominent. As Rahman and Mamun (2016, p. 807) pointed out "an absence of international trade variable (export plus import) in a growth model may underestimate or overestimate the effect of energy consumption on macroeconomic growth." Both the trade-led growth (where the assumption is that exports and imports are conducive to the country's economic growth) and the growth-led trade (where the assumption is the economic growth boosts the trade performance of the country) hypotheses are evaluated extensively. In the literature, four possible directions of the relationship were detected in the literature: (1) Transfer of academic and technical knowledge and technology as well as foreign direct investment (FDI) flows can motivate economic growth; (2) economic growth might create the conditions for a country to have a competitive advantage in the trade with other countries; (3) Trade and economic growth are independent; (4) There is a negative relationship between trade and economic growth, which is usually attributed to imports being higher than exports that affect economic growth negatively.

In the energy literature, the issue of international trade started gaining significance recently among other factors. As was explained in Katircioglu et al. (2016) and Sadorsky (2012), trade, economic growth, and energy usage patterns are moving together over time, and hence, possible causal relationships should firstly be examined not only to comprehend the implied dynamics better but also to assist with trade policy designing and implementation. A few channels of this tripartite relationship are presented next:

- When economic growth is being promoted by export activities, then the exporting sectors require higher amounts and better quality of factors of production to increase their output, and hence, their demand for energy input increases;
- When policy makers focus on the reduction of energy consumption for environmental reasons or in cases (as has occurred in many developing countries recently), with frequent power interruptions, extra constraints are put from an input-availability point of view in the production of goods and services both domestically (and hence, economic growth is pressured) and for trade purposes (and hence, exports are pressured);
- The more straightforward connection among the three is the following; when real economic output increases, subsequently there is an upward push in energy capacity and consumption. Based on fundamental economic thinking, this usually results in an increase in imports in the following periods. The effect is more intense for energy importing countries;
- A general increase in energy consumption has high potential to boost the country's economic activity of the country and the household incomes which will result in higher trading opportunities both from an export and import point of view;
- Energy conservation policies that aim at mitigation of greenhouse gas emissions would decrease the volume of international trade, when trading activities are greatly associated with energy consumption (Sadorsky, 2012). Such activities include, for example, transportation that is energy-intensive and particularly dependent on oil.

Trade can be proxied in the quantitative analysis by exports (total volume or percentage to GDP), or imports (total volume or percentage to GDP), or openness. Trade openness can be defined as trade shares (sum of exports and imports divided by GDP); or trade barriers (taxes on international trade or indices of tariff and nontariff barriers); or exchange rates and comparative prices, as a signal or an indication of the competitive advantages of the country.

Quantitative analysts must carefully select the proxy for energy, as the causal relationship depends on whether the energy is considered from the supply or demand point of view (Abidin et al., 2015). In addition to this, the type of energy plays also a particularly different role in the establishment of the relationship with economic growth and trade: Amri (2017) explains that specifically renewable energy "helps the integration of the developing and developed countries into international trade."

All things considered, the role of international trade on energy should not be neglected when designing and planning energy and environmental policies. To achieve the desired sustainability through policy making, a well-coordinated effort is necessary to balance the dynamics of energy use and international trade. Trade policies should also consider the sustainability and energy consequences (Rahman and Mamun, 2016).

2.3 Financial Development

As per Karanfil (2009) suggestions, variables that capture a country's financial development have a potential to influence the relationship between energy consumption and economic growth. More specifically, the promotion of the activities such as FDI, stock market activity, and banking activities is what the important aspect included in the term coined as *financial development*. All these activities promote economic growth and will have an effect on energy demand (Sadorsky, 2010). Theoretically, financial institutions that differ in quality and quantity are positive contributors to the growth path of a country. Fung (2009) described the two main channels, through which financial development can promote economic growth: (1) through factor productivity (asymmetries in information availability are reduced due to financial system improvements, leading to more efficient monitoring of investment projects. Also, the lower cost of equity will result in rising economic growth); and (2) through factor accumulation (the spread and reorganization of financial systems have the potential to increase the efficiency and allocation of resources.

Financial development can promote the economic efficiency of a country's financial system. It can also stimulate changes such as the reduction of risks and costs of borrowing, and facilitate access to capital and investment flows. Following all these possible developments, consumers and producers will acquire access to modern and advanced energy efficient products and technologies.

Generally, two main ways are identified, through which financial sector growth affects energy demand through either individual cross-sectoral development or indirect impacts from increases in access to financial development leading to higher demand for human capital, resulting in higher incomes and demand for goods, services, and hence energy consumption. More specifically, the channels through which financial development affects energy are as follows:

- Access to financial products and markets enables consumers go for borrowing to buy more extensive and more energy-intensive items such as houses, cars, and big-household appliances. Such purchases by the consumers increase the energy directly as automobiles cannot be utilized without—nowadays still—oil and housing have also high energy needs in heating and cooling (Sadorsky, 2010). Thus the satisfaction of consumers' needs, through goods and services, leads to the escalation of the energy consumption.
- Businesses also get benefit from a developed financial sector because it provides access to financial capital that enables them to expand the existing operations (infrastructure development, more employees, and acquisition of more machinery and equipment) or build new ones. Understandably, these activities intensify the energy requirements of the economy.
- A well-developed and well-established stock market can provide reasonable suggestion to investors and businesses of upcoming economic growth, which subsequently boosts

consumer and business confidence. That, in turn promotes demand for energy-intensive products.

- By developing further the stock market allows businesses access to equity financing in addition to debt financing, and increases risk diversification for end-users and business to create wealth.

These channels can be active only if there is an effective and well-operating banking and financial system improving the functioning of stock market. Through this, the energy consumption will keep increasing as a result. Policy makers should take the impact of financial development into consideration when designing energy and environmental policies.

On the other side, access to financial products provides capital to invest on energy-efficient technologies and products, which contrary to all the other channels, create the potential to reduce energy usage and promote energy efficiency. In addition to that, financial development and access to finances create more opportunities for the research and the development of more energy-efficient technologies, whose benefits will be borne by the society in its entirety (Abidin et al., 2015). If this channel is promoted, the increasing demand due to the previously discussed channels can be controlled with policies and incentives that promote the adoption of energy-efficient technologies and products.

To quantify the impact of financial development, the literature has proposed a number of variables, such as

- *FDI*: It is the most commonly used proxy for the financial development in studies. Stimulation of economic growth can be achieved through technology transfer, knowledge spillovers, and introduction of new technologies, processes, products, and productivity gains. It can be measured both as total net inflows and even as a percentage to GDP. Barro and Sala-i-Martin (1995) confirmed that FDI is one of the main reasons toward modernization of an economy, and Abidin et al. (2015, p. 842) explained that particularly FDI can increase energy consumption "through expansion of industrial, logistic, and the manufacturing sector development, whereas energy is essential to support the industrialized process."
- *The ratio of stock market capitalization to GDP or the stock market turnover ratio, or the stock market value traded to GDP*: as Minier (2009) described, there are two channels, through which higher stock market activity can improve economic growth. First, with the continuous development of stock markets, more regulations are established leading to higher investor confidence (both locally and at a foreign level). Second, diversification and liquidity increase with the rising of stock market activities leading to investments with higher profitability.
- *And the ratio of deposit money bank assets to GDP* (Tamazian et al., 2009): the higher this variable is, the more funding is available for loaning out to boost consumption, economic growth, and energy consumption.

2.4 Pricing Structures

Energy is a commercialized product and as such, its demand is highly connected not only with the economic output and household income, but also very importantly, to its cost. In many energy studies, the energy price has been proven to be the most important factor that could potentially lower the magnitude of the relationship between energy consumption and economic growth. Policy makers and investors can take advantage of the understanding of such behavioral responses as being able to adhere to consumers' reaction when planning infrastructure development, price restructuring, and environmental policies (Blignaut et al., 2015).

Borrowing from basic economic analysis, the concept of *price elasticity*, to measure the consumers' reaction in electricity consumption, subject to price changes, is defined as follows:

$$e_d = \frac{\Delta Q\%}{\Delta P\%} = \frac{\dfrac{Q_1 - Q_0}{Q_0}}{\dfrac{P_1 - P_0}{P_0}} \tag{4.2}$$

where e_d is the price elasticity of electricity demand, Q is the electricity quantity demanded, and P is the price of electricity (where Δ denotes the change between two periods while 1 is the latest time period and 0 is the initial time period).

- $e_d = 0 \rightarrow$ the demand for electricity is perfect inelastic: the consumers do not change their quantity consumed, when the price of electricity fluctuates, *ceteris paribus*.
- $-1 < e_d < 0 \rightarrow$ the demand for electricity is relatively inelastic: the consumers' electricity demand change (in percentage) is smaller than the percentage change in price, *ceteris paribus*.
- $-\infty < e_d < -1 \rightarrow$ the demand for electricity is relatively elastic: the consumers' electricity demand change (in percentage) is higher than the percentage change in price, *ceteris paribus*.

The general expectations from the literature are that: (1) developed countries have less negative elasticity than developing countries (this is more evident for the residential sector in the developed countries); (2) less negative for richer consumers than poorer ones; and (3) elasticities are less negative in the short run than in the long run. Residential consumers also have varying elasticities during different time of the day, in case different time of use tariffs structuring is implemented (Deloitte, 2009).

In econometric terms, the quantification of the elasticity of electricity demand can be modeled as follows:

$$lnQ_t = a + \beta_1 lnX_t + \beta_2 lnW_t + \beta_3 lnP_t \tag{4.3}$$

where Q is the electricity consumed; X is the economic output; W includes all the other exogenous factors; P is the price of electricity; and β are all the coefficients/elasticities with the subscript t referring to the time period under investigation.

The significance of the price elasticity is not static; it can fluctuate depending on the price levels, the conditions of the energy market and the socioeconomic conditions of the country (Inglesi-Lotz, 2011). Also, across economic sectors, the direction and size of the behavioral response to price change vary substantially, depending on the availability of substitutes and the energy intensity of the production (Blignaut et al., 2015). Hence, from a policy making point of view, it might be beneficial to implement sectorally differentiated pricing structures. Thus by increasing the electricity prices of energy-intensive sectors and users, the policy makers can influence the energy usage and saving by various sectors (Blignaut et al., 2015). Even with the same elasticities though, the extent of potential adoption of energy-efficiency measures varies (Thollander et al., 2005).

Due to the 2008 financial crisis, the literature also dealt with bubble-like behavior for prices in the financial sector, the housing market, as well as commodity and energy prices. Not taking into account, the possible consequences of the aggravated bubble might create serious problems in the market: the suppliers may react by increasing investment in capital. "However, policy makers should be cautious before they take drastic measures to deflate bubbles. A policy should meet three requirements: the bubbles should be accurately identified; the policy should improve macroeconomic stability; and finally, the policy should be tested with results in effective deflation of bubbles" (Gupta and Inglesi-Lotz, 2016).

Except for the effect of the prices on energy consumption and demand, the cost of energy plays an important role to the economy in its entirety. The level of energy prices has an impact on the behavior of consumers and suppliers in the economy: if they are low, then they reduce the cost for almost all products, increasing thus both the supply and the demand and hence, economic growth; while if they are higher, they put downward pressure on economic growth (with the exception of energy-exporting countries, such as Nigeria that benefits from higher oil prices). In addition, with lower (higher) energy tariffs, the real disposable income of the households can be increased (decreased) and allow or restrict consumers to (from) spending in other ways (WEF and IHS CERA, 2012).

2.5 Economic Structures

The structure of the economies is not fixed and constant over time. According to Rostow theory (Rostow, 1960), there are five basic stages in the development of an economy:

1. Traditional society: agriculture, hunting, and fishing are the main economic activities with limited technology and capital utilization;
2. Preconditions to takeoff: development and commercialization of the already established sectors, investment and better exploitation of natural resources, initial stage of product share and exchange, and trade;

3. Take-off: industrialization and urbanization processes initiate, secondary sectors expand;
4. Drive to maturity: diversification of manufacturing goods, general and social infrastructure, and transportation systems develop;
5. Age of mass consumption: the manufacturing sector dominates the domestic production, general consumption of necessity and luxury goods and services, generally an urban society.

Although, Rostow supports that the passing through each of the five stages is fairly linear, the length of the stages though is not predefined and certain.

Not only all these changes are characterized by a general movement toward more energy-intensive industries, but also the higher incomes of the population at more developed stages have generally more needs for goods and services (e.g., transportation) that also move the energy demand upwards. Even the service industries and the financial sector, which are dominant contributors at the later stages of economic development, need higher amounts and quality of energy resources to function; although, the production is not considered to be tangible, the building and maintenance of the infrastructure, equipment and buildings where production takes place, are resource and capital-intensive.

The research proposed hypothesis such as the Environmental Kuznets Curve which advocates that the environmental degradation and the energy intensity/usage worsen, as economic development improves, until it reaches a certain threshold point, from which the environmental quality and energy efficiency improve. The reasoning behind this trend is that, the economic structures have changed within the developmental timelines and thus more capital is available to invest on energy-efficient technologies.

However, the counterargument suggests that the ratio of energy to GDP has lowered through the years, not due to the output mix changes, but rather to the energy mix changes (Stern, 2004), as we discuss next.

2.6 Different Energy Types and Fuels

As the economies evolve through different developmental stages, it is observed that the energy consumption mix changes and with these, also changes the relationships between energy and economic activity (Toman and Jemelkova, 2003). The well-known "ladder effect," as defined by Barnes and Floor (1996) is not necessarily a monotonic linear transition from one type or combination of energies to another. Households and other energy users do not move on the ladder only in an upward direction (vertically), but in many cases, depending on the opportunity and real costs as well as the household incomes (horizontally).

Generally, at lower levels of socioeconomic development, economies are characterized by traditional production activities and rely mostly on primary economic sectors such as the agriculture. Hence, energy comes from easily accessible sources, such as wood and dung (which are mostly renewable and cleaner forms of energy; Tahvonen and Salo, 2001).

In the next stages of development, the economy moves toward a higher contribution of the manufacturing sector, and thus, energy is sourced from some commercial fossil energy, animal powers, and biofuels (charcoal). Finally, at most industrialized levels of socioeconomic development, commercial fossil fuels, and electricity become the dominant source of energy.

However, the concept of energy quality is important in understanding the advantages and disadvantages of substitution, as well as the different effect of various fuels on economic production. Stern (2004, p. 46) defines energy quality as "the relative economic usefulness per heat equivalent unit of different fuels and electricity. One way of measuring energy quality is the marginal product of the fuel, which is the marginal increase in the quantity of a good or service produced by the use of one additional heat unit per fuel." Some energy carriers can be utilized in a variety of activities and sectors comparatively to others. For instance, electricity is used widely by all sectors and energy users, while coal cannot be utilized directly, but only through power generation. There are certain characteristics of the fuels that influence their marginal product: density, environmental friendliness, safety, accessibility, cost of conversion, storage capacity, and natural scarcity. Apart from the physical attributes, there are economic traits such as the cost and quantity of the production factors required for its generation and usage, and its requirement for various applications and sectors. Energy quality is also not constant over time; it can fluctuate depending on new technologies implemented, as well as the relative costs per unit of energy. Nowadays, according to the prices and marginal products, electricity is considered to be the highest quality type of energy (it should still be noted that electricity generation however, is still connected to other fuels quantity) followed by natural gas, fossil fuels, and wood. As far back as the decade of 1960s (Schurr and Netschert, 1960), the importance of quality was appreciated, noting that a shift to higher quality of fuels (as per the energy ladder) will decrease the energy requirement of the economy to produce a unit of GDP.

2.7 Population and Urbanization

Increasing population and population growth, especially in developing countries, as well as the rising amount of households that have now access to energy services compared to the past years, have rendered the population and population growth significant factors to a country's energy demand. Although it might be expected to observe a proportional increase of population and energy consumption, this is not always captured in quantitative studies. With access to energy, as well as to economic development, increasing numbers of people have higher incomes and thus, needs in goods and services, thus increasing energy demand more. The population's indirect influence on energy consumption, through specific characteristics such as population density, age structure, household size, and urbanization movements, is certainly important and should be explored (Liddle, 2014).

The dynamics of population density, age structure, and household size are relatively straightforward. In the literature, studies have proven the existence of a relationship between population density and factors such as lower electricity consumption in buildings and transport energy demand. With regard to age structure, life-cycle behaviors are examined: younger adults have positive elasticity in contrast with older adults that present negative elasticities of energy demand (Liddle and Lung, 2010). As Liddle (2014) explains, age structure is important because:

1. Population in different life stages and age groups have different behaviors, habits, levels of income, and affluence, that increase energy demand accordingly;
2. The size of household and the age of the head of the household affect the energy consumption behaviors: for example, bigger households tend to have lower per capita road energy use (Liddle, 2004).

Over the centuries, the migration of populations in different geographical positions either by choice or due to unforeseeable or unavoidable circumstances is not something new in the history of humanity. Recently, the population of rural areas seek more opportunities both in the job market, but also—and maybe more importantly—in the access to basic services and infrastructure in main cities and generally, urban areas. Urbanization is, thus, "a dynamic moderation phenomenon on social and economic capability originating from rural areas and is based on the agriculture economy toward urban areas with industries and services sectors" (Shahbaz et al., 2015). Poumanyvong and Kaneko (2010) forecast that by 2030, the population that stays in urban areas will reach 4.6 billion—almost three times higher than in 1970. However, the impact of these population movements toward urban areas is difficult to estimate and forecast, because it can be twofold: on the one side, urbanization boosts economic activity as consumption and production become more concentrated, while at the same time it results in economies of scale and opportunities for energy efficiency and cleaner energy types (Sadorsky, 2013).

Particularly in developing economies, economic transition often passes through rapid urbanization that transforms the "picture" of their energy markets, and changes economic patterns of resource employment (Parikh and Shukla, 1995). Urbanization is often considered responsible for higher energy demand and decline in environmental quality. IEA (2012) reported that more than 50% of the global population exist in urban areas, where more than half of the global energy consumption takes place and more than 60% of CO_2 emissions are produced. High tendency of economic activities in economic sectors such as manufacturing and transport make urban areas more energy-intensive through the years and comparatively to rural areas. Moreover, urban populations have higher living standards because access to modern and advanced infrastructure, water and energy supplies, transport and telecommunications, is easier. The succeeding changes as Ji and Chen (2017) explained are that "…their consumption patterns gradually shift from 'survival mode' to 'development mode' and even to 'enjoyment mode' which may directly or indirectly push up urban energy use." Apart from these, increasing housing needs and investment growth are also considered to be increasing contributors to energy demand. As Jones (1991) observed, urban citizens rely more on energy-intensive electric appliances and tend to use private transport modes more, comparatively to rural citizens.

On the other side, the energy-urbanization literature discusses the benefits of urbanization toward decreasing energy use and emissions. This negative association of urbanization with energy consumption can be explained by the urban-compaction theory. Economies of scale can be exploited: This exploitation of economies of scale ensures a decrease in vehicle usage, distance traveled, as well as reduced distribution losses in electricity supply. These reductions subsequently reduce overall energy consumption. This effect can primarily be experienced in developed countries where financing infrastructure is more accessible.

Poumanyvong and Kaneko (2010) found a negative relationship between urbanization and energy consumption in low-income group countries. This relationship appears to be counter intuitive as low-income countries have less efficient energy infrastructure and urban planning than high-income countries, and should therefore exhibit a positive relationship between energy consumption and urbanization. Studies in the literature seem to be against the urban-compaction theory, arguing that population increases in urban areas are highly linked with increased air pollution and traffic congestion. Poumanyvong and Kaneko (2010) counterargue that the effects of modernization resulting in more modern and more efficient uses and ways of using fuels are introduced to urbanized populations. This thinking path and results are also confirmed in Pachauri (2004) and Pachauri and Jiang (2008) who showed that fuel substitution from inefficient and "dirty" solid fuels such as wood to cleaner and more efficient energies in urban areas results in lower per household energy consumption and surely, less emissions.

Urbanization creates pressure not only on the receiving end of population, but also on the agricultural sector, from which a significant amount of labor leaves behind. Kalnay and Cal (2005) explained that the agricultural sector is required to overproduce with less human capital, creating thus an impact on extra land use and a consequent increase in the energy demand. Simultaneously, urbanization has been found to have an effect on industrialization and commercialization which will intensify the demand not only for electricity, but also for consumer goods, which in their turn will rise the energy use during their production (Mishra et al., 2009; York, 2007). All things considered, it seems that urbanization movements cause structural changes in numerous economic sectors and the economy in its entirety.

The actual impact and specific impact of urbanization depend on the level of affluence of the country (Ji and Chen, 2017) and hence, the magnitude of the effect is not the same for all the countries and during all the time periods. The positive relationship between the urbanization and energy consumption is more intense in middle-income countries; while the poorer countries although characterized by rapid urbanization, the low financing opportunities do not favor infrastructure building in the new expanded urban areas in any case. On the other hand, the already advanced, industrialized, high-income countries have reached maximum urbanization levels, so the smaller variance in urbanization trends will not alter the already established infrastructure as well as the average income per capita is already relatively high and hence, the consumption patterns will not change drastically.

Summarizing the potential channels of impact of urbanization on energy and growth, Sadorsky (2013) has identified four particular paths:

1. Through the effect of urbanization to economic production: economic activity is concentrated in cities and metropolitan areas. That assists with economies of scale in production, especially as one of the main factors of production—the labor—is now in closer proximity. However, through urbanization, energy-intensive sectors such as manufacturing are substituting for the less energy-intensive agricultural activities.
2. Through the increases in transport use: the traffic inside and outside of the urban areas increase, resulting in higher energy demand. With the development of public transport, some of the traffic congestions (and increased energy demand) can be controlled.
3. Through the rising demand for infrastructure and services: with the aforementioned example of the transport demand, it is understandable that especially during the initial stages of urbanization and growth of the urban areas, energy demand will be on the rise in conjunction with the building of the infrastructure (and later maintenance) needed.
4. Through the changes in household preferences and consumption baskets: the moving population's incomes tend to increase and with that their consumption choices and preferences alter. Also, the way the energy is used for the household purposes changes with urbanization: for example, for cooking and food preparation purposes, more energy-intensive appliances are used, such as refrigerators etc.

From a policy perspective, the city governments are usually the ones responsible first to absorb the negative effects of urbanization and second, to make sure the transition can be as energy-saving and green as it can be. Ji and Chen (2017, p. 7) suggested the following:

1. *Focus on the "incremental reform," supplemented by the improvement of current stock* Stock-based reform is the unplanned reform of old cities. Local governments reconstruct old cities infrastructure. However, only the rebuilding itself may cause bigger problems, especially in the short run than it tries to solve. Also, city planning based on traditional approaches might also intensify the energy and water limitations. On the other hand, incremental reform refers to newly designed municipalities or towns. This approach is preferred because it takes into consideration the needs of production and of the populations. By encouraging the most appropriate layouts for living as well as industrial areas, the negative effect caused by the increasing demand transport on energy resources can be reduced. Finally, the correct positioning and geographical location of new medical

centers, schools, and other basic services will also assist in the energy efficiency of the newly urbanized population.

2. *Advocate the "active" bottom-up urbanization instead of the "passive" government-led and top-down urbanization*
   In the process of active urbanization, structures and infrastructures are prepared to accommodate the newly urbanized citizens. Following that approach, and avoiding the "passive" one, the policy makers can ensure urbanization quality. To do so, infrastructure and public services should be encouraged and promoted to exploit the benefits of the growing population.

3. *Increase the efficiency of rigid demand and eradicate the nonrigid demand*
   Given the scarcity in energy resources and the development evolution of most countries toward industrialization, the efficiency of the rigid energy consumption is subject to future improvements. Policy makers are responsible to make sure that the controlling of overconsumption will not affect the well-being of the citizens; one way to do so is the correct management of the energy pricing system.

2.8 Miscellaneous Factors

2.8.1 Militarization and national defense

International efforts to reduce energy consumption, improve energy efficiency, and consequently, reduce CO_2 emissions, have focused primarily on the behavior and the energy profiles of the industrial and residential sectors. One of the mostly neglected sectors, which however, receive a significant share of the countries budgets, is the national military sector. The literature has pointed to the fact that the impact of the defense dimension to the energy consumption and environmental degradation should be explored (Clark et al., 2010), particularly due to the wars (directly or indirectly) for dominance over oil and other natural resources (Klare, 2002). Wars that are energy-intensive and hence increase energy consumption, also need strong military forces. Moreover, countries with more advanced military forces take advantage of that position to achieve access to natural resources (more modern military forces boost the country's strength into pursuing disproportionate, sometimes, access to global natural resources).

Industrially advanced countries direct more expenditure toward militarization due to increasing concerns for their defense in recent years (Bildirici, 2016). Particularly after the World War II, military operations have become more capital—and technology—intensive, consuming oil at a higher proportion than other forms of energy. Furthermore, the building and the equipping of new military bases were considered necessary recently (that usually are exempted from environmental laws and regulations (Gould, 2007) which does not incentivize efforts for energy efficiency efforts). As Clark et al. (2010, p. 27) point out: "The military authorities requires bases being in close proximity to potential theaters of war to properly meet the energy needs of modern equipment and the personnel needs of soldiers."

All these increase the needs for energy in the defense sector: all activities and related technologies utilize immense amounts, primarily, of fossil fuel energies. "The expansion and development of high-tech equipment and vehicles have increased the energy demands of the military, as enormous quantities of fossil fuels are required to operate the planes, ships, tanks, helicopters, and vehicles of the armed forces" (Clark et al., 2010, p. 38).

Additionally, the sector relies heavily on new and advanced technologies that are constantly improved and hence, research and development, as well as training of troops, also increase the needs for energy.

During peacetime, the military functions, and hence, expenses and energy use, do not fade. Particularly, high-tech and modern militaries consume energy for maintenance of the equipment and infrastructure even without actively using them (Clark et al., 2010; Roberts et al., 2003). Also, the subsistence of the military force is necessary: troops and administrative personnel need to be trained, housed, armed, and clothed, as well as transported. Furthermore, the sector needs to stock on raw materials, munitions, spare parts, and other ancillary equipment. All things considered, this military infrastructure entails and increases energy usage—without high contribution to the country's economic growth.

Observing trends across countries overtime, we understand that the energy consumption is higher in the more advanced developed countries that anyway have higher military spending. Hence, a simple correlation analysis might be misleading. To quantify and evaluate the impact of militarization to the relationship between energy consumption and economic growth, the literature has used various military measures that can be used in isolation or in combination, to capture the holistic effect: (1) Total military expenditures or per soldier (this usually includes all capital expenditures on the armed forces, the defense ministries, and administration, for operation, maintenance, R&D, remunerations, social services, and military aid); and (2) Military personnel (this usually includes both active duty military personnel as well as paramilitary forces).

2.8.2 Tourism development

Tourism (both international and domestic) is considered to be one of the highly important sectors for the economic development of a country, as well as the industrialization of an economy. It also contributes to the economic growth by foreign exchange inflows, movements of capital, creation of job opportunities, and investments on infrastructure development (Tang and Abosedra, 2014). All these activities, however, have an impact on the energy demands of a country. Hence, the role of the tourism to a country's aggregate energy consumption as well as the role of energy consumption within the sector should not be neglected in the literature.

The goal of sustainability within the sector calls for environmental consideration and energy efficiency to all types of tourism—both in its more traditional form and the more specific sectors such as ecotourism—and requires the efficient allocation of natural resources, the consideration of the environment, and the respect for local populations, communities, and businesses. As per Tiwari et al. (2013), tourism has the potential to play an even more important role in the future when it comes to climate change and energy efficiency, as it is characterized by rapidly evolving and innovative resources and processes.

The tourism sector's development contributes to total energy consumption (and climate change) through various channels: energy consumption increases with different functions such as transport, accommodation, catering, and operation of facilities, irrigation, and entertainment. Also, the energy demand of the sector will increase already during the construction period of hotels and other facilities.

The most important contributor to the sector's energy usage is linked with the transport modes that tourists use to travel to and from host destinations (Gossling et al., 2002). The specific energy usage for transport depends highly on the mode of transport and the distance of the tourism attraction and facilities from main stations and urban centers. Thus in remote and island destinations, the majority of tourists arrive by air modes increasing the energy consumption of the sector (Kelly and Williams, 2007).

Accommodation (or lodging) facilities are the main contributor to the sector's energy consumption, as there are two types of users: on the one hand, the tourist guests make use of small appliances during their stay and on the other hand, the operation of accommodation facilities requires certain amounts of energy (Becken et al., 2001). "Hotels use two types of energy namely electricity and thermal energy. Electricity is mainly used (disregarding the nature of the source) for illumination and to power motor driven equipment and electronic devices. Example of electricity includes air conditioning units, fans and air-handlers, such as humidifiers, lighting fixtures, refrigeration equipment, water pumps, large appliances (e.g., clothes and dish washing machines), small appliances (e.g., toasters, microwave ovens, and hair dryers), electronic devices (e.g., television sets, stereos, and computers), and communications equipment (e.g., cellular telephones and computers). Thermal energy is used (disregarding the nature of the source) as a source of energy in heating applications, for example space heaters, water heaters, cooking equipment (such as stoves and ovens), and laundry dryers" (Tiwari et al., 2013). However, the energy consumption depends not only on the type of lodging, for example hotels, bed and breakfast accommodation, backpackers and camping areas, but also, on the existence of energy-intensive facilities namely restaurants, laundries, swimming pools, gyms, etc. Becken et al. (2001) concluded that hotels have generally the higher energy demand per visitor. The visitors' country of origin should also be considered when evaluating their energy usage behavior; studies have shown that tourists with home destinations characterized by the highest energy awareness and efficiency levels tend to choose accommodation solutions that promote energy-saving infrastructures and use of renewable energy sources (Tsagarakis et al., 2011).

Katircioglu (2014) noted that the tourism sector is not very energy-intensive comparatively to others. Future development of the sector will intensify the needs for energy certainly; but electricity and energy can be sourced through cleaner and more environmental friendly energies, such as photovoltaic panels. Moreover, promoting energy and environmental targets for all new developments of facilities and the redevelopment of new projects will have a significant impact on the sector's energy usage and efficiency while simultaneously, allowing the sector to increase its economic output and contribution to the national economy. More specifically, all these interventions should be characterized by innovations in planning and designing as well as taking into account proximity to energy supply infrastructures.

2.8.3 Energy security

The concept of energy security attracts increasing attention among the circles of academics and policy makers, when exploring changes in the sector. The European Commission has included energy security as a part of the priority triangle of its energy policies, together with energy efficiency and sustainability, while the promotion of energy security is manifested in the promotion through a concerted effort against power cuts by the International Energy Agency (IEA).

The first definition of energy security can be found back in 1985, and was proved by the IEA which simply stated that energy security is "an adequate supply of energy at a reasonable cost." The European Commission defines the concept more thoroughly in a Green paper (European Commission, 2000) as "the uninterrupted physical availability of energy products on the market, at a price that is affordable for all consumers (private and industrial), while respecting environmental concerns and looking towards sustainable development." The same Green Paper makes an effort to clarify some of the issues in the definition by identifying a few sources of risk: (1) physical risks "distinguishing between permanent disruption (due to stoppages in energy production or to exhaustion of energy resources) and temporary disruptions (due to geopolitical crisis or natural disasters)" (Labandeira and Manzano, 2012); (2) economic risks; (3) regulatory risks; (4) social risks; and (5) environmental risks.

To quantify the concept of energy security, and to be able to utilize it in modeling applications, Scheepers et al. (2007) suggest two proxies: (1) the ratio of supply to demand (information can be accessed at the national energy balances); and (2) the Crisis Capability Index (managing capability to control and minimize the effects of temporary power disruptions).

Negative externalities related to energy security are concerned first, with the oil imports (the market power of exporting activities are conducive to negative effects related to oil imports that create market failures) and second, with price volatility (especially as the adjustment is slow and hence, the factor and product markets are not strong enough to bounce back, resulting in higher economic costs) (Bohi and Toman, 1993).

2.8.4 Political considerations

The energy system is not different than any other sector where political and policy-making decision-making processes influence the relationship between natural (energy) and economic resources.

As Jacobsson and Lauber (2006) explain, the policy-making process is not "rational" from an economic point of view, but rather more technocratic and politicized. It depends highly on historical and cultural influences and traditions, but most importantly, on the opposition and/or support of several pressure groups. This fact is particularly obvious in energy markets where the energy suppliers are state-owned monopolies and also, the location of distribution and transmission lines was not decided based on economic optimality but rather, to win the favor of support groups. However, recently, the energy sector is characterized by the regulatory reforms, high financial flows, and concentration of the decision-making.

As mentioned, the majority of power suppliers traditionally were owned by the national governments. For political reasons, hence, prices were kept at low levels. At the same time, national power suppliers employ large numbers of skilled and unskilled employees, while the power generation facilities should be always maintained and expanded. The combination of low revenue and high costs leads to unplanned financial losses for the utilities. The losses have to be covered by government revenue, resulting in some cases in poor provision of energy services due to the lack of funds from the government budget.

Among the *political* issues dealt in the energy sector, energy poverty or lack of access to energy goods and services is also one among many. Barnett (2014) explained further that the fundamental truth is that the low-income population does not have access to energy because they cannot afford it. Greater efforts by policy makers have been made recently to ensure provision to modern energy services aiming to uplift the citizens' living standards and the countries' overall economic growth and development. "Decentralized energy solutions (both with diesels and renewables) offer possibilities in remote locations, or where the utility is dysfunctional. But it is unlikely that these systems can provide connections to electrical power (rather than milliamps for lighting and mobile phones) unless through effective grid extension" (Barnett, 2014).

3 Regime Switching and Time-Varying Models

3.1 Regime Switching Approach

Models that assume a switching mechanism have attracted attention in the applied energy econometric literature due to their ability to capture the unique dynamics and trends in demand and electricity spot prices (examples can be found in Bierbrauer et al. (2007), Karakatsani and Bunn (2008), Alizadeh et al. (2008), Haldrup et al. (2009), and Rahman et al. (2017)). Energy and electricity have distinctive characteristics that do not allow the researchers to use models developed for normal financial and commodity markets. Energy does not have high storage capacities (with the current technologies) and so, the transfer to users should be direct and immediate. The end users show high cyclical traits attributed to weather fluctuations and business cycle dependence. All of these reasons contribute to high seasonal variability (at various frequencies: annual, monthly, and even daily), volatility, and extreme spikes and falls in electricity spot prices. These increases also have a tendency to gather in clusters (Janczura and Weron, 2010).

The most commonly used modeling approach when studies consider regime-switching is the Markov regime-switching models. The first papers that considered Markov regime-switching models (first generation Markov regime-switching models) were published at the end of 1990s: Ethier and Mount (1998) and Deng (1998). Within a derivatives pricing context, three spot price models were suggested (Deng, 1998): for instance, "a 2-state regime-switching specification for the log-prices in which the base regime was driven by an autoregressive process of order one, i.e. AR(1), and the spike regime by the same AR(1) process (i.e. with the same parameters) shifted by an exponentially distributed random variable (spike/jump size)" (Janczura and Weron, 2010, p. 7). Later on, Huisman and Mahieu (2003) used a regime-switching model with three possible states "where the initial jump regime was followed by the reverse one and moved back to the base."

First generation models for regime-switching have two common characteristics: (1) the log-prices were considered; and (2) a mean-reverting process drives the base regime. These are the traits that second generation models came to improve on. They are classified into two categories: (1) those that use basic exogenous information to improve the model (Anderson and Davidson, 2008); and (2) those that aim at improving the statistical aspects to improve the models' capturing of reality (Lucheroni, 2010).

Markov regime-switching models denote the "observed stochastic behavior of a specific time series by two (or more) separate states or regimes with different underlying stochastic processes" (Janczura and Weron, 2012, p. 386). This capability of the Markov switching models is important considering they can capture consecutive spikes, as in reality. Furthermore, the Markov switching models enable the prices to return to the "normal regime" after a spike, as well as "temporary dependence within the regimes, in particular, for mean reversion" (Janczura and Weron, 2012, p. 387) which describes one of the main characteristics of electricity prices.

After selecting the specific model class, there are two aspects that need to be defined: the type of dependence between regimes, the number of regimes chosen, the calibration to market data. With regard to the first, aiming for computational simplicity, the assumption is suggested to assume regimes that are characterized by dependency to each other but with the same random noise. However, aiming for greater flexibility of the model, interdependent regimes are suggested, which are closer to reality taking into account the volatile behavior of spot electricity prices.

3.2 The Time-Varying Approach

One of the main modeling approaches within the EGN is the cointegration to establish long-run relationships. Although this approach assumes stationarity in the series and constant parameters (elasticities) over time, these characteristics do not necessarily capture the reality. Hunt et al. (2003) propose methodologies with stochastically time-varying parameters which has the capacity to present more robust estimation.

The Kalman filter methodology covers all the above-mentioned requirements in modeling and can be an ideal estimation with variables whose elasticities to electricity demand vary over time. Originally, the estimation state-space models, where the Kalman filter is based, were developed for chemistry and engineering (Kalman, 1960). Only in 1980s, these type of estimations were adopted by the economics field.

According to the basic representation of the Kalman filter, two types of models are the unobservable components and the time-varying parameter models.

Case Study: Time-Varying Price Elasticitiy of Demand for Energy

The following description of a Kalman filter application for an aggregate electricity demand model is adopted by Inglesi-Lotz (2011, p. 3691).

First, the formal representation of a dynamic system written in state-space form suitable for the Kalman filter should be described. The following system of equations presents the state-space model of the dynamics of an nX1 vector, y_t

$$\text{Observation (or measurement) equation: } y_t = Ax_t + H\xi_t + w_t \tag{4.4}$$

$$\text{State (or transition) equation: } \xi_{t+1} = F\xi_t + v_{t+1} \tag{4.5}$$

where A, H, and F are matrices of parameters of dimension (nXk), (nXr), and (rXr), respectively, and x_t is the a(kX1) vector of exogenous or predetermined variables, ξ_t is the a(rX1) vector of possibly unobserved state variables, known as the state vector.

The following two equations represent the characteristics of the disturbance vectors w_t and v_t, which are assumed to be independent white noise.

$$E\left(v_\tau v_\tau'\right) = \left\{ \begin{array}{ll} Q, & \text{for } t = \tau \\ 0, & \text{otherwise} \end{array} \right\} \tag{4.6}$$

$$E\left(w_\tau w_\tau'\right) = \left\{ \begin{array}{ll} R, & \text{for } t = \tau \\ 0, & \text{otherwise} \end{array} \right\} \tag{4.7}$$

As shown in the following two equations, the disturbances v_t and w_t are uncorrelated at all lags:

$$E\left(v_\tau w_\tau'\right) = 0 \quad \text{for all } t \text{ and } \tau \tag{4.8}$$

In the observation equation, the factor x_t is considered to be predetermined or exogenous which does not provide information about ξ_{t+s} or w_{t+s} for $s = 0,1,2,....$, beyond what is given by the sequence $y_{t-1}, y_{t-2}, ... y_1$. Thus x_t could include lagged values of y or variables which are uncorrelated with x_t and w_t for all t. The overall system of equations is used to explain a finite series of observations $\{y_1, y_2, y, y_t\}$ for which assumptions about the initial value of the state vector x_t are needed. With the assumption that the parameter matrices (F, Q, A, H, or R) are functions of time, then the state-space representation (Eqs. (4.4) and (4.5)) becomes:

$$Y_t = \alpha(x_t) + \left[H(x_t) \right]' \xi_t + w_t \tag{4.9}$$

$$\xi_{t+1} = F(x_t) \xi_t + v_{t+1} \tag{4.10}$$

where $F(x_t)$ is the a(rXr) matrix whose elements are functions of x_t, a(x_t) is the a(nX1) vector-valued function and $H(x_t)$ is a(rXn) matrix-valued function. Eqs. (4.9) and (4.10) allow for stochastically varying parameters, but are more restrictive in the sense that a Gaussian distribution is assumed.

Eq. (4.11) includes standard variables, used in international and local literature such as prices of electricity and output of the economy, to explain the electricity consumption.

$$lneleccons_t = \alpha * lnelecprice_t + \beta * lnoutput_t + \varepsilon \tag{4.11}$$

where eleccons is the electricity consumption; elecprice is the price of electricity and output is the gross domestic product of the economy in time t. All variables are in their natural logs, as indicated.

The estimation of this equation would result in a constant coefficient a, representing the price elasticity of electricity and a constant coefficient b, representing the income elasticity of electricity. However, in this study, by applying a Kalman filter estimation, the coefficients a and b are time varying; hence, the equation to be estimated looks as follows:

$$lneleccons_t = \alpha_t * lnelecprice_t + \beta_t * lnoutput_t + \varepsilon_t \tag{4.12}$$

To estimate this, the model contains four equations based on the notation of Eviews software to allow for time varying coefficients:

$$lneleccons_t = sv1 * lnelecprice_t + sv2 lnoutput_t + sv3 \tag{4.13}$$

$$Sv1 = sv1(-1) \tag{4.14}$$

$$Sv2 = sv2(-2) \tag{4.15}$$

$$Sv3 = c(2)sv3(-1) + \left[var = \exp(c1) \right] \tag{4.16}$$

Eqs. (4.14) and (4.15) show that the time varying coefficients evolve through time according to a random walk process. All variables are integrated of order 1.

4 Concluding Remarks

This chapter dealt with a battery of factors that can influence and determine not only the relationship between energy and economic growth, such as production factors, international trade, financial development, population growth, and urbanization, but also some less popular ones in the energy economics literature, such as militarization and tourism development. Last, the chapter made a primary introduction to two approaches that attract attention in the EGN: regime switching modeling and time-varying estimations.

The chapter examined the various dynamics and channels of impact. Based on the literature, it also proposed, proxies to quantify these determinants in modeling exercises. The usual bivariate linkage between energy consumption and economic growth is estimated in the literature; however, the policy recommendations for promoting energy savings through economic growth or vice versa are restricted to energy and environmental policies. "Additionally, electricity is a national or regional market. Power cannot be stored or transported on a global scale. These facts have crucial implications for policy decisions. High-energy prices can reduce consumption and investment at the household, business and industrial level. Capital is mobile, but electricity supply, generally, is not. Maximizing the number of jobs in the electricity sector is not likely to be an efficient way to maximize employment in the economy as a whole (WEF and IHS CERA, 2012)." Confirming and quantifying the impact of other factors provide the policy makers with alternative proposed solutions. For example, promoting financial access to small businesses aiming at improving their technological development may result in energy efficiencies and increased economic output.

To make an effort to converge the findings of various studies, future research should adopt and employ advanced methods (even borrow from other fields of research) and particularly use multivariate approaches. As Ozturk (2010) confirms, multivariate models are preferable including case-specific factors for different sectors, countries, or regions.

The obvious question for future research is which are the most influential of all these factors. There is no straightforward answer to this question; there is no one model-fits-all. As in all econometric and other quantitative analyses, the specific case (country or sector) should be examined thoroughly. Only the changing of time periods and the replication of past studies without any methodological improvement offer no contribution to the discussion of the EGN. When data are permitting this, studies should exploit high frequency data to explore both short- and long-run dynamics; determinants such as technological progress and innovation would affect the dynamics of the relationship with a slower and through a lagged reaction, as new technologies are absorbed in the economic production in the long run. Finally, the choice of the most appropriate variables to include depends highly on the developmental stage of countries. For example, the importance and the impact of urbanization to the dynamics of energy consumption are lower in the already industrialized countries where infrastructure is already existent.

References

Abidin, I.S., Haseeb, M., Azam, M., Islam, R., 2015. Foreign direct investment, financial development, international trade and energy consumption: panel data evidence from selected ASEAN countries. Int. J. Energy Econ. Policy 5, 841–850.

Alizadeh, A.H., Nomikos, N.K., Pouliasis, P.K., 2008. A Markov regime switching approach for hedging energy commodities. J. Banking Finance 32, 1970–1983.

Amri, F., 2017. Intercourse across economic growth, trade and renewable energy consumption in developing and developed countries. Renew. Sustainable Energy Rev. 69, 527–534.

Anderson, C.L., Davidson, M., 2008. A hybrid system-econometric model for electricity spot prices: considering spike sensitivity to forced outage distributions. IEEE Trans. Power Syst. 23, 927–937.

Barnes, D., Floor, W.M., 1996. Rural energy in developing countries: a challenge for economic development. Annu. Rev. Energy Environ. 21, 497–530.

Barnett, A., 2014. Evidence on demand [Online]. Available from: http://www.evidenceondemand.info/political-considerations-relevant-to-energy-andeconomic-growth.

Barro, R.J., Sala-i-Martin, X., 1995. Economic Growth. McGraw-Hill, Cambridge.

Becken, F., Frampton, C., Simmons, D., 2001. Energy consumption patterns in the accommodation sector – the New Zealand case. Ecol. Econ. 39, 371–386.

Bierbrauer, M., Menn, C., Rachev, S.T., Truck, S., 2007. Spot and derivative pricing in the EEX power market. J. Banking Finance 31, 3462–3485.

Bildirici, M.E., 2016. The causal link among militarization, economic growth, CO_2 emissions, and energy consumption. Environ. Sci. Pollut. Res.doi: 10.1007/s11356-016-8158-z.

Blignaut, J.N., Inglesi-Lotz, R., Weideman, J.P., 2015. Sectoral electricity elasticities in South Africa: before and after the supply crisis of 2008. S. Afr. J. Sci. 9 (10), 1–7.

Bohi, D.R., Toman, M.A., 1993. Energy security: externalities and policies. Energy Policy 21, 1093–1109.

Clark, B., Jorgenson, A.K., Kentor, J., 2010. Militarization and energy consumption. Int. J. Sociol. 40 (2), 23–43.

Deloitte, 2009. Estimating the Price Elasticity of the Demand for Electricity in South Africa. Deloitte, Pretoria.

Deng, S.-J., 1998. Stochastic Models of Energy Commodity Prices and Their Applications: Mean-Reversion With Jumps and Spikes. PSerc, Punjab, Working paper 98-28.

Ethier, R., Mount, T., 1998. Estimating the Volatility of Spot Prices in Restructured Electricity Markets and the Implications for Option Values. PSerc, Punjab, Working paper 98-31.

European Commission, 2000. Towards a European Strategy for the Security of Energy Supply. European Commission, Paris.

Fung, M.K., 2009. Fianancial development and economic growth: convergence or divergence? J. Int. Money Finance 28, 56–67.

Ghali, K.H., El-Sakka, M.I., 2004. Energy use and output growth in Canada: a multivariate cointegration analysis. Energy Econ. 26, 225–238.

Gossling, S., Hansson, C.B., Horstmeier, O., Saggel, S., 2002. Ecological footprint analysis as a tool to assess tourism sustainability. Ecol. Econ. 43, 199–211.

Gould, K., 2007. The ecological costs of militarization. Peace Rev. J. Social Justice 19, 331–334.

Gupta, R., Inglesi-Lotz, R., 2016. Detection of multiple bubbles in South African prices. Energy Sources B Econ. Plann. Policy 11, 637–642.

Haldrup, N., Nielsen, F.S., Nielsen, M.O., 2009. A Vector Autoregressive Model for Electricity Prices Subject to Long Memory and Regime Switching. Queen's Economics Department, Leibniz, Working Paper, No. 1211.

Howarth, R.B., 1997. Energy efficiency and economic growth. Contemp. Econ. Policy 25, 1–9.

Huisman, R., Mahieu, R., 2003. Regime jump in electricity prices. Energy Econ. 25, 425–434.

Hunt, L.C., Judge, G., Ninomiya, Y., 2003. Underlying trends and seasonality in UK energy demand: a sectoral analysis. Energy Econ. 25, 93–118.

IEA, 2012. World Energy Outlook. International Energy Agency, Paris.

Inglesi-Lotz, R., 2011. The evolution of price elasticity of electricity demand in South Africa: a Kalman filter application. Energy Policy 39 (6), 3690–3696.

Inglesi-Lotz, R., Pouris, A., 2016. On the causality and determinants of energy and electricity demand in South Africa: a review. Energy Sources B Econ. Plann. Policy 11, 626–636.

Jacobsson, S., Lauber, V., 2006. The politics and policy of energy system transformation – explaining the German diffusion of renewable energy technology. Energy Policy 34, 256–276.

Janczura, J., Weron, R., 2010. An empirical comparison of alternate regime-switching models for electricity spot prices. Energy Econ. 32, 1059–1073.

Janczura, J., Weron, R., 2012. Efficient estimation of Markov regime-switching models: an application to electricity sport prices. AStA – Adv. Stat. Anal. 96, 385–407.

Ji, X., Chen, B., 2017. Assessing the energy-saving effect of urbanization in China based on stochastic impacts by regression on population, affluence and technology (STIRPAT) model. J. Cleaner Prod. 163, 306–314.

Jones, D.W., 1991. How urbanisation affects energy-use in developing countries. Energy Policy 12, 621–629.

Kalman, R.E., 1960. A new approach to linear filtering and prediction problems. J. Basic Eng. 82, 35–45.

Kalnay, E., Cal, M., 2005. Impact of urbanization and land-use change on climate. Lett. Nat. 423, 528–531.

Karakatsani, N.V., Bunn, D.W., 2008. Intra-day and regime-switching dynamics in electricity price formation. Energy Econ. 30, 1776–1797.

Karanfil, F., 2009. How many times again will we examine the energy-income using a limited range of traditional econometric tools. Energy Policy 37, 1191–1194.

Katircioglu, S.T., 2014. International tourism, energy consumption, and environmental pollution: the case of Turkey. Renew. Sustainable Energy Rev. 36, 180–187.

Katircioglu, S., Fethi, S., Kalmaz, D.B., Caglar, D., 2016. Interactions between energy consumption, international trade, and real income in Canada: an empirical investigation from a new version of the Solow growth model. Int. J. Green Energy 13, 1059–1074.

Kelly, J., Williams, P.W., 2007. Modelling tourism destination energy consumption and greenhouse gas emissions: Whistler, British Columbia, Canada. J. Sustainable Tourism 15 (1), 67–90.

Klare, M., 2002. Resource Wars. Henry Holt, New York.

Labandeira, X., Manzano, B., 2012. EforEnergy [Online]. Available from: file:///C:/Users/user/Downloads/WP092012.pdf.

Liddle, B., 2004. Demographic dynamics and per capita environmental impact: using panel regressions and household decomposition to examine population and transport. Popul. Environ. 26, 23–39.

Liddle, B., 2014. Impact of population, age structure, and urbanization on carbon emissions/energy consumption: evidence from macro-level, cross-country analyses. Popul. Environ. 35, 286–304.

Liddle, B., Lung, S., 2010. Age structure, urbanization, and climate change in developed countries: revisiting STIRPAT for disaggregated population and conusmption-related environmental impacts. Popul. Environ. 31, 317–343.

Lucheroni, C., 2010. Stochastic models of resonating markets. J. Econ. Interact. Coord. 5, 77.

Minier, J., 2009. Opening a stock exchange. J. Dev. Econ. 90, 135–143.

Mishra, V., Smyth, R., Sharma, S., 2009. The energy-GDP nexus: evidence from a panel of Pacific Island countries. Res. Energy Econ. 31, 210–220.

Ozturk, I., 2010. A literature survey on energy-growth nexus. Energy Policy 38, 340–349.

Pachauri, S., 2004. An analysis of cross-sectional variations in total household energy requirements in India using micro survey data. Energy Policy 32, 1723–1735.

Pachauri, S., Jiang, L., 2008. The household energy transition in India and China. Energy Policy 36, 4022–4035.

Parikh, J., Shukla, V., 1995. Energy use and greenhouse effects in economic development: results from a cross-country study of developing economies. Global Environ. Change 5, 87–103.

Poumanyvong, P., Kaneko, S., 2010. Does urbanisation lead to less energy use and lower CO_2 emissions? A cross country analysis. Ecological Econ. 70, 434–444.

Rahman, M.M., Mamun, S.A., 2016. Energy use, international trade and economic growth nexus in Australia: new evidence from an extended growth model. Renew. Sustainable Energy Rev. 64, 806–816.

Rahman, M.S., Shahari, E., Rahman, M., Noman, A.H.M., 2017. The interdependent relationship between sectoral productivity and disaggregated energy consumption in Malaysia: a Markov switching approach. Renew. Sustainable Energy Rev. 67, 752–759.

Roberts, T.J., Grimes, P., Manale, J., 2003. Social roots of global environmental change: a world-systems analysis of carbon dioxide emissions. J. World Syst. Res. 9, 277–315.

Rostow, W.W., 1960. The Stages of Economic Growth: A Non-communist Manifesto. Cambridge University Press, Cambridge.

Sadorsky, P., 2010. The impact of financial development on energy consumption in emerging economies. Energy Policy 38, 2528–2535.

Sadorsky, P., 2012. Energy, output, and trade in South America. Energy Econ. 33, 476–488.

Sadorsky, P., 2013. Do urbanization and industrialization affect energy intensity in developing countries? Energy Econ. 37, 52–59.

Scheepers, M., Seebregts, A., de Jong, J., Maters, H., 2007. EU standards for energy security of supply. ECN/Clignendael International Energy Programme, s.l.

Schurr, S., Netschert, B., 1960. Energy and the American Economy, 1850–1975. Johns Hopkins University Press, Baltimore.

Shahbaz, M., Sbia, R., Nanthakumar, L., Afza, T., 2015. The effect of urbanization, affluence and trade openness on energy consumption: a time series analysis in Malaysia. Renew. Sustainable Energy Rev. 47, 683–693.

Stern, D.I., 1993. Energy and economic growth in the USA: a multivariate approach. Energy Econ. 15, 137–150.

Stern, D.I., 2004. Economic growth and Energy. Encycl. Energy 2, 35–51.

Stern, D.I., 2011. The role of energy in economic growth. Ann. N. Y. Acad. Sci. 1219, 26–51.

Tahvonen, O., Salo, S., 2001. Economic growth and transitions between renewable and nonrenewable energy resources. Eur. Econ. Rev. 45, 1379–1398.

Tamazian, A., Chousa, J.P., Vadlamannati, C., 2009. Does higher economic and financial development lead to environmental degradation: evidence from BRIC countries. Energy Policy 37, 246–253.

Tang, C.F., Abosedra, S., 2014. The impacts of tourism, energy consumption and political instability on economic growth in the MENA countries. Energy Policy 68, 458–464.

Thollander, P., Karlsson, M., Soderstrom, M., Creutz, D., 2005. Reducing industrial energy costs through energy costs through energy efficient measures in a liberalized European electricity market: case study of a Swedish iron foundry. Appl. Energy 81, 115–126.

Tiwari, A.K., Ozturk, I., Aruna, M., 2013. Tourism, energy consumption and climate change in OECD countries. Int. J. Energy Econ. Policy 3 (3), 247–261.

Toman, M., Jemelkova, B., 2003. Energy and Economic Development: An assessment of the State of Knowledge. Resources for the Future, Washington, DC.

Tsagarakis, K.P., et al., 2011. Tourists' attitudes for selecting accommodation with investments in renewable energy and energy saving systems. Renew. Sustainable Energy Rev. 15, 1335–1342.

WEF and IHS CERA, 2012. Energy for Economic Growth: Energy Vision Update 2012. World Economic Forum, Geneva.

York, M., 2007. Demographic trends and energy consumption in European Union nations, 1960–2025. Social Sci. Res. 29, 855–872.

CHAPTER 5

Critical Issues to Be Answered in the Energy-Growth Nexus (EGN) Research Field

Angeliki N. Menegaki*,‡, Stella Tsani**,†

*Hellenic Open University, Patras, Greece; **Athens University of Economics and Business, Athens, Greece; †International Centre for Research on the Environment and the Economy, Athens, Greece; ‡TEI STEREAS ELLADAS, University of Applied Sciences, Lamia, Greece

Chapter Outline

1 Introduction

Energy policies are associated with today's environmental problems: greenhouse effect, deforestation and acid precipitation that damages forests and biodiversity, and the extreme water withdrawals used for energy production. Climate change bears a large social dimension, because the extreme weather phenomena it causes and the alterations it affects to farmland are the reasons for the relocation of many populations. The increase in global temperature may cause such changes that will disrupt economies. Thus the economy and the environment (not less the society) are inextricably woven together, and the problems of one cannot be solved in isolation from the other.

Besides, rapid population growth and urbanization, which cause a tremendous impact on energy and environmental problems, constitute both economic and social problems. Pollution problems and natural resource degradation are not national problems but international ones, as their effects can be felt by more countries. Despite the nonexistence of a common worldwide blueprint for energy use and sustainable development, the EGN offers plenty of food for thought to policy makers.

Countries need to propose long-term policies not only with respect to energy use, but also to take short-run measures to adapt and direct themselves toward the long-run goals. Given that energy problems are interlinked to social and economic ones, energy problems should be widely acknowledged within the cadre of sustainable development. Governments need to evaluate the benefits of energy conservation and pass the corresponding price signals to consumers. Energy efficiency emerges as the cutting edge of energy policy making, because tremendous benefits can be derived from it.

As a result of that, the questions that the research on EGN attempts to answer are regarding the support of one of the four known hypotheses: Growth, Conservation, Feedback, and Neutrality. The reason this is so is because policy makers are keen to know the effects that energy conservation measures will, or will not, have on economic growth, as represented by the GDP indicator.

Energy conservation stems both from the need to preserve and safeguard natural resources for future generations and from the need to reduce carbon emissions, thereby stopping or reversing climate change and environmental pollution in general. The need to embark on energy conservation activities stems from not only common logic and altruistic attitudes toward the environment, but it is also dictated by major international environmental agreements. Some of them are binding, while some are not. Some have been ratified by a lot of countries, while some have not.

The rest of this chapter is organized as follows. Section 2 deals with the most important international environmental agreements and recognizes that very few studies of the EGN research attempt to answer whether the agreement goals will be fulfilled; the section continues with the research opportunities offered by those agreements. Section 3 highlights comparison difficulties among different types of energy-growth studies. Section 4 describes recent shifts in the EGN such as sustainability and energy efficiency issues and presents additional shifts such as the food-energy-water nexus and the energy-water nexus. Section 5 suggests literature segregation paradigms. Section 6 concludes.

2 Environmental Agreements and Research Opportunities in the EGN

In 1983, the Brundtland Commission or the World Commission on Environment and Development was established independent of the United Nations to focus on sustainability and coordinate international strategies on the matter of harmonic coexistence of economic welfare with ecology. The Brundtland Commission published in 1987 its main report entitled "Our Common future" (United Nations, 1987), which greatly influenced the Rio Summit some years later.

In 1992, the Rio Earth Summit generated the most important international agreement on climate change—the United Nations Framework Convention on Climate Change (UNFCCC). As its aftermath, in 1997, the Kyoto Protocol was adopted, and in 2005, it came into effect. This agreement is the first commitment period (2008–2012) of the UNFCCC and sets legally binding targets in terms of greenhouse emission reduction. In 2013, the Doha Amendment to Kyoto Protocol came into the foreground, but has not come into effect, because it has not been ratified by the required number of countries. This concerned the second commitment period (2013–2020), whereby countries would agree to reduce emissions by at least 18% below 1990 levels (United Nations, 2016).

Unfortunately, large economies of the world, such as the United States, have never entered the Kyoto Protocol agreement or have retreated, such as Canada, while Japan and Russia have not signed the Doha Amendment at all. In 2015, the Paris Agreement was generated, according to which, countries agreed to reduce global warming by at least 2°C from 2020 and thereafter. The measures each country announces to take for the combat of climate change have been termed as "Intended Nationally Determined Contributions" (INDCs). The Paris Agreement entered into force on November 4, 2016, and up to date, only 75 countries have ratified it. For reasons of comparison, we report that the Kyoto Protocol had been ratified by 83 countries from 193 existent parties (United Nations, 2016), while the Doha Amendment has been ratified by 65 countries, but entering into force required the ratification from at least 144 countries.

The EGN research provides useful feedback for the formulation of the INDCs, because it is based on the sensitivity of economic growth to energy conservation measures, and thus policy makers in participating countries would be benefited if they took them into consideration. The fact that the countries, which ratify the Paris Agreement, will be asked to update their INDCs every 5 years (World Resources Institute, 2016) underlines the necessity for ongoing and more sophisticated future research in the EGN. Moreover, the EGN can provide useful input to the Intergovernmental Panel on Climate Change (IPCC), which provides policy makers with research results (derived from scientific, technological, and socioeconomic information) on climate change so that they make informed policies.

Until 2017, the remedy tools recommended by the Kyoto Protocol (or other environmental agreement or strategy) have not been explicitly hosted in the EGN research. The most important tools suggested by the Kyoto Protocol agreement had been carbon trading, reforestation, land improvement, and joint implementation opportunities. Nevertheless, the EGN research, as exercised today, is blind to the methods each country opts to reach its Kyoto Protocol goals (or other environmental goals). For example, a country may invest on becoming more energy efficient or develop more renewable energy sources and thus may reduce fossil energy consumption; however, at the same time, if it has a large land surface and embarks on huge reforestation programs, this will outweigh the greenhouse emitted by its economy. Thus different reduction needs and targets will be present for energy conservation, which aim at the reduction of emissions. Not recognizing the environmental agreements a country has signed may lead to a misspecified EGN model ignoring interdependencies that will produce unrealistic results both for the causality relationship itself and the quantification of causality.

In addition, the "joint implementation" tool of the Kyoto Protocol is a concept whose effects can be studied if countries are studied within panels that make sense, not only from a purely geographical, economic development or data availability point of view, but also from an environmental commitment point of view. According to the joint implementation tool, countries can gain emission reduction units (one unit per CO_2 tone equivalent) for themselves if they contribute to another country's reduction of emissions. However, both the two hypothetical countries must be embracing Kyoto Protocol agreement. Again, the EGN relationship in a panel of countries in such a situation is misspecified if that information is not enclosed.

Last, the EGN does not make a tradeoff between the costs of carbon trading with the cost of building renewable energy infrastructure. Not taking into account the level capital and technology each country is found at, or the level of its international indebtedness, again causes a knowledge gap that cuts crucial information out. If we witness causality from renewable energy to economic growth, this does not guarantee that the correct recipe for the continued economic growth in this country is to increase renewable energy development, if the country is indebted. In that case, the renewable energy expansion could take place through international funds.

2.1 What Do International Agreements and National Efforts to Reduce Greenhouse Gas (GHG) (Emissions) Mean for the EGN Literature Development?

The majority of the EGN studies investigate a relationship between energy consumption and economic growth (with various control variables). The study titles reflect the perused variables, the method, or main findings, and more rarely, titles contain the reason to study the EGN in the particular form the authors imagine or suggest. Very few examples of EGN

Table 5.1: Examples of EGN studies which claim to answer a concrete policy question.

	Studies	EGN Relationships	Policy Targets and Questions	Answers
1	Rafindadi and Ozturk (2015)	Natural gas consumption = f(Y, K, Ex, L)	Is the 10th Malaysian plan attainable (2011–2020)? → Achieving a gross national income per capita of US$12,140 in 2020.	Natural gas consumption does not Granger cause economic growth in the long run. Support of the Feedback hypothesis in the short run. 1% increase in economic growth in Malaysia will lead to the decline of natural gas consumption by 0.1304%.
2	Śmiech and Papież (2014)	Level of compliance with EU policy: Y = f (E)	Climate and energy package for 2020 (20% reduction in CO_2 emissions, 20% increase in renewable energies and 20% increase in energy efficiency).	Neutrality hypothesis and the strength of this result depends on the level of compliance.
3	Soytas and Sari (2009)	Y = f(CO_2, K, L, E)	Reduction of emissions by 60%–80% by 2050 as compared to 1990. Energy reduction by 20% by 2020.	Energy consumption does not Granger cause income in the long run. Unidirectional causality from emissions to energy use.

CO_2, Carbon emissions; *Ex*, exports; *K*, capital; *L*, labor; *Y*, income.

paper titles specifically claim to be answering a particular policy question, and these are shown in Table 5.1.

To the best of our knowledge, there are only three papers that claim to be studying the feasibility or compliance toward an energy or environmental target.

In Study 1 (Rafindadi and Ozturk, 2015): The question was rather meant to be "What would be the contribution of natural gas to the 10th Malaysian plan?" The 10th Malaysian plan aims at the achievement of a certain level of economic growth and thus any possible drivers toward that should be investigated. The study investigates the consumption of natural gas to this respect but does not make a clear projection for 2020. The study uses data from 1971 to 2012 to estimate the long-run and short-run relationships, but it does not make a clear-cut projection to 2020 economic growth, based on the variables in the model.

Based on causality results from this study, natural gas consumption cannot lead to economic growth in Malaysia. However, natural gas consumption has indirect effects on economic growth through its direct effect on exports and capital by 42.29% and 20.04%, respectively. Additional policy conclusions suggest that the very low contribution of labor should be strengthened so that the country will not need foreign expert labor.

In Study 2 (Śmiech and Papież, 2014): This study claims to be studying the energy-growth causality with respect to compliance to EU (European Union) policies. It uses data from 1993 to 2011 and reaches the conclusion that energy does not affect economic growth and hence, on average, all 2020 plans could be materialized, *ceteris paribus*. The study does not make projections about specific future years and the compliance accomplishment made in each one of them.

In Study 3 (Soytas and Sari, 2009): This study applies data from 1960 to 2000 for Turkey and uses the Toda–Yamamoto procedure to find out whether energy and emissions can be cut down in this country to comply with the EU strategies (20% reduction in energy by 2020, and reduction of greenhouse gases by 60%–80% by 2050). The emissions' causal relationship to energy use in the absence of long-run causality from energy use to growth can give very general insights; However, these tell little about what could be done on a year-by-year basis until 2020.

Overall, it is our opinion that the EGN studies should take one additional step after identifying cointegration and causality relationships. And that, an additional step is the forecasting and exact quantification of the intertemporal causal effect of one variable on the other. Notwithstanding robustness tests with stability and impulse response functions, a necessary step to make papers essential for policy making is to perform dynamic forecasting analysis and employ h-causality tests, which very few studies until now have utilized. Conversely, a static prediction would mean to forecast one-step ahead. Unless a forecasting model can be secured, it is very unlikely to express safe statements about future relationships and whether deadlines and policy goals can be fulfilled.

2.2 An Application with Intertemporal Causality

Tiwari et al. (2016) had conducted an empirical study on the causal relationship between CO_2 emissions, GDP per capita, energy consumption, domestic credit, exports, and money supply in Saudi Arabia from 1971 to 2014. Based on their results, CO_2 emissions can be curtailed in Saudi Arabia without the economy slowing down. The perused data are as follows: GDP is measured by GDP per capita (current US\$), EC represents fossil fuel energy consumption (as % of total), EXP is measured by exports of goods and services (as %GDP), DCTOP is measured by domestic credit to private sector (as %GDP), C is measured by CO_2 emissions (metric tons per capita), and MS is measured by Money and quasi-Money M2 to total reserves ratio. From a methodological point of view, they have perused an autoregressive distributed lag (ARDL) model and an h-horizon causality approach, which proved to be more informative than the bounds-test approach causality results. Their results revealed that CO_2 emissions do not cause any other variables, and they are only caused by GDP growth at h-horizon equal to 2 (Table 5.2).

Observing the direction of causalities between energy consumption and each of the variables, we observe no causality whatsoever starting from energy consumption to any of the rest of the variables. Thus energy consumption in Saudi Arabia does not cause economic growth to

Table 5.2: H-step or infinite step Granger noncausality.

	h = 1	h = 2	h = 3	h = 4	h = 5	h = 6	h = 7
LNDGPPC → LNEC	0.070**	0.322	0.165	0.485	0.398	0.138	0.531
LNDGPPC → LNEXP	0.329	0.124	0.512	0.240	0.188	0.238	0.271
LNDGPPC → LNDCTOP	0.364	0.340	0.317	0.262	0.080**	0.118	0.359
LNDGPPC → $LNCO_2$	0.373	0.064**	0.214	0.497	0.444	0.388	0.549
LNDGPPC → LNMS	0.255	0.433	0.276	0.158	0.212	0.289	0.568
LNEC → LNDGPPC	0.421	0.435	0.302	0.494	0.418	0.284	0.197
LNEC → LNEXP	0.340	0.482	0.493	0.476	0.487	0.339	0.051**
LNEC → LNDCTOP	0.195	0.445	0.403	0.326	0.509	0.436	0.222
LNEC → $LNCO_2$	0.307	0.183	0.213	0.156	0.155	0.106	0.086**
LNEC → LNMS	0.378	0.224	0.298	0.347	0.556	0.571	0.581
LNEXP → LNDGPPC	0.005*	0.192	0.459	0.625	0.255	0.105	0.607
LNEXP → LNEC	0.342	0.128	0.23	0.465	0.495	0.337	0.559
LNEXP → LNDCTOP	0.047*	0.486	0.339	0.377	0.548	0.234	0.637
LNEXP → $LNCO_2$	0.124	0.504	0.245	0.398	0.622	0.637	0.668
LNEXP → LNMS	0.032*	0.047*	0.108	0.091**	0.227	0.147	0.284
LNDCTOP → LNDGPPC	0.018*	0.209	0.431	0.375	0.015*	0.156	0.500
LNDCTOP → LNEC	0.263	0.366	0.060**	0.467	0.414	0.479	0.336
LNDCTOP → LNEXP	0.437	0.129	0.457	0.249	0.005*	0.085**	0.458
LNDCTOP → $LNCO_2$	0.229	0.239	0.478	0.464	0.479	0.326	0.275
LNDCTOP → LNMS	0.036*	0.169	0.399	0.450	0.411	0.260	0.490
$LNCO_2$ → LNDGPPC	0.125	0.254	0.21	0.144	0.187	0.369	0.59
$LNCO_2$ → LNEC	0.289	0.314	0.148	0.435	0.344	0.457	0.281
$LNCO_2$ → LNEXP	0.322	0.204	0.128	0.200	0.378	0.387	0.196
$LNCO_2$ → LNDCTOP	0.223	0.470	0.489	0.254	0.485	0.524	0.470
$LNCO_2$ → LNMS	0.331	0.247	0.144	0.261	0.357	0.486	0.596
LNMS → LNDGPPC	0.314	0.462	0.439	0.359	0.014*	<0.000*	0.24
LNMS → LNEC	0.353	0.488	0.357	0.357	0.446	0.47	0.278
LNMS → LNEXP	0.332	0.498	0.324	0.378	0.090**	0.013*	0.284
LNMS → LNDCTOP	0.490	0.439	0.379	0.534	0.117	0.004*	0.155
LNMS → $LNCO_2$	0.480	0.248	0.472	0.425	0.012*	0.174	0.347

Figures reported are *p*-values and h denotes horizons to which Granger-causality is tested. The determination of h is based on formula h = number of variables in the VARXlags+1.

* denotes significance at 5%. ** denote significance at 10%.

Source: Adapted from Tiwari, A.K., Mensi, W. et al., 2016. CO_2 emissions, energy consumption & economic growth: fresh evidence from Saudi Arabia using multi-horizon, time varying and Fourier-based Granger causality approaches, Working Paper.

any of the rest of the variables in the short run, with the exception of exports and emissions. However, we observe future causality, namely at horizon ($h = 7$), from energy consumption to exports and CO_2 emissions.

Moreover, evidence of causality at horizon ($h = 1$) exists from exports to GDP, to domestic credit and money supply. The latter finding is strong and persistent across horizons 1, 2, and 5. Domestic credit drives causalities to GDP and money supply at $h = 1$, toward energy consumption at $h = 3$, and toward exports at $h = 5$ and 6, but there is no causality toward CO_2

emissions. The latter, on the other hand, drive no causality toward any of the variables either in the short horizons or the longer ones.

Last, the money supply drives the causality toward almost all variables except for energy consumption, and this causality occurs at longer horizons, namely at $h = 5$ or 6. Infinite-step causality and time-varying Granger causality are significant too for causality starting from money supply toward domestic credit and emissions.

3 Difference Between EGN Results from a Single Country and EGN Results from Panel Countries Which Contain the Single Country

Results from EGN studies based on a large sample of countries are used for the identification of global relationships that can form the basis for the information of theory, but must be treated with caution when relied to perform any policy making concerning individual countries, particularly when these countries are small in population and size of economy (e.g., low income). Thus single-country analyses should be considered as equally important as multiple-country analyses and at the moment policy makers should consult them in a complementary and synthetic way. Furthermore, sometimes, depending on the perused econometric method, results for individual countries can be reached even within multiple-country studies such as in the study by Omri et al. (2015) in which dynamic simultaneous equation analysis was employed. To further support our argument, we provide later an illustrative example.

3.1 Application: Single Country Versus Multiple Countries

The seminal paper by Apergis and Payne (2012) investigated the relationship between renewable (RE) and nonrenewable energy consumption (NRE) and growth for 80 countries in the period 1990–2007. Besides the already mentioned energy variables, it used real gross fixed capital formation (K) and labor force (L). The detailed countries of their sample are provided in the first column of Table 5.3. The paper refers to no particular reason why these 80 countries were selected. We assume that this choice was based on data availability by an as-large-as possible number of countries, and that the study aimed to produce a global relationship.

Based on causality results, both renewable and nonrenewable energies affect growth and growth affects both types of energies too (RE↔Y in the short run and non-RE↔Y in the long run, also RE↔non-RE; the arrows show the direction of causality). Thus there is evidence of the feedback hypothesis. Support for the same hypothesis applies between renewable and nonrenewable energies.

If, however, we try collecting studies for the individual countries composing the sample in Apergis and Payne (2012), we observe that the causality results are not confirmed in all

Table 5.3: Identification of causality results from single countries encompassed in a global study.

Countries [1]	Studies [2]	Time Spans [3]	Relationship [4]	Causality Results for Energy [5]	Agree/Disagree with Global Relationship [6]
Argentina	Omri et al. (2015)	1990–2011	Y = f(RE, L)	Y→RE	Disagree
Australia	Azad et al. (2015)	1976–2013	Y = f(RE, NRE, K, L, CO_2)	NRE↔Y RE↔CO_2	Agree (NRE)
Belgium	Omri et al. (2015)	1990–2011	Y = f(RE, L)	RE↔Y	Agree (RE)
Brazil	Omri et al. (2015)	1990–2011	Y = f(RE, L)	RE~Y	Disagree
Bulgaria	Omri et al. (2015)	1990–2011	Y = f(RE, L)	RE↔Y	Agree (RE)
	Koçak and Şarkgüneşi (2017)	1990–2012	Y = f(RE, L, K)	RE→Y	Disagree
Canada	Omri et al. (2015)	1990–2011	Y = f(RE, L)	RE↔Y	Agree (RE)
China	Lin and Moubarak (2014)	1977–2011	Y = f(RE, L, K, CO_2)	RE↔Y	Agree (RE)
Denmark	Irandoust (2016)	1975–2012	RE = f(Y, CO_2, Tech)	Y→RE RE→CO2	Disagree
Finland	Omri et al. (2015)	1990–2011	Y = f(RE, L)	RE~Y	Disagree
	Irandoust (2016)	1975–2012	RE = f(Y, CO_2, Tech)	Y→RE RE→CO2	Disagree
France	Omri et al. (2015)	1990–2011	Y = f(RE, L)	RE↔Y	Agree (RE)
Greece	Koçak and Şarkgüneşi (2017)	1990–2012	Y = f(RE, L, K)	RE→Y	Disagree
Hungary	Omri et al. (2015)	1990–2011	Y = f(RE, L)	RE→Y	Disagree
India	Omri et al. (2015)	1990–2011	Y = f(RE, L)	RE→Y	Disagree
Japan	Omri et al. (2015)	1990–2011	Y = f(RE, L)	RE→Y	Disagree
NRL	Omri et al. (2015)	1990–2011	Y = f(RE, L)	RE→Y	Disagree
Norway	Irandoust (2016)	1975–2012	RE = f(Y, CO_2, Tech)	Y→RE RE↔CO2	Disagree
Pakistan	Omri et al. (2015)	1990–2011	Y = f(RE, L)	RE↔Y	Agree (RE)
	Shahbaz et al. (2015)	1972–2011	Y = f(RE, L, K)	RE↔Y	Agree (RE)
Romania	Koçak and Şarkgüneşi (2017)	1990–2012	Y = f(RE, L, K)	RE↔Y	Agree (RE)
Spain	Omri et al. (2015)	1990–2011	Y = f(RE, L)	Y→RE	Disagree
Sweden	Omri et al. (2015)	1990–2011	Y = f(RE, L)	RE→Y	Disagree
	Irandoust (2016)	1975–2012	RE = f(Y, CO_2, Tech)	Y→RE RE↔CO2	Disagree
Switzerland	Omri et al. (2015)	1990–2011	Y = f(RE, L)	RE~Y	Disagree
Turkey	Koçak and Şarkgüneşi (2017)	1990–2012	Y = f(RE, L, K)	RE↔Y	Agree (RE)
	Dogan (2016)		Y = f(RE, L, K)	Y→RE (Short run) RE→Y (Long run)	Disagree
USA	Omri et al. (2015)	1990–2011	Y = f(RE, L)	RE↔Y	Partly Agree (RE)

NA stands for not available. Column [1] refers to the countries in the sample from Apergis and Payne (2012). Column [2] refers to the studies on individual countries. Column [3] refers to the time span of the study in Column [2]. Column [4] refers to the relationship through which renewable energy and growth appear in the study of Column [2]. Column [6] refers to whether results from individual studies in Column [2] agree and up to what degree, with results from principal study, namely by Apergis and Payne (2012).

individual studies, and hence caution is suggested at policy making. The identification of the individual studies was implemented based on whether the paper studied the renewable EGN. Studies where only total energy is used have not been included in Table 5.3.

Apparently, only for 18 countries individual studies on the renewable EGN were identified. For 5 countries, we have identified at least two studies. For the rest of the countries, namely for 57 countries, individual studies do not exist, to the best of our knowledge. For 73% of the individual studies, causality results between renewable energy and growth are different from the principal study, namely the study by Apergis and Payne (2012). The principal study supports evidence of the feedback hypothesis, while 73% of the individual studies find support of all the other three hypotheses.

Thus the usefulness of these studies is for the benefit of science progress and development. They may unveil important empirical relationships and give a boost to the development of the research field. Practical importance is sometimes little. The latter can be produced mostly from single-country studies or from small panels of countries with similar geographical, economic, and social characteristics. The G7, the MENA countries, the PIGST, the BRICS, Nordic countries, or European Union countries are groups that make some sense historically, politically, geographically, and economically. On the other hand, data availability is sometimes a key factor for many of the country samples we meet in these studies. Albeit irrelevant the countries may be, results are better than nothing in an attempt to empirically recognize a new relationship or update an existing one.

4 Recent Shifts in the EGN: Sustainability

Sustainability refers to the ability of an economy to maintain at least the same (or a higher) level of economic, environmental, and social sources across generations. According to the United Nations (United Nations, 2016), there are 17 goals for sustainable development: the eradication of poverty, food security and sustainable agriculture, good health and well-being, quality education, gender equality, clean water and sanitation, affordable and clean energy, decent employment for all and economic growth, industry, innovation and infrastructure, reduced inequalities, sustainable cities and communities, responsible consumption and production, climate action, life below water, life on land (ecosystems, biodiversity, and prevention of land degradation), peace, justice and strong institutions, and last, partnerships for the goals.

To the best of our knowledge, the EGN does not take into account sustainability except for the degree of renewable or alternative energies as part of full energy consumption. Only one study by Zaman et al. (2016) makes explicit reference to sustainability and includes independent variables: some environmental variables, energy variables, and health variables proxied by a fertility rate and an infant mortality rate. Of course, we cannot oversee the fact that the conservation of energy or energy efficiency is by themselves movements toward sustainability too. Hence, when we investigate the effect of energy conservation to economic growth, we practically investigate the effect of energy sustainability on economic growth.

Recently, a new tendency in the EGN literature incorporates sustainability in an additional way. Instead of GDP as an indicator for economic welfare, which is typically used in almost all EGN studies, this new literature (Menegaki and Tiwari, 2017; Menegaki and Tugcu, 2016; Menegaki and Tugcu, 2017) uses the Index of Sustainable Economic Welfare (ISEW) in place of the GDP. Different results apply when the ISEW is used in place of the GDP, which means that when economies focus on sustainability, the best thing is to use the two indicators in different models and compare them. This is so because policy makers should be at least equally concerned with the effects of energy conservation on sustainable GDP rather than solely on the effects on the conventional GDP.

The GDP has been criticized for not being able to measure genuine progress (Kubiszewski et al., 2013), because it is not able to distinguish between welfare-improving and welfare-reducing activities (Talbreth et al., 2007). There are many indicators for sustainable development such as the Human Development Index (HDI), but only the ISEW is all inclusive as it contains all possible dimensions that form the three spheres of sustainability: economy, environment, and society. The ISEW was first applied and later improved by Daly and Cobb (1989) and Cobb and Cobb (1994), respectively. It has passed various stages of evolutionary development from the Measure of Economic Welfare by Nordhaus and Tobin (1972) to the Genuine Progress Index, which is nowadays used alternatively with the ISEW (Lawn, 2003).

The calculation of the ISEW would be straightforward if all its components were available. Currently, while the economy part of the ISEW can be calculated with relative safety and easiness, the environment and society parts are difficult to calculate because of lack or incompleteness of environmental or social data (Fig. 5.1).

The ISEW has been calculated for few countries worldwide; for example, Brazil (Torras, 2005), Chile (Castañeda, 1999), China (Cheng et al., 2005), Indonesia (Torras, 2005), Thailand (Clarke, 2004), France (Nourry, 2008), Italy (Pulselli et al., 2012), USA (Bagstad et al., 2014), and Greece (Menegaki and Tsagarakis, 2015).

Since the ISEW is not a magnitude readily provided by major international databases, it must first be calculated before using in the EGN empirical projects. However, because not all the ISEW components are publicly available, a reasonable approach would be to calculate the ISEW in as much depth and width as it is allowed by available data for each country. Menegaki and Tugcu (2016) for sub-Saharan African countries, Menegaki and Tsagarakis (2015) for Greece, Menegaki and Tiwari (2017) for American countries, Menegaki and Tugcu (2017) for G7 countries, Menegaki and Tugcu (2016) for emerging economies and Menegaki and Tiwari (2017) for Europe have used the ISEW components that are shown in Table 5.1.

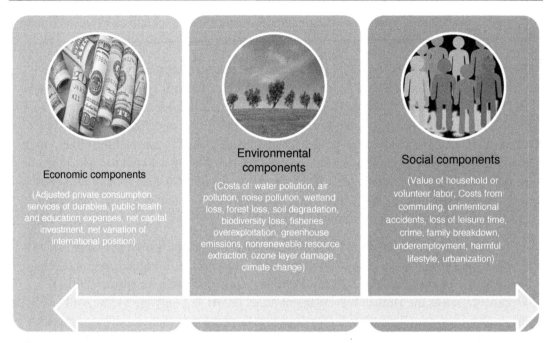

Figure 5.1: The Components of the Index of Sustainable Economic Welfare.
Adapted from Beça, P., Santos, R., 2010. Measuring sustainable welfare: a new approach to the ISEW. Ecol. Econ. 69(4), 810–819 and Bagstad, K.J., Berik, G. et al., 2014. Methodological developments in US state-level genuine progress indicators: toward GPI 2.0. Ecol. Indic. 45, 474–485.

The components shown in Table 5.4 are based on Eq. (5.1):

$$ISEW = C_w + G_{eh} + K_n + S + N + C_s \qquad (5.1)$$

where C_w is the weighted consumption, G_{eh} stands for non-defensive public expenditure, K_n is the net capital growth, S is the unpaid work benefit, N is the depletion of natural environment, and C_s is the cost from social problems, which is the most difficult part to be included in the calculations due to lack of data. Hence, it has been omitted from Table 5.4.

5 Recent Shifts in the EGN: Energy Intensity and Energy Efficiency

5.1 Definitions and Scope

Minimizing energy intensity is a goal that needs to be pursued together with the constraint to maximize economic growth and development, environmental protection, and energy security for all countries and economic sectors. Energy intensity measures energy efficiency. Namely, it measures the capability to "do more with little," and to produce more output using less inputs. This requires a unique combination of the inputs, together with the adoption of best and leanest technology and management. Eurostat measures energy efficiency as the

Table 5.4: ISEW components, sign, calculation methods, and data sources.

Components	Signs	Calculation Methods	Source/Available From
1. Adjusted personal consumption with durables (C_w)	+	Multiplied personal consumption and durables' expenditure (PC) with Gini coefficient (G) and poverty index (P) as: PC $\times (1-G) \times (1-P)$	PC: http://data.worldbank.org/indicator/NE.CON.PRVT.CDT.CD Gini coefficient: http://data.worldbank.org/indicator/SI.POV.GINI Poverty index (headcount ratio): http://data.worldbank.org/indicator/SI.POV.2DAY
2. Education expenditure (G_{eh})	+	Public expenditure on education (current operating expenditures in education, including wages and salaries and excluding capital investments in buildings and equipment). Assuming that half of it is defensive, we multiply this amount by 50%.	http://data.worldbank.org/indicator/NY.ADJ.AEDU.CD
3. Health expenditure (G_{eh})	+	Public health expenditure is also multiplied with 50% for the same reason as above.	http://data.worldbank.org/indicator/SH.XPD.PUBL
4. Net capital growth (K_n)	±	Use of data on fixed capital accumulation (FCA). Subtract consumption of fixed capital (CFC) to find the net capital and then calculate its growth rate.	FCA: http://data.worldbank.org/indicator/NE.GDI.TOTL.CD CFC: http://data.worldbank.org/indicator/NY.ADJ.DKAP.CD
5. Mineral depletion (N_1)	−	Mineral depletion is the ratio of the value of the stock of mineral resources to the remaining reserve lifetime (capped at 25 years). It covers tin, gold, lead, zinc, iron, copper, nickel, silver, bauxite, and phosphate.	http://data.worldbank.org/indicator/NY.ADJ.DMIN.CD
6. Energy depletion (N_2)	−	It is the ratio of the value of the stock of energy resources to the remaining reserve lifetime (capped at 25 years). It covers coal, crude oil, and natural gas.	http://data.worldbank.org/indicator/NY.ADJ.DNGY.CD
7. Forest depletion (N_3)	−	Net forest depletion is calculated as the product of unit resource rents and the excess of roundwood harvest over natural growth.	http://data.worldbank.org/indicator/NY.ADJ.DFOR.CD
8. Damage from CO_2 emissions (climate change-long-run environmental damage) (N_4)	−	It is estimated to be $20 per ton of carbon dioxide (the unit damage in 1995 US dollars) times the number of tons of carbon dioxide emitted. World bank estimations are based on Samuel Fankhauser's "Valuing Climate Change: The Economics of the Greenhouse" (1995). No other greenhouse gases are included.	http://data.worldbank.org/indicator/NY.ADJ.DCO2.CD

Source: Menegaki, A.N. and Tiwari, A.K., 2017. The index of sustainable economic welfare in the energy-growth nexus for American countries. Ecol. Indic. 72, 494–509, Menegaki, A.N. and Tsagarakis, K.P., 2015. More indebted than we know? Informing fiscal policy with an index of sustainable welfare for Greece. Ecol. Indic. 57, 159–163, Menegaki and Tugcu (2016), and Menegaki, A.N. and Tugcu, C.T., 2017. Energy consumption and sustainable economic welfare in G7 countries: a comparison with the conventional nexus. Renew. Sustainable Energy Rev., forthcoming, 10.1016/j.rser.2016.11.133.

ratio of gross inland consumption of energy over GDP. Thus it is counted as kilogram of oil equivalent per 1000 € (Eurostat, 2016). Energy consumption is in the numerator of this ratio and includes renewables too. This means that energy intensity is not concerned with energy cleanness but rather with energy leanness (quantity). Energy intensity can also be regarded as the inverse of energy productivity.

Unavoidably, a high-energy intensity causes additional and redundant emissions, particularly if it is about fossil energy consumption. High-energy intensity produced from renewables is surely clean, but not lean. Therefore, energy intensity reveals the level of energy leanness and energy smartness of an economy. Energy intensity does not include energy imported from other countries, but only primary energy supply and the energy exported to foreign markets (The Conference Board of Canada, 2016). This indicator reveals to what degree growth is decoupled from energy consumption. Thus energy intensity can be lowered either by energy conservation measures or through the adoption of energy efficient technologies or by combining these two solutions.

Usually, countries with extreme climates are also more energy-intensive (due to the heating and/or cooling demand for building space). In addition, countries with a vast industrial sector are more energy-intensive, because the industry is six times more energy-intensive than the services sector (World Energy Council, 2013). The latter is particularly evidenced for low-value industrial sectors. Energy exporters are also energy-intensive, because their GDP production part is connected with high-energy production. Another factor that contributes to high-energy intensity is a slow-growth economy whereby economic growth increases at a lower pace than energy consumption and results in low efficiency in the electricity sector, which is a major energy user in all economies. Low prices additionally play a role in driving high-energy intensity as well as the lack of orientation to sustainable energy practices and advanced technology (Research & Development). Last, capital and equipment obsolescence (Gómez et al., 2014) are crucial in building modern energy efficiency opportunities. Energy policies and population dynamics together with urbanization trends also play a role in determining the energy intensity. A wealthy population with ample access to credit opportunities is more likely to drive energy intensity high. In addition, noteworthy is the fact that about 20% of end-use efficiency improvements are offset by conversion losses (World Energy Council, 2013). Moreover, the power sector is to be blamed in cases when it had a slow progress in thermal power generation, which is a means toward efficient production of electricity.

Given the high economic recession, in whose whirl many countries worldwide are currently found, a rather turbulent period for energy intensity might be expected, because economies may give higher priorities to large-scale infrastructure developments to restart economies. According to the World Energy Council (2013), 86% of the countries worldwide had set quantitative energy saving targets in 2012. Some targets are expressed as energy-use reduction, others as energy efficiency increase, energy intensity reduction, or conventional-lamps replacement. Half of the countries worldwide have an energy efficiency law, and 75% of countries have an energy efficiency agency. Europe moves on a plan dictated by the demands of the Energy Efficiency Directive 2012/27/EU, which sets the goal for 20% energy savings in primary energy for the year 2020 (European Commission, 2012a). Similar targets are set by the Energy Roadmap 2015, according to which, an 80%–95% reduction in greenhouse emissions is required by 2050 (European Commission, 2012b).

To this end, countries have put forward building codes, appliance standards, product labeling, tax and other incentives, urban planning such as zoning and traffic design, research and development such as battery technology, and energy audits for industries (Doris et al., 2009). Countries differ in their aims and targets because of their primary energy mix, their climate, their existing level of development and lifestyles, and the level of their industrial activities (World Energy Council, 2013).

5.2 Stylized Facts About the Energy Intensity of Major Economies and Major Industries of the World

It is useful at this point to have an overview of energy intensity facts and numbers worldwide. Major economies do pave the track and are the basic actors in the international economic and energy scene. Hence, starting from Canada, this country has reduced its energy intensity by 39% since 1971, but it still is a more energy-intense country, above a 17-country average of 0.15 toe per US$1000 of GDP (The Conference Board of Canada, 2016). During the same period, the United States have lowered their energy intensity by 54%. An annual increase of 0.8% in global intensity was noted between 1981 and 2010 with a peaked change in 2010. Global energy intensity has increased by 1.35% in 2010 despite a decline in the previous 30-year period, and this is largely caused by the emerging economies and industrialized countries that have not paid enough attention to sustainable development (Yoder, 2013). The federal Energy Policy Act of 2005 (EPAct 2005) established energy-saving goals for buildings in the United States (U.S. Department of Energy, 2016). Moreover, they have established a wide set of initiatives and regulation for the transport industry and other sectors encompassed in the Action Plan 2007 (U.S. Government, 2008), which places the adoption of cost-effective energy efficiency as a high-priority resource and indicates the mechanisms through which this can be delivered, measured, and evaluated.

Other major countries such as China have increased productivity and moved to higher-value products, thus lowering energy intensity. Indonesia is a major coal exporter with almost 90% of the production of coal in Southeast Asia from this country (International Energy Agency, 2015). According to the same source, Asian countries have introduced mandatory minimum-energy performance and fuel standards for appliances and vehicles, respectively. Various countries have set different targets for the reduction of their energy intensity or for increasing their energy-saving performance by different percentages and by different target dates.

With respect to African countries, they had planned to achieve the following energy use reduction by sector: economy wide (12%), industry and mining (15%), power generation (15%), commercial (15%), residential (10%), and transport 9% (Househam, undated). Deindustrialization and trade pattern changes as well as technological transfer through Foreign Direct Investments have induced energy intensity in South Africa to decline (Adom, 2015). Africa's energy demand is expected to grow by 85% until 2040 due to rapid urbanization and the provision of access to electricity by a larger population (Pielli et al., 2016). MENA countries have the third largest growth of carbon emissions originating mainly from the oil producing countries such as Saudi Arabia. There is a lot of scope for energy efficiency (Worldbank, 2010).

Based on Table 5.5, world energy intensity has been gradually falling from 1995 to 2005 and this is also evidenced for most of the countries or regions in Table 5.5; however, MENA countries are an exception. Energy intensity has been increasing with an average annual increase of 0.6%. As far as Australia is concerned (which is not contained in Table 5.5), energy intensity has declined yearly by 1.4% from 1980 to 2013 characterized by a structural shift of the Australian economy from industry to a services one (Australian Government, 2015).

As far as the energy intensity in major energy industries is concerned, based on data from the World Energy Council (2013) and starting from the industry sector, which is a major economic pillar (it consumes about 33% of global energy), most emphasis has been laid on energy saving rather than energy intensity reduction. For example, Middle East countries, Latin America, Russian Commonwealth countries, and Asia (OECD members) have set no targets with respect to energy intensity reduction. The rest of the countries have set only a minor part of measures concerning energy intensity reduction (almost less than one triple the

Table 5.5: Total energy intensity for selected countries, 1995–2005.

	1995	2000	2004	2005	Average Annual Change, 1995–2005	Per Capita Gross Inland Energy Consumption in 2005 (TOE per Inhabitant)
World	100.0	94.0	94.6	94.1	−0.6%	1.8
EU-27	100.0	90.0	88.7	87.4	−1.3%	3.7
MENA	100.0	102.4	105.2	106.4	0.6%	2.0
USA	100.0	91.1	83.5	81.0	−2.1%	7.8
China	100.0	68.3	68.1	67.2	−3.9%	1.3
Russia	100.0	103.5	85.4	80.9	−2.1%	4.5

Source: Adapted from European Environment Agency, 2005. EN17 Total primary Energy Intensity. Available from: www.eea.europa.eu/.../ EN17-CS128.

quantity of energy saving targets). Furthermore, about half of the targets are financial and the rest are fiscal, relative, and of voluntary agreement type. The North American industry sector has no targets of the aforementioned kind.

As far as transport is concerned (it consumes 29% of total energy), all countries have set targets for cars except for the Russian Commonwealth countries, which have set measures on other types of transport and transport companies. Financial measures have been set in all countries except for the Middle East, where only regulative measures apply. Most fiscal measures apply for Europe and North American countries. Regions where energy intensity reduction has not made any progress in the years from 1990 to 2011 are the Middle East, Latin America, and Africa.

Regarding the residential sector (it is about 35% of total energy), it has the largest potential of energy intensity improvement through proper insulation, the usage of resilient energy materials, and energy economizing electric appliances and lamps. Europe and Asia (OECD) have set energy intensity targets for existing buildings, while only Latin America and Asia (non-OECD) have posed targets on electric appliances. The Middle East and Russian Commonwealth countries have placed energy intensity targets only on lighting. The largest part of financial measures in the domestic sector is provided in North American countries.

Last, with respect to agriculture, the lowest energy intensity (World Energy Council, 2016) is held by Niger and the highest by Denmark. Very poor countries use manual labor in agriculture (for tilling, harvesting, and processing) and no artificial fertilizers. Thus the energy input to agriculture is almost zero. The opposite holds for highly developed countries, where corporate agriculture is the norm and most work is automated.

5.3 Energy Efficiency and the EGN

Energy efficiency is highly involved in the EGN, but it is rarely discussed as such. We need to remember that all four major hypotheses in the EGN bear an "if part," which results in the reaction of the economy (through its GDP) to possible energy conservation. A concept very relevant to energy conservation is energy efficiency. The two terms are relevant, but they are not the same concept. Energy conservation refers to the reduction of energy consumption, while energy efficiency refers to producing more output with the same quantity of energy consumption. It is not clearly straightforward whether energy conservation under the known "Four Hypotheses" in the EGN literature includes energy efficiency therein, or they refer to mere energy reduction per se. The US Energy Information Administration (US Energy Information Administration, 2016) clarifies that energy conservation means using less (e.g., turn off the lights when leaving a room), while efficiency means using wisely (e.g., make use of technology that requires less energy to perform the same function). If energy conservation referred to in the four hypotheses of the EGN and does include energy efficiency, then things are more straightforward. If, however, energy efficiency is not included in the energy conservation meant in the "if" part of the four hypotheses of the EGN, we can imagine how energy efficiency will affect the four hypotheses as follows:

1. The Conservation Hypothesis with the existence of energy efficiency: According to this hypothesis definition, energy conservation does not hinder growth. With the presence of energy efficiency as well, the higher the energy efficiency, the further this outcome (energy conservation) is facilitated.
2. The Growth Hypothesis with the existence of energy efficiency: According to this hypothesis definition, energy conservation impedes growth. With the additional presence of energy efficiency, the higher the energy efficiency, the more it will impede growth.
3. The Feedback Hypothesis with the existence of energy efficiency: According to this hypothesis definition, energy conservation impedes growth, and the lower growth causes lower energy consumption too. As above, the higher energy efficiency impedes growth, and the consequent impeded or lower growth cannot produce more efficiency, which is an unavoidable consequence when investment on technology, which improves efficiency, does not take place.
4. The Neutrality Hypothesis with the existence of energy efficiency: According to this hypothesis definition, energy conservation does not affect growth, and these two magnitudes do not appear as being related to each other. Energy efficiency does not affect the relationship either.

Typically, but indirectly, EGN studies include energy intensity measures in their models and analyses, but they make no particular discussion on this matter. What is meant by this is that EGN studies do not use a variable verbatim called "energy intensity," but most of them use energy use per capita as an independent variable, which is actually one measure of energy intensity of an economy. That being said, we could state that the energy economics literature has studied extensively the relationship of energy use per capita and in this sense the energy intensity, with economic growth and the causality among energy intensity, economic growth, and a multitude of other variables that describe the productive functions of economies (e.g., employment, capital, emissions, financial development, management). In this sense, any measure, index, or variable that can make the usage of energy comparable across countries, or across people, can constitute an energy intensity measure.

To the best of our knowledge, the EGN has not perused a variable measuring the energy efficiency of a country or a sector. There is one piece of research that uses energy intensity as an independent variable, but this is rather proxied by renewable energy and thus does not contain any measure of energy efficiency (Riti and Shu, 2016). In addition to this, recently Csereklyei et al. (2016) recognizing the neglected importance of energy intensity, suggesting two new hypotheses in the EGN literature: the strong decoupling hypothesis and the weak decoupling hypothesis. While the former is concerned with the investigation and contribution of economic growth on energy consumption, the latter is concerned with testing the intertemporal coevolution of economic growth and energy intensity. A simultaneous examination of the two provides a deeper understanding of the role energy consumption plays in the formation of economic growth.

Besides the above, there is an inherent way to measure energy efficiency after the EGN model has been estimated. Apparently, in the EGN models that contain, among others, both CO_2 emissions and energy use as independent variables, the ratio of the coefficients of emissions over energy use might serve as a good indication of energy efficiency. However, the simultaneous use of emissions and energy consumption may introduce collinearity, and most studies avoid the simultaneous use of these two variables (Fig. 5.2).

Gradually, the EGN research has started incorporating energy use in two different types of energy: the renewable and the nonrenewable. Besides the variables that are estimated in the nexus models and may directly represent energy efficiency, there are other indirect ways that could be employed to make results across studies comparable. The ratio of the coefficient of renewable energy use over the coefficient of nonrenewable energy use could serve as a measure of energy efficiency, given that renewable energy can proxy energy efficiency. The higher the numerator and the lower the denominator, the higher this index will be. This ratio can be used for comparison across different studies and economies, both for the long and the short run.

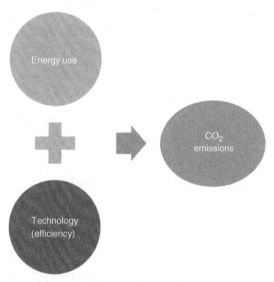

Figure 5.2: The Level of Emissions Is Determined by the Amount of Energy Used and the Technology Perused.

5.3.1 An example of energy efficiency measurement in the EGN

Next, we provide three applications for the measurement of energy efficiency in three estimated EGN models through the ratio of the coefficient of renewable energy over the coefficient of nonrenewable energy sources.

Example 1. The study by Dogan (2016) investigates the relationship between renewable and nonrenewable energy consumption with growth for Turkey with an ARDL approach. The aforementioned ratio is $-0.04/0.75 = -0.053$ for the long run and $0.03/0.54 = 0.056$ for the short run (Table 5.6). This ratio is lower in the long run. The minus sign indicates that the two types of energies affect growth differently, one by contributing to its increase and the other contributing to its decrease.

Example 2. The study by Kahia et al. (2016) investigates the relationship between renewable and nonrenewable energy consumption with growth for MENA countries with panel FMOLS estimation. The aforementioned ratio is $0.058/0.772 = 0.075$ (Table 5.7). This is higher in MENA countries than the ratio estimated for Turkey only by Dogan (Dogan, 2016). If we divide the ratio by 13, which denotes the number of countries involved, then this turns out to be equal to 0.006, which is the lowest ratio in all samples. As the paper provides an analogous model variant, if the ratio is calculated based only on the five countries that are the highest producers of renewables, the ratio equals to $0.110/0.241 = 0.456$, and if it is divided by 5, this reduces to 0.091.

Table 5.6: Results from the long-run and short-run estimates for renewable and nonrenewable EGN for Turkey.

Variables	Coefficients
Long-run Results	
Renewable energy	−0.04 (0.25)
Nonrenewable energy	0.75*** (0.00)
Capital	0.10*** (0.01)
Labor	−0.18* (0.07)
Constant	1.38*** (0.00)
D_{2003} structural break	0.01** (0.05)
Short-run Results	
Renewable energy	0.03 (0.48)
Nonrenewable energy	0.54*** (0.00)
Capital	0.14*** (0.00)
Labor	−0.17* (0.09)
ECT (−1)	−0.95*** (0.00)
F-statistic	83.54*** (0.00)
R^2	0.96

Three (***), two (**) and one (*) asterisks denote significance at 1%, 5%, and 10% respectively
The model ARDL (2.1.2.1.0) is selected based on the Akaike Information Criterion (AIC).
Source: Adapted from Dogan, E., 2016. Analyzing the linkage between renewable and non-renewable energy consumption and economic growth by considering structural break in time-series data. Renew. Energy 99, 1126–1136.

Table 5.7: Parameter estimation using Fully Modified OLS (FMOLS) for renewable energy, nonrenewable energy and growth in MENA countries.

Variables	All MENA Countries Coefficient	Selected MENA Countries Coefficient
Renewable energy	0.058 (0.000)***	0.110 (0.000)***
Nonrenewable energy	0.772 (0.000)***	0.241 (0.000)***
Capital	0.548 (0.009)**	0.877 (0.000)***
Labor	0.479 (0.000)***	0.896 (0.030)**
White's Heteroskedasticity test	2.56 (0.32)	17.93 (0.21)
Ramsey's RESET test	0.74 (0.52)	1.16 (0.32)
R^2	0.996	0.895
Adj. R^2	0.993	0.892
Durbin–Watson test for serial correlation	2.081	0.961

Three (***) and two (**) asterisks denote significance at 1% and 5% respectively.

Source: Kahia, M., Ben Aïssa, M.S. et al., 2016. Impact of renewable and non-renewable energy consumption on economic growth: new evidence from the MENA Net Oil Exporting Countries (NOECs). Energy 116, 102–115.

Example 3. The study by Apergis and Payne (2012) investigates the relationship between renewable and nonrenewable energy consumption with growth for 80 countries with panel FMOLS estimation. The aforementioned ratio is 0.371/0.384 = 0.966 (Table 5.8). This is the highest from the earlier two examples, because it contains heterogeneous countries, one with high penetration of renewable and other with lower. If the ratio is divided by 80, then this turns out to be 0.01.

If the ratio is high, it is in support for higher efficiency and vice versa. There is no sacred price of the ratio, but it is an indication and provides a comparison across countries and sectors, under the precondition that the compared studies peruse similar variables.

Table 5.8: FMOLS long-run parameter estimates.

Variables	Coefficients
Renewable energy	0.371 (9.72)*
Nonrenewable energy	0.384 (22.3)*
Capital	0.388 (41.5)*
Labor	0.493 (52.5)*
Constant	0.139 (5.19)*
White's Heteroskedasticity test	1.66 (0.29)
Ramsey's RESET test	1.28 (0.40)
Lagrange multiplier test for serial correlation	0.53 (0.62)
Adj. R^2	0.74

Figures in parentheses are t-values and all variables are significant at 1%.

Source: Adapted from Apergis, N., Payne, J.E., 2012. Renewable and non-renewable energy consumption-growth nexus: evidence from a panel error correction model. Energy Econ. 34(3), 733–738.

5.4 Suggested Shifts in the EGN: The Energy-Water Nexus and the Water- Energy-Food (WEF) Nexus

The WEF nexus is becoming increasingly popular in environmental and resource economics and management, because there is a growing concern due to water, energy, and food shortages, resource depletion and interlinkages between these three sources that cannot be neglected, if we aim to pursue sustainable development. This may be interpreted in other words as by Bazilian et al. (2011), who identified three spheres that add to the existence of the WEF nexus: the accessibility toward food, water, and energy, the environmental degradation that is caused through their exploitation-production-consumption, and their price volatility.

The urging need to investigate the WEF nexus stems from the somewhat threatening numbers: 60% more food will be required up to 2050 to feed the world (FAO, 2009), not to mention that one billion people are currently already suffering from hunger (OECD, 2016). Energy demand will increase by 50% by 2035 (IEA, 2010). Water demand for irrigation will increase by 10% by 2050 (FAO, 2011). Food production is the highest user of energy, and this supports the need for further investigation with respect to this sector.

Continuous population growth (an estimation is that global population will reach 9.2 billion people by 2050) (UN-ESCAP, 2013), urbanization, industrialization, consumerism, and climate change, all mean that the demand for food, water, and energy will increase, despite technological progress that enables more accurate irrigation application, the penetration of renewable energies, and the development of hybrid and smart cultivations that demand less water or the increased mechanization of crops reaping, planting, and so on. Thus since food, water, and energy are interconnected in multiple ways, policy making should be all-inclusive.

Regarding the interrelationships between WEF the production of food requires the usage of water (80% of freshwater goes to agriculture) and the consumption of energy for irrigation, cultivation (growing and planting), and transportation of the food production. This is also the reason why food prices are so much connected with fuel prices. On the other hand, water production requires energy consumption for water to be extracted, as well as for the implementation of sanitation facilities (electrification of wastewater plants). Moreover, energy production requires the use of water (e.g., for hydropower, the cultivation of biofuel plants that need water to grow, thermoelectric power generation that demands water for cooling, technologies such as carbon sequestration in the future which might also absorb large quantities of water). Energy consumption is used for food production, not only for the pumping and distribution of water (8% of energy is used for this purpose), but also for the distribution of agricultural inputs and products along the agrifood supply chain (30% of energy consumption is devoted to this direction).

There are additional synergies and trade-offs between food, energy, and water; for example, water consumption for irrigation increases water withdrawals and can thus jeopardize the water cycle. Establishing efficient irrigation systems might eventually cost more in energy terms (UN-Water, 2014). In addition, cultivating biofuel plants reduces the available land for food production. Thus the trade-offs are intriguing and need careful balancing.

Quoting from Howarth and Monasterolo (2016): "The nexus represents a multi-dimensional means of scientific enquiry, which seeks to describe the complex and non-linear interactions between water, energy, food with the climate and further understand wider implications for society." The nexus thinking can safeguard water, energy, and food security. This statement is different from what economists think and have done up-to-date for the EGN, wherein they study most of the time, the direction, strength, and quantity of causality as well as its time horizon.

The studies in the newly developed WEF nexus are recent and very heterogeneous in methods and logic. For example, Leung Pah Hang et al. (2016) use the WEF nexus with resource accounting, process engineering tools, and energy for designing local production systems. Al-Saidi and Elagib (2017) state that the WEF paradigm hosts integration in three forms: incorporation, cross-linking, and assimilation. The former refers to embedding of one component into the other, but each one keeps its initial integrity; cross-linking refers to mutual relationships; and assimilation refers to one resource transforming into the other. Cairns and Krzywoszynska (2016) state that the term nexus (with respect to the water) was coined in the World Economic Forum (WEF) in 2008, where it was recognized that water affects growth through a nexus of issues. The FEW nexus describes the complex and interrelated nature of our global resource systems (FAO, 2009).

It is unavoidable to think that as demand for resources grows, higher competition will be placed among resources and the need to observe and forecast the effect of conservation measures for them. Will conservation measures reduce food production, or will the latter be left unaffected? In cases where conservation will not have a negative effect on food growth, it means there is room for a rationalization of resource use without impeding food growth.

The nexus approach forecasts tradeoffs, synergies, and helps policy makers to prioritize their options (FAO, 2009). According to Yang et al. (2016), nexus thinking identifies situations where development paths are in conflict; for example, development in one sector contradicts development in another sector. Rasul (2016) underlines that the FEW nexus perspective adds to policy coherence and sectoral coordination, because there are examples of countries (usually developing ones) that subsidize water and energy to increase cereal production, but this, in turn, causes environmental degradation and leads to underinvestment in all resource-related infrastructure or does not allow markets to clear rationally and the comparative advantage principle of the international trade theory to take over.

FEW studies nowadays are more or less descriptions of biophysical processes and input-output analyses (Fang and Chang, 2016). A somewhat different approach is followed by

Ozturk (2015), who peruses the Kuznets curve to study sustainability in the FEW among BRICs countries. Other approaches are a life cycle assessment approach by Al-Ansari et al. (2015), or a social-ecological network analysis (Spiegelberg et al., 2017) combing qualitative and quantitative information. Sometimes analysis takes place at water basin level, because this is a more realistic segregation when water analysis is also involved. Thus there are studies that set up scenarios and possible outcome-trajectories depending on precipitation and physically based hydrological models (Yang et al., 2016).

Under the suggested WEF nexus, economic modeling could possibly work as in the long-established EGN. With the dependent variable being food production, independent variables could be set to be: water withdrawals, energy consumption, fertilizers and pesticides, employment in agriculture, agricultural machinery, and land. Cointegration and causality analysis could be based on the following known hypotheses, applicable in EGN, but also transferable in the WEF nexus. Various assumptions can be made about the production technology too.

1. **The WEF Growth hypothesis:** Energy and/or water usage decrease and food production decreases too. The size of decrease also depends on whether a year is a "good year" or "a bad year" with a generally low production, which has taken place not only because of systemic reasons, but also because of unforeseen weather conditions or other circumstances.
2. **The WEF Conservation hypothesis:** Energy and/or water usage decrease, but food production increases. It is evident that the technology applied had not been very efficient, and there is room for improvement and reduction of squandering in resources.
3. **The WEF Feedback hypothesis:** Energy and/or water usage decrease, and this reduction is transferred to food production, while the latter causes additional reduction in the demand for energy and/or water.
4. **The WEF Neutrality hypothesis:** Energy consumption and/or water withdrawals have no causal relationship to and from food production. This could be applicable for arid cultivations that do not demand the involvement of machinery or other energy-consuming entities. Ideal climatic conditions, manual labor, and very small agricultural plots are the reasons for not perusing energy.

Noteworthy is the fact that water and energy might not co-move at all times. These two magnitudes bear a degree of correlation, which is variable on technology. In the particular case of the energy-water (E-W) nexus, the focus of investigation is on these two variables, but various control variables could be added depending on the study framework. Thus the four known hypotheses (from the EGN) could be adapted for the E-W nexus as follows:

1. **The E-W Growth hypothesis:** An energy decrease will also decrease water withdrawal. This would be particularly felt for regions where water is not abundant locally and has to be transferred from long distances, or the ground morphology is such that it requires water to be pumped to high altitudes to enable flow.

2. **The E-W Conservation hypothesis:** An energy decrease will increase water withdrawal. In this case, water withdrawals are not due to energy by a large extent. In addition, efficient technologies have replaced less efficient ones.
3. **The E-W Feedback hypothesis:** An energy decrease will have a decreased effect on water withdrawal, and the latter will have a decreased effect on energy too.
4. **The E-W Neutrality hypothesis:** Energy and water have no causality relationship. This situation could well apply when renewable energy is abundant, namely when an energy generation technology is present that requires no water withdrawal and vice versa. Although it would be rare to have water pumped with no energy, this situation applies when there is no need for water pumping, namely when water is so abundant and located at such a convenient place from an altitude point of view that pumping costs are practically zero.

6 Literature Segregation Paradigms

One aspect that needs to be addressed in the EGN papers is the presentation of literature. It appears that authors, when they write literature review for their papers, do not focus solely on papers of the same type as theirs, but they provide a mixed presentation of literature which, due to its miscellaneous nature, cannot be directly comparable with the results that a new paper produces.

Recently, Menegaki and Tiwari (2017) have identified $2 \times 4 \times 4 = 32$ combinations of energy-growth papers and thus suggest the following meta-analytic energy-growth literature separation as shown in Fig. 5.3.

It appears that despite the plethora of EGN papers, little consensus with regard to the results has been reached despite some attempts to systemize literature (Ozturk, 2010) or even to meta-analyze it (Kalimeris et al., 2014; Menegaki, 2014).

Based on Fig. 5.3, we observe that separation of the EGN literature can take place at three levels. The first level is the separation between aggregated energy consumption and disaggregated energy consumption. Studies that investigate the energy consumption of the economy as a whole are not directly comparable with studies that investigate energy consumption in particular sectors of the economy or those that investigate a particular type of energy or fuel consumption.

A second level of literature separation is by the hypothesis evidence (growth, conservation, feedback, and neutrality) it brings forth. The third level of literature separation is based on whether studies are single-country studies or multiple country studies. "Groups I" studies refer to studies that encompass a certain geographical region, while "Groups II" studies refer to studies with larger country sets that are not dictated by geography or other grouping. In a similar vein, "Single I" studies refer to absolutely single countries, while "Single II" studies

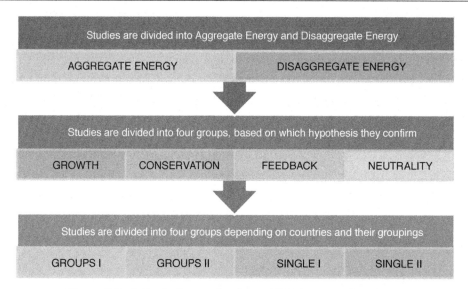

Figure 5.3: A Fresh Categorization on EGN Nexus Literature.
From Menegaki, A.N. and Tiwari, A.K., 2017. The index of sustainable economic welfare in the energy-growth nexus for American countries. Ecol. Indic. 72, 494–509.

refer to studies that encompass the single country in a wider group of countries that makes geographical sense or economic sense, for example the G7 countries.

The three levels are not necessarily in this turn. Additional separations of literature could be added not only based on the perused econometric method analysis or the control variables, but also whether the analysis is production function or demand function oriented. Examples of literature presentation as suggested by Fig. 5.3 are provided in Tables 5.9, 5.10, and 5.11. Table 5.12 presents EGN studies on Asian countries.

Furthermore, current literature reviews on EGN studies present some literature highlights but the choice of these highlights—papers to be reviewed—is rather subjective and mixed. Thus we recommend that the search yardstick of the literature review be reported at the beginning of the literature review section. More specifically, the paper should report the search database, the search keys, and the number of papers finally included as well as the reason for the noninclusion of some others.

7 Conclusions

This chapter deals with critical issues in the EGN that need to be addressed so that the nexus keeps up with developments and current demands for sustainable development. The critical issues are related not only to the content and orientation of the studies, but also to their comparability and presentation issues. Thus first, the chapter reveals that EGN studies are not particularly verbose to whether international energy strategies and targets can be

Table 5.9: Studies on the EGN for Sub-Saharan countries.

Studies [1]	Countries [2]	Aggregate/ Disaggregate [3]	Hypotheses [4]	Types [5]
Saidi and Hammami (2015)	13 SSA countries	Agg.	N	1
Alam and Paramati (2015)	18 developing countries (South Africa)	Dis.(oil)	F	1
Khan et al. (2014)	World (South Africa)	Agg.	C	1
Tiwari et al. (2015)	12 SSA countries	Agg.	C, G	1
Karanfil and Li (2015)	160 countries worldwide	Dis. (electricity)	N	1
Bildirici and Bakirtas (2014)	BRICTS countries (South Africa)	Dis. (Oil)	F	1
Akkemik and Göksal (2012)	79 countries worldwide (Cote d' Ivoire and Senegal)	Agg.	G	1
Eggoh et al. (2011)	21 African countries (15 SSA)	Agg.	F	1
Ozturk et al. (2010)	51 low- and middle-income countries	Agg.	C, F	1
Ouedraogo (2013)	15 West African countries (Economic Community of West African States-ECOWAS)	Agg.	C, G	1
Lee (2005)	18 developing countries (Kenya and Ghana)	Agg.	G	1
Mahadevan and Asafu-Adjaye (2007)	20 energy importer-exporter countries (Ghana, Senegal, and South Africa)	Agg.	C, G	1
Iyke (2015)	Nigeria	Dis. (electricity)	G	2
Ozturk and Bilgili (2015)	51 SSA countries	Dis. (biomass)	G	2
Ali et al. (2015)	Nigeria		C	2
Kumar et al. (2015)	South Africa	Agg.	G	2
Michieka (2015)	5 SSA countries	Dis. (electricity)	C, N	2
Albiman et al. (2015)	Tanzania	Dis. (electricity)	C	2
Dlamini et al. (2015)	South Africa	Dis. (electricity)	N	2
Dogan (2014)	4 low-income SSA countries	Agg.	C, N	2
Fatai (2014)	SSA countries (East and Southern region)	Agg.	G, N	2
Gao and Zhang (2014)	14 SSA countries	Dis. (electricity)	G, F	2
Richard (2012)	12 SSA countries	Agg.	C, G	2
Kahsai et al. (2012)	SSA (middle-income countries)	Agg.	F, N	2
Kouakou (2011)	Cote d'Ivoire	Dis. (electricity)	G	2
Esso (2010)	7 SSA countries	Agg.	C, F	2
Odhiambo (2010)	3 SSA countries	Agg.	C, G	2
Odhiambo (2009)	South Africa	Dis. (electricity)	F	2
Ziramba (2009)	South Africa	Dis. (oil)	F	2
Akinlo (2008)	11 SSA countries	Agg.	G, F, N	2
Behmiri and Manso (2013)	23 SSA countries	Dis. (oil)	G, F	2
Esso (2010)	7 SSA countries	Agg.	C, F	2
Ebohon (1996)	2 SSA countries	Agg.	F	2

In Column [2], the parenthesized countries are SSA countries contained in the study sample, in Column [3] the term "Agg." stands for aggregate energy consumption, "Dis." Stands for disaggregate energy consumption, for example renewable energy, oil, natural gas. Column [4] reports the verified hypotheses, namely C: Conservation, G: Growth, F: Feedback, N: Neutrality. In Column [5], number 1 stands for studies where sub-Saharan countries are within groups of countries and sub-panels while number 2 stands for single countries or sub-Saharan country panels.
Source: Menegaki and Tugcu (2016).

Table 5.10: Literature review of EGN in emerging countries.

	Total Energy Consumption and Economic Growth (Group 1)				Partial Energy Consumption and Economic Growth (Group 2)			
	Conservation (G→E)	Growth (E→G)	Feedback (E↔G)	Neutrality (E~G)	Conservation (G→E)	Growth (E→G)	Feedback (E↔G)	Neutrality (E~G)
Brazil	Pao and Fu (2013a)	Menegaki and Tugcu (2016)	Menegaki and Tugcu (2016)		Pao and Fu (2013b), nonrenewable/ renewable	Yoo and Kwak (2010), electricity Pao and Fu (2013b), nonhydro renewable energy Bildirici et al. (2012), electricity	Omri et al. (2015), renewable Pao and Fu (2013b), renewable	Cowan et al. (2014), electricity
Chile		Joo et al. (2015)	Menegaki and Tugcu (2016)					
China		Menegaki and Tugcu (2016)	Menegaki and Tugcu (2016)			Yoo and Kwak (2010), electricity Bildirici et al. (2012), electricity		Cowan et al. (2014), electricity
Colombia		Menegaki and Tugcu (2016)	Menegaki and Tugcu (2016)			Yoo and Kwak (2010), electricity		
Hungary		Menegaki and Tugcu (2016)	Ozturk and Acaravci (2010) Caraiani et al. (2015)	Menegaki and Tugcu (2016)	Caraiani et al. (2015), gas/ renewable	Omri et al. (2015), renewable		Omri et al. (2015), nuclear
India		Chang et al. (2013)	Menegaki and Tugcu (2016)		Bildirici et al. (2012), electricity	Omri et al. (2015), renewable		Cowan et al. (2014), electricity
Indonesia	Hwang and Yoo (2014) Soares et al. (2014) Yildirim et al. (2014) Shahbaz et al. (2013) Aslan and Kum (2010)	Pao et al. (2014) Asafu-Adjaye (2000) Chiou-Wei et al. (2008) Menegaki and Tugcu (2016)	Menegaki and Tugcu (2016)	Soares et al. (2014) Razzaqi et al. (2011) Asafu-Adjaye (2000)	Yoo (2006), electricity	Pao et al. (2014), renewable Chandran and Tang (2013), transport energy	Pao et al. (2014), nuclear	

Country								
Malaysia	Yildirim et al. (2014) Saboori and Sulaiman (2013) Islam et al. (2013) Ang (2008)	Aslan and Kum (2010) Chiou-Wei et al. (2008)	Tang (2008) Tang and Tan (2014) Menegaki and Tugcu (2016)	Saboori and Sulaiman (2013)		Chandran and Tang (2013), transport energy Chandran et al. (2010), electricity	Yoo (2006), electricity Tang and Tan (2013), electricity Tang (2008), electricity	
Mexico	Bozoklu and Yilanci (2013) Galindo, (Galindo 2005)	Pao et al. (2014)	Pao et al. (2014)	Menegaki and Tugcu (2016)		Pao et al. (2014), renewable	Pao et al. (2014), nuclear	
Morocco		Raheem and Yusuf (2015)		Issa Shahateet (2014) Menegaki and Tugcu (2016)				
Philippines	Yildirim et al. (2014) Aslan and Kum (2010) Chiou-Wei et al. (2008) Menegaki and Tugcu (2016)	Chang et al. (2013)	Asafu-Adjaye (2000)					
Poland	Lach (2015) Menegaki and Tugcu (2016)	Lach (2015) Menegaki and Tugcu (2016)		Wolde-Rufael (2014), Menegaki and Tugcu (2016)	Caraiani et al. (2015), gas	Bildirici and Özaksoy (2013), biomass Caraiani et al. (2015), primary energy		Gurgul and Lach (2011), coal

Table 5.10: Literature review of EGN in emerging countries. (*cont.*)

	Total Energy Consumption and Economic Growth (Group 1)				Partial Energy Consumption and Economic Growth (Group 2)			
	Conservation (G→E)	Growth (E→G)	Feedback (E↔G)	Neutrality (E~G)	Conservation (G→E)	Growth (E→G)	Feedback (E↔G)	Neutrality (E~G)
South Africa	Esso (2010a) Esso (2010b)	Odhiambo (2012) Raheem and Yusuf (2015) Menegaki and Tugcu (2016)	Menegaki and Tugcu (2016)	Liu (2013) Menegaki and Tugcu (2016)	Cowan et al. (2014), electricity Bildirici et al. (2012), electricity Yoo (2006), electricity		Bildirici et al. (2012), electricity	
Thailand		Aslan and Kum (2010) Menegaki and Tugcu (2016)	Asafu-Adjaye (2000) Yildirim et al. (2014) Chang et al. (2013)	Chiou-Wei et al. (2008)				
Turkey	Ozun and Cifter, Ozturk et al. (2013) Menegaki and Tugcu (2016)	Pao et al. (2014)	Ozun and Cifter (2007) Fuinhas and Marques (2012)	Nazlioglu et al. (2014)	Bildirici et al. (2012), electricity Ocal and Aslan (2013), Ocal and Aslan (2013), renewable Caraiani et al. (2015), electricity		Aslan (2014), electricity Aytaç and Güran (2011), electricity	Ocal et al. (2013), coal
Total number of studies	20	15	11	9	11	15	10	6

Source: Adapted by Menegaki, A.N., Tugcu, C.T., 2016. Rethinking the energy-growth nexus: proposing an index of sustainable economic welfare for Sub-Saharan Africa. Energy Res. Soc. Sci. 17, 147–159.

Table 5.11: Studies on the EGN for American countries.

Studies [1]	Types [2]	Countries [3]	Agg./Disagg. [4]	Hypothesis [5]
Akarca and Long (1980)	Groups I	7 South American countries	Disagg. (electricity)	G, F, N
Cheng (1995)	Groups I	6 Central American countries	Disagg. (RES)	G, F
Payne and Taylor (2010)	Groups I	6 Central American countries	Disagg. (nuclear)	F
Payne (2009)	Groups I	5 Latin American countries	Disagg. (Hydro)	G
Apergis and Payne (2011)	Groups I	6 Central American countries	Disagg. (RES)	F
Yoo and Kwak (2010)	Groups I	7 South American countries	Disagg.(electricity)	G, F, N
Cheng (1997)	Groups I	3 Latin countries (Brazil, Mexico, and Venezuela)	Agg.	G, N
Bowden and Payne (2010)	Groups II	20 American countries (in a group of 79 worldwide)	Disagg. (electricity)	G, F
Karanfil and Li (2015)	Groups II	North American countries (in a panel of 160 countries)	Disagg. (electricity)	N
Lee (2005)	Groups II	6 Latin American countries (in a group of 18 developing countries)	Agg.	G
Mehrara (2007)	Groups II	3 Central and South American countries (in a group of 11 of oil exporting countries)	Agg.	C
Lee and Chang (2007a)	Groups II	6 Latin countries (in a group of 22 developed countries)	Agg.	C, G
Mahadevan and Asafu-Adjaye (2007)	Groups II	3 American countries—Argentina, USA, and Venezuela (in a group of 20 energy importing or exporting countries)	Agg.	G, F
Huang et al. (2008)	Groups II	18 American countries (in a group of 82 countries)	Agg.	C, N
Saunoris and Sheridan (2013)	Single I	USA	Disagg. (electricity)	C, G
Akarca and Long (1980)	Single I	USA	Agg.	N
Yu and Hwang (1984)	Single I	USA	Agg.	N
Yu and Jin (1992)	Single I	USA	Agg.	N
Cheng (1995)	Single I	USA	Agg.	N
Payne and Taylor (2010)	Single I	USA	Disagg. (nuclear)	N
Menyah and Wolde-Rufael (2010)	Single I	USA	Disagg. (nuclear)	N
Sari et al. (2008)	Single I	USA	Disagg.(coal, fossil fuels, hydro, solar, wind, natural gas, wood, and waste)	C
Payne (2009)	Single I	USA	Disagg. (RES)	N
Menyah and Wolde-Rufael (2010)	Single I	USA	Disagg. (RES)	C
Bowden and Payne (2010)	Single I	USA	Disagg. (RES)	N
Payne (2011)	Single I	USA	Disagg. (biomass)	G
Yildirim et al. (2012)	Single I	USA	Disagg. (RES)	N

(Continued)

Table 5.11: Studies on the EGN for American countries. (*cont.*)

Studies [1]	Types [2]	Countries [3]	Agg./Disagg. [4]	Hypothesis [5]
Pao and Fu (2013a)	Single I	Brazil	Disagg. (RES)	F
Thoma (2004)	Single I	USA	Disagg. (electricity)	F
Kraft and Kraft (1978)	Single I	USA	Agg.	C
Erol and Yu (1987)	Single I	USA	Agg.	N
Menyah and Wolde-Rufael (2010)	Single II	Canada and Chile (in a group of 26 OECD countries)	Disagg. (electricity)	G
Yildirim et al. (2012)	Single II	USA and Canada (in a group of G7 countries)	Disagg. (biomass)	G
Bhattacharya et al. (2016)	Single II	USA (in a group of 38 countries)	Disagg. (RES)	C
Chang et al. (2015)	Single II	USA and Canada (in a group of a G7 countries)	Disagg. (RES)	N
Yu and Hwang (1984)	Single II	Mexico (in a group of 11 countries)	Agg.	N
Yoo and Ku (2009)	Single II	Argentina (in a group of 6 countries)	Disagg. (nuclear)	N
Wolde-Rufael and Menyah (2010)	Single II	Canada and USA (in a group of 9 developed countries)	Disagg. (nuclear)	C, F
Chu and Chang (2012)	Single II	USA and Canada (in a group of G6 countries)	Disagg. (nuclear)	G, N
Tugcu et al. (2012)	Single II	Canada and USA (in G7 countries)	Disagg. (RES)	N
Lee and Chiu (2011)	Single II	USA (in a group of 6 industrialized countries)	Disagg. (nuclear)	N
Omri et al. (2015)	Single II	Argentina and USA (in a group of 17 developed and developing countries)	Disagg. (nuclear)	G, F, N
Cowan et al. (2014)	Single II	Brazil (in a group of BRICS countries)	Disagg. (electricity)	N
Yu and Choi (1985)	Single II	USA (in a group of 5 countries)	Agg.	N
Narayan and Smyth (2008)	Single II	USA and Canada (in a group of G7 countries)	Agg.	G
Ozturk et al. (2010)	Single II	17 American countries (in a group of 51 countries)	Agg.	C, F
Soytas and Sari (2003)	Single II	3 American countries— Argentina, Canada, and USA (in a group of G7 countries)	Agg.	G

Column [2]: Group 1 encompasses studies that focus on American countries groups solely, Group 2 encompasses studies where American groups of countries are one group among many other countries or groups of countries, Column [4]: Agg = aggregate energy, Disagg.= disaggregate energy, Column [5]: *G* for Growth, *F* for Feedback, *C* for Conservation, and *N* for Neutrality.

Source: Menegaki, A.N., Tiwari, A.K., 2017. The index of sustainable economic welfare in the energy-growth nexus for American countries. Ecol. Indic. 72, 494–509.

Table 5.12: EGN studies on Asian countries.

Ref [1]	Studies [2]	Country type [3]	Countries [4]	Years [5]	Methods [6]	Variables [7]	Agg./Dis. [8]	G [9]	C [10]	N [11]	F [12]
1	Raza et al. (2016)	2	4 South Asian countries (Pakistan, Sri Lanka, India, and Bangladesh)	1980–2010	Random effects modeling	E, L, C	2: Electricity	Yes			
2	Fang and Chang (2016)	2	16 Asian Pacific countries	1970–2011	Continuously updated fully modified estimation	C, L, HC	1		Yes		Yes
3	Furuoka (2016)	2	China and Japan	1980–2012	ARDL	E, C, Ex	2: Natural gas	Yes			Yes
4	Shakeel and Iqbal (2015)	2	4 South Asian (Pakistan, Sri Lanka, India, and Bangladesh)	1980–2009	VECM causality	E, Ex, C, L	1	Yes			Yes
5	Furuoka (2015)	2	12 Asian countries (east and south)	1971–2011	PANIC ANALYSIS	E	2: Electricity	Yes	Yes		
6	Vidyarthi (2015)	2	5 South Asian countries (India, Pakistan, Bangladesh, Sri Lanka, and Nepal)	1971–2010	Pedroni cointegration and Granger Causality	E, C, L	1	Yes			Yes
7	Mahdi Ziaei (2015)	3	East Asia	1989–2011	Panel Vector Autoregression model	E, CO$_2$, C, S	1	Yes			Yes
8	Yildirim et al. (2014)	2	Indonesia, Malaysia, Philippines, Singapore, and Thailand	1971–2009	Panel data analysis	I, L	1		Yes	Yes	Yes
9	Zeshan and Ahmed (2013)	2	Bangladesh, India, Pakistan, Sri Lanka, and Nepal	1980–2010	Kao's panel cointegration	E, CO$_2$	1			Yes	Yes
10	Abbas and Choudhury (2013)	2	India and Pakistan	1972–2008	Johansen and VECM	E	2: Electricity	Yes	Yes		Yes
11	Hossain and Saeki (2011)	2	India, Nepal, and Pakistan	1971–2007	Johansen	E	2: Electricity	Yes	Yes		Yes
12	Chandran et al. (2010)	1	Malaysia	1971–2003	ARDL	E	2: Electricity	Yes			

(Continued)

Table 5.12: EGN studies on Asian countries. (cont.)

Ref [1]	Studies [2]	Country type [3]	Countries [4]	Years [5]	Methods [6]	Variables [7]	Agg./Dis. [8]	G [9]	C [10]	N [11]	F [12]
13	Wolde-Rufael (2010)	1	India	1969–2006	Toda-Yamamoto and ARDL	E	2: Nuclear energy	Yes			
14	Ghosh (2009)	1	India	1970–2006	Toda-Yamamoto and ARDL	E, L	2: Electricity				Yes
15	Yuan et al. (2008)	1	China	1953–2005	Johansen	E	1, 2: Coal, Oil, Electricity	Yes			
16	Narayan and Prasad (2008)	3	30 OECD countries: Korea	1965–2003	Bootstrapped causality approach	E	2: Electricity			Yes	
17	Lee and Chang (2007b)	1	Taiwan	1955–2003	Threshold regression	E	1	Yes			
18	Shahbaz et al. (2012)	1	Pakistan	1972–2011	ARDL	E, L, C	2: RES, non-RES				Yes
19	Chen et al. (2007)	2	10 Asian countries (China, Hong-Kong, Indonesia, India, Korea, Malaysia, Philippines, Singapore, Taiwan, and Thailand)	1971–2001	Pedroni's cointegration and Granger causality	E	2: Electricity	Yes			Yes
20	Ghosh (2002)	1	India	1950–2001	Johansen cointegration and Granger causality	E	2: Electricity	Yes			
21	Morimoto and Hope (2004)	1	Sri Lanka	1960–1998	Yang's method	E	2: Electricity				Yes
22	Mozumder and Marathe (2007)	1	Bangladesh	1971–1999	Johansen and VECM	E	2: Electricity		Yes		
23	Oh and Lee (2004)	1	Korea	1981–2000 (quarterly)	Johansen and VECM	E, C, L, P[a]	1		Yes		

#	Study	[3]	Countries	Period	Method	Variables	[8]			
24	Ajmi et al. (2013)	3	Japan	1960–2010	Nonlinear tests	E	1			Yes
25	Yoo (2005)	1	Korea	1970–2002	Granger causality	E	2: Electricity			Yes
26	Yoo (2006)	2	4 ASEAN countries (Indonesia, Malaysia, Singapore, and Thailand)	1971–2002	Granger causality	E	2: Electricity		Yes	Yes
27	Huang et al. (2008)	3	82 countries worldwide: 12 Asian (low, middle and high income) countries	1972–2002	GMM estimation	E, CO_2, Ind.	1	Yes	Yes	Yes
28	Al-Mulali (2014)	3	30 worldwide countries: Korea, Japan, and Pakistan	1990–2000	Granger causality	CO_2[a], L, I, U	2: Electricity from coal, natural gas, and nuclear energy	Yes		
29	Glasure and Lee (1998)	2	South Korea and Singapore	1961–1990	Granger causality	E	1	Yes		
30	Fatai et al. (2004)	3	India, Indonesia, Philippines, and Thailand	1960–1999	Toda–Yamamoto approach	E	2: Coal, natural gas, oil, and electricity	Yes		Yes
31	Squalli (2007)	3	OPEC countries: Indonesia	1980–2003	ARDL, Toda–Yamamoto approach	E	2: Electricity	Yes		
32	Chiou-Wei et al. (2008)	3	Newly industrialized countries (8 Asian)	1954–2006	Granger causality	E	1	Yes	Yes	Yes
33	Ang (2008)	1	Malaysia	1971–1999	Johansen and VECM	E, CO_2	1	Yes	Yes	
34	Bildirici and Bakirtas (2014)	3	BRICS (India and China)	1980–2011	ARDL	E	2: Coal, natural gas, and oil			Yes

C, Domestic credit to private sector over GDP; I, Investment; Ind, value added in industry; S, stock traded turnover ratio; U, urbanization.
Column [3] is interpreted as follows: 1: Single country, 2: Multiple Asian countries, 3: Multiple countries also containing Asian countries, Column [8]: 1 stands for aggregate, 2 stands for disaggregate.
[a]A second model has been estimated with dependent variable being CO_2. They also employ a demand side model.

fulfilled, although the studies are motivated by them and they derive their usefulness by them. Even studies that claim in their title to be answering policy problems and assessments of the possibility to fulfill international agreements and stipulations do not in the end reveal much about that question. In addition, the EGN studies do not make distant future projections and forecasting.

Second, the EGN studies that involve voluminous numbers of countries serve mostly for the discovery of global relationships that aim to build and evolve theory, but cannot provide single countries with insightful policy making results. To this end, we have taken, as example, a principal study based on a large number of countries, and then we have tried collecting studies for each of the individual countries constituting the sample of the principal study. Results of individual countries reveal a large heterogeneity with respect to the hypothesis they support.

Third, we have suggested and discussed several new directions to which the EGN could be shifted and enriched accordingly. Those directions are sustainable development, energy efficiency, and the set-up of new nexuses related to energy such as the EGN and WEF nexuses. The chapter extends the four known hypotheses in the EGN and the two new nexuses. In addition, examples are provided to compare energy efficiency across studies and examples of incorporation of the notion of sustainability in the EGN.

Last, the chapter recommends a format of literature presentation that will contribute toward putting some order in the papers and will help highlight the differences of new papers from the already existing ones. The literature format will help identify the exact type of EGN papers with each new paper to be compared.

References

Abbas, F., Choudhury, N., 2013. Electricity consumption-economic growth Nexus: an aggregated and disaggregated causality analysis in India and Pakistan. J. Policy Model. 35 (4), 538–553.

Adom, P.K., 2015. Determinants of energy intensity in South Africa: testing for structural effects in parameters. Energy 89, 334–346.

Ajmi, A.N., El Montasser, G., Nguyen, D.K., 2013. Testing the relationships between energy consumption and income in G7 countries with nonlinear causality tests. Econ. Model. 35, 126–133.

Akarca, A.T., Long, T.V., 1980. On the relationship between energy and GNP: a reexamination. J. Energy Dev. 5, 326–331.

Akinlo, A.E., 2008. Energy consumption and economic growth: evidence from 11 Sub-Sahara African countries. Energy Econ. 30 (5), 2391–2400.

Akkemik, K.A., Göksal, K., 2012. Energy consumption-GDP nexus: heterogeneous panel causality analysis. Energy Econ. 34 (4), 865–873.

Al-Ansari, T., Korre, A., Nie, Z., Shah, N., 2015. Development of a life cycle assessment tool for the assessment of food production systems within the energy, water and food nexus. Sustainable Prod. Consumption 2, 52–66.

Al-Mulali, U., 2014. Investigating the impact of nuclear energy consumption on GDP growth and CO_2 emission: a panel data analysis. Prog. Nucl. Energy 73, 172–178.

Al-Saidi, M., Elagib, N.A., 2017. Towards understanding the integrative approach of the water, energy and food nexus. Sci. Total Environ. 574, 1131–1139.

Alam, M.S., Paramati, S.R., 2015. Do oil consumption and economic growth intensify environmental degradation? Evidence from developing economies. Appl. Econ. 47 (48), 5186–5203.

Albiman, M.M., Suleiman, N.N., et al., 2015. The relationship between energy consumption CO_2 emissions and economic growth in Tanzania. Int. J. Energy Sect. Manage. 9 (3), 361–375.

Ali, H.S., Hook, L.S., et al., 2015. Financial development and energy consumption nexus in Nigeria: an application of autoregressive distributed lag bound testing approach. Int. J. Energy Econ. Policy 5 (3), 816–821.

Ang, J.B., 2008. Economic development, pollutant emissions and energy consumption in Malaysia. J. Policy Model. 30 (2), 271–278.

Apergis, N., Payne, J., 2011. The renewable energy consumption-growth nexus in Central America. Appl. Energy 88, 343–347.

Apergis, N., Payne, J.E., 2012. Renewable and non-renewable energy consumption-growth nexus: evidence from a panel error correction model. Energy Econ. 34 (3), 733–738.

Asafu-Adjaye, J., 2000. The relationship between energy consumption, energy prices and economic growth: time series evidence from Asian developing countries. Energy Econ. 22 (6), 615–625.

Aslan, A., 2014. Causality between electricity consumption and economic growth in Turkey: an ARDL bounds testing approach. Energy Sources, Part B 9 (1), 25–31.

Aslan, A., Kum, H., 2010. An investigation of cointegration between energy consumption and economic growth: dynamic evidence from East Asian countries. Global Econ. Rev. 39 (4), 431–439.

Australian Government (2015). End-use energy intensity in Australia. Available from: www.industry.gov.au.

Aytaç, D., Güran, M.C., 2011. The relationship between electricity consumption, electricity price and economic growth in Turkey: 1984–2007. Argum. Oecon. 27 (2), 101–123.

Azad, A.K., Rasul, M.G., et al., 2015. Study on Australian energy policy, socio-economic, and environment issues. J. Renew. Sustainable Energy 7 (6).

Bagstad, K.J., Berik, G., et al., 2014. Methodological developments in US state-level genuine progress indicators: toward GPI 2.0. Ecol. Indic. 45, 474–485.

Bazilian, M., Rogner, H., et al., 2011. Considering the energy, water and food nexus: towards an integrated modelling approach. Energy Policy 39 (12), 7896–7906.

Behmiri, N.B., Manso, J.R.P., 2013. How crude oil consumption impacts on economic growth of Sub-Saharan Africa? Energy 54, 74–83.

Bhattacharya, M., Paramati, S.R., et al., 2016. The effect of renewable energy consumption on economic growth: evidence from top 38 countries. Appl. Energy 162, 733–741.

Bildirici, M.E., Bakirtas, T., 2014. The relationship among oil, natural gas and coal consumption and economic growth in BRICTS (Brazil, Russian, India, China, Turkey and South Africa) countries. Energy 65, 134–144.

Bildirici, M.E., Bakirtas, T., et al., 2012. Economic growth and electricity consumption: auto regressive distributed lag analysis. J. Energy South. Afr. 23 (4), 29–45.

Bildirici, M.E., Özaksoy, F., 2013. The relationship between economic growth and biomass energy consumption in some European countries. J. Renew. Sustainable Energy 5 (2).

Bowden, N., Payne, J., 2010. Sectoral analysis of the causal relationship between renewable and non-renewable energy consumption and real output in the US. Energy Source Part B 5, 400–408.

Bozoklu, S., Yilanci, V., 2013. Energy consumption and economic growth for selected OECD countries: further evidence from the Granger causality test in the frequency domain. Energy Policy 63, 877–881.

Cairns, R., Krzywoszynska, A., 2016. Anatomy of a buzzword: the emergence of "the water-energy-food nexus" in UK natural resource debates. Environ. Sci. Pol. 64, 164–170.

Caraiani, C., Lungu, C.I., et al., 2015. Energy consumption and GDP causality: a three-step analysis for emerging European countries. Renew. Sustainable Energy Rev. 44, 198–210.

Castañeda, B.E., 1999. An index of sustainable economic welfare (ISEW) for Chile. Ecol. Econ. 28 (2), 231–244.

Chandran, V.G.R., Sharma, S., et al., 2010. Electricity consumption-growth nexus: the case of Malaysia. Energy Policy 38 (1), 606–612.

Chandran, V.G.R., Tang, C.F., 2013. The impacts of transport energy consumption, foreign direct investment and income on CO_2 emissions in ASEAN-5 economies. Renew. Sustainable Energy Rev. 24, 445–453.

Chang, T., Chu, H.P., et al., 2013. Energy consumption and economic growth in 12 Asian countries: panel data analysis. Appl. Econ. Lett. 20 (3), 282–287.

Chang, T., Gupta, R., et al., 2015. Renewable energy and growth: evidence from heterogeneous panel of G7 countries using Granger causality. Renew. Sustainable Energy Rev. 52, 1405–1412.

Chen, S.T., Kuo, H.I., et al., 2007. The relationship between GDP and electricity consumption in 10 Asian countries. Energy Policy 35 (4), 2611–2621.

Cheng, B.S., 1995. An investigation of cointegration and causality between energy consumption and economic growth. J. Energy Dev. 21, 73–84.

Cheng, B.S., 1997. Energy consumption and economic growth in Brazil, Mexico and Venezuela: a time series analysis. Appl. Econ. Lett. 4 (11), 671–674.

Cheng, G., Xu, Z., et al., 2005. Vision of integrated happiness accounting system in China. Acta Geogr. Sin. 60 (6), 883–893.

Chiou-Wei, S.Z., Chen, C.F., et al., 2008. Economic growth and energy consumption revisited: evidence from linear and nonlinear Granger causality. Energy Econ. 30 (6), 3063–3076.

Chu, H.P., Chang, T., 2012. Nuclear energy consumption, oil consumption and economic growth in G-6 countries: bootstrap panel causality test. Energy Policy 48, 762–769.

Clarke, M., 2004. Widening development prescriptions: policy implications of an Index of Sustainable Economic Welfare (ISEW) for Thailand. Int. J. Environ. Sustainable Dev. 3 (3–4), 262–275.

Cobb, C.W., Cobb, J.J., 1994. The Green National Product: A Proposed Index of Sustainable Economic Welfare. University Press of America, Lanham, MD.

Cowan, W.N., Chang, T., et al., 2014. The nexus of electricity consumption, economic growth and CO_2 emissions in the BRICS countries. Energy Policy 66, 359–368.

Csereklyei, Z., Rubio-Varas, M., et al., 2016. Energy and economic growth: the stylized facts. Energy J. 37 (2), 223–255.

Daly, H., Cobb, J.J., 1989. For the Common Good. Beacon Press, Boston, MA.

Dlamini, J., Balcilar, M., et al., 2015. Revisiting the causality between electricity consumption and economic growth in South Africa: a bootstrap rolling-window approach. Int. J. Econ. Policy Emerging Econ. 8 (2), 169–190.

Dogan, E., 2014. Energy consumption and economic growth: evidence from low-income countries in Sub-Saharan Africa. Int. J. Energy Econ. Policy 4 (2), 154–162.

Dogan, E., 2016. Analyzing the linkage between renewable and non-renewable energy consumption and economic growth sby considering structural break in time-series data. Renew. Energy 99, 1126–1136.

Doris, E., Cochran, J., et al., 2009. Energy efficiency policy in the United States: overview of trends at different levels of government. Available from: www.nrel.gov/docs.

Ebohon, O.J., 1996. Energy, economic growth and causality in developing countries: a case study of Tanzania and Nigeria. Energy Policy 24 (5), 447–453.

Eggoh, J.C., Bangake, C., et al., 2011. Energy consumption and economic growth revisited in African countries. Energy Policy 39 (11), 7408–7421.

Erol, U., Yu, E.S.H., 1987. On the causal relationship between energy and income for industrialised countries. J. Energy Dev. 9, 75–89.

Esso, J.L., 2010. The energy consumption-growth nexus in seven sub-saharan African countries. Econ. Bull. 30 (2), 1191–1209.

Esso, J.L., 2010a. The energy consumption-growth nexus in seven sub-Saharan African countries. Econ. Bull. 30 (2), 1191–1209.

Esso, L.J., 2010b. Threshold cointegration and causality relationship between energy use and growth in seven African countries. Energy Econ. 32 (6), 1383–1391.

European Commission, 2012a. Energy efficiency directive. Available from: ec.europa.eu/energy/en/topics/energy-efficiency/energy-efficiency-directive.

European Commission, 2012b. Energy roadmap 2015. Available from: ec.europa.eu/energy/sites/ener/files.

Eurostat, 2016. Energy intensity of the economy. Available from: ec.eurostat/web/products-datasets/-tsdec360.

Fang, Z., Chang, Y., 2016. Energy, human capital and economic growth in Asia Pacific countries: evidence from a panel cointegration and causality analysis. Energy Econ. 56, 177–184.

FAO, 2009. How to feed the world by 2050. Available from: http://www.fao.org/fileadmin/templates/wsfs/docs/expert_paper/How_to_Feed_the_World_in_2050.pdf.

FAO, 2011. The state of the world's land and water resources for food and agriculture. Available from: http://www.fao.org/docrep/017/i1688e/i1688e.pdf.

Fatai, B.O., 2014. Energy consumption and economic growth nexus: panel co-integration and causality tests for Sub-Saharan Africa. J. Energy South. Afr. 25 (4), 93–100.

Fatai, K., Oxley, L., et al., 2004. Modelling the causal relationship between energy consumption and GDP in New Zealand, Australia, India, Indonesia, the Philippines and Thailand. Math. Comput. Simul. 64 (3–4), 431–445.

Fuinhas, J.A., Marques, A.C., 2012. Energy consumption and economic growth nexus in Portugal, Italy, Greece, Spain and Turkey: an ARDL bounds test approach (1965–2009). Energy Econ. 34 (2), 511–517.

Furuoka, F., 2015. Financial development and energy consumption: evidence from a heterogeneous panel of Asian countries. Renew. Sustainable Energy Rev. 52, 430–444.

Furuoka, F., 2016. Natural gas consumption and economic development in China and Japan: an empirical examination of the Asian context. Renew. Sustainable Energy Rev. 56, 100–115.

Galindo, L.M., 2005. Short- and long-run demand for energy in Mexico: a cointegration approach. Energy Policy 33 (9), 1179–1185.

Gao, J., Zhang, L., 2014. Electricity consumption-economic growth-CO_2 emissions nexus in Sub-Saharan Africa: evidence from panel cointegration. Afr. Dev. Rev. 26 (2), 359–371.

Ghosh, S., 2002. Electricity consumption and economic growth in India. Energy Policy 30 (2), 125–129.

Ghosh, S., 2009. Electricity supply, employment and real GDP in India: evidence from cointegration and Granger-causality tests. Energy Policy 37 (8), 2926–2929.

Glasure, Y.U., Lee, A.R., 1998. Cointegration, error-correction, and the relationship between GDP and energy: the case of South Korea and Singapore. Resour. Energy Econ. 20 (1), 17–25.

Gómez, A., Dopazo, C., et al., 2014. The causes of the high energy intensity of the Kazakh economy: a characterization of its energy system. Energy 71, 556–568.

Gurgul, H., Lach, Ł., 2011. The role of coal consumption in the economic growth of the Polish economy in transition. Energy Policy 39 (4), 2088–2099.

Hossain, M.S., Saeki, C., 2011. Does electricity consumption panel granger cause economic growth in South Asia? Evidence from Bangladesh, India, Iran, Nepal, Pakistan and Sri-Lanka. Euro. J. Social Sci. 25 (3), 316–328.

Househam, I. (undated). "Energy Efficiency Metrics". Available from: www.oecd.org/sti/inno.

Howarth, C., Monasterolo, I., 2016. Understanding barriers to decision making in the UK energy-food-water nexus: the added value of interdisciplinary approaches. Environ. Sci. Policy 61, 53–60.

Huang, B.N., Hwang, M.J., et al., 2008. Causal relationship between energy consumption and GDP growth revisited: a dynamic panel data approach. Ecol. Econ. 67, 41–54.

Hwang, J.H., Yoo, S.H., 2014. Energy consumption, CO_2 emissions, and economic growth: evidence from Indonesia. Qual. Quant. 48 (1), 63–73.

International Energy Agency (IEA), 2010. World energy outlook 2010, executive summary. Available from: http://www.iea.org/Textbase/npsum/weo2010sum.pdf.

International Energy Agency, 2015. Southeast Asia Energy Outlook. Available from: www.iea.org.

Irandoust, M., 2016. The renewable energy-growth nexus with carbon emissions and technological innovation: evidence from the Nordic countries. Ecol. Indic. 69, 118–125.

Islam, F., Shahbaz, M., et al., 2013. Financial development and energy consumption nexus in Malaysia: a multivariate time series analysis. Econ. Model. 30 (1), 435–441.

Issa Shahateet, M., 2014. Modeling economic growth and energy consumption in Arab countries: cointegration and causality analysis. Int. J. Energy Econ. Policy 4 (3), 349–359.

Iyke, B.N., 2015. Electricity consumption and economic growth in Nigeria: a revisit of the energy-growth debate. Energy Econ. 51, 166–176.

Joo, Y.J., Kim, C.S., et al., 2015. Energy consumption, CO_2 emission, and economic growth: evidence from Chile. Int. J. Green Energy 12 (5), 543–550.

Kahia, M., Ben Aïssa, M.S., et al., 2016. Impact of renewable and non-renewable energy consumption on economic growth: new evidence from the MENA Net Oil Exporting Countries (NOECs). Energy 116, 102–115.

Kahsai, M.S., Nondo, C., et al., 2012. Income level and the energy consumption-GDP nexus: evidence from Sub-Saharan Africa. Energy Econ. 34 (3), 739–746.

Kalimeris, P., Richardson, C., et al., 2014. A meta-analysis investigation of the direction of the energy-GDP causal relationship: implications for the growth-degrowth dialogue. J. Clean. Prod. 67, 1–13.

Karanfil, F., Li, Y., 2015. Electricity consumption and economic growth: exploring panel-specific differences. Energy Policy 82 (1), 264–277.

Khan, M.A., Khan, M.Z., et al., 2014. Global estimates of energy-growth nexus: application of seemingly unrelated regressions. Renew. Sustainable Energy Rev. 29, 63–71.

Koçak, E., Şarkgüneşi, A., 2017. The renewable energy and economic growth nexus in Black Sea and Balkan countries. Energy Policy 100, 51–57.

Kouakou, A.K., 2011. Economic growth and electricity consumption in Cote d'Ivoire: evidence from time series analysis. Energy Policy 39 (6), 3638–3644.

Kraft, J., Kraft, A., 1978. On the relationship between energy and GNP. J. Energy Dev. 3, 401–403.

Kubiszewski, I., Costanza, R., et al., 2013. Beyond GDP: measuring and achieving global genuine progress. Ecol. Econ. 93, 57–68.

Kumar, R.R., Stauvermann, P.J., et al., 2015. Exploring the role of energy, trade and financial development in explaining economic growth in South Africa: a revisit. Renew. Sustainable Energy Rev. 52, 1300–1311.

Lach, L., 2015. Oil usage, gas consumption and economic growth: evidence from Poland. Energy Sources Part B 10 (3), 223–232.

Lawn, P.A., 2003. A theoretical foundation to support the Index of Sustainable Economic Welfare (ISEW), Genuine Progress Indicator (GPI), and other related indexes. Ecol. Econ. 44 (1), 105–118.

Lee, C., 2005. Energy consumption and GDP in developing countries: a cointegrated panel analysis. Energy Econ. 27 (3), 415–427.

Lee, C., Chang, C., 2007a. Energy consumption and GDP revisited: a panel analysis of developed and developing countries. Energy Econ. 29 (6), 1206–1223.

Lee, C.C., Chang, C.P., 2007b. The impact of energy consumption on economic growth: evidence from linear and nonlinear models in Taiwan. Energy 32 (12), 2282–2294.

Lee, C.Y., Chiu, Y., 2011. Nuclear energy consumption, oil prices, and economic growth: evidence from highly industrialized countries. Energy Econ. 33, 236–248.

Leung Pah Hang, M.Y., Martinez-Hernandez, E., et al., 2016. Designing integrated local production systems: a study on the food-energy-water nexus. J. Clean. Prod. 135, 1065–1084.

Lin, B., Moubarak, M., 2014. Renewable energy consumption: economic growth nexus for China. Renew. Sustainable Energy Rev. 40, 111–117.

Liu, W.C., 2013. The relationship between energy consumption and output: a frequency domain approach. Rom. J. Econ. Forecast. 16 (4), 44–55.

Mahadevan, R., Asafu-Adjaye, J., 2007. Energy consumption, economic growth and prices: a reassessment using panel VECM for developed and developing countries. Energy Policy 35 (4), 2481–2490.

Mahdi Ziaei, S., 2015. Effects of financial development indicators on energy consumption and CO_2 emission of European, East Asian and Oceania countries. Renew. Sustainable Energy Rev. 42, 752–759.

Mehrara, M., 2007. Energy consumption and economic growth: the case of oil exporting countries. Energy Policy 35 (5), 2939–2945.

Menegaki, A.N., 2014. On energy consumption and GDP studies: a meta-analysis of the last two decades. Renew. Sustainable Energy Rev. 29, 31–36.

Menegaki, A.N., Tiwari, A.K., 2017. The index of sustainable economic welfare in the energy-growth nexus for American countries. Ecol. Indic. 72, 494–509.

Menegaki, A.N., Tsagarakis, K.P., 2015. More indebted than we know? Informing fiscal policy with an index of sustainable welfare for Greece. Ecol. Indic. 57, 159–163.

Menegaki, A.N., Tugcu, C.T., 2016. Rethinking the energy-growth nexus: proposing an index of sustainable economic welfare for Sub-Saharan Africa. Energy Res. Soc. Sci. 17, 147–159.

Menegaki, A.N., Tugcu, C.T., 2017. Energy consumption and sustainable economic welfare in G7 countries: a comparison with the conventional nexus. Renew. Sustainable Energy Rev. 69, 892–901.

Menyah, K., Wolde-Rufael, Y., 2010. CO_2 emissions, nuclear energy, renewable energy and economic growth in the US. Energy Policy 38, 2911–2915.

Michieka, N.M., 2015. Short- and long-run analysis of factors affecting electricity consumption in Sub-Saharan africa. Int. J. Energy Econ. Policy 5 (3), 639–646.

Morimoto, R., Hope, C., 2004. The impact of electricity supply on economic growth in Sri Lanka. Energy Econ. 26 (1), 77–85.

Mozumder, P., Marathe, A., 2007. Causality relationship between electricity consumption and GDP in Bangladesh. Energy Policy 35 (1), 395–402.

Narayan, P.K., Prasad, A., 2008. Electricity consumption-real GDP causality nexus: evidence from a bootstrapped causality test for 30 OECD countries. Energy Policy 36 (2), 910–918.

Narayan, P.K., Smyth, R., 2008. Energy consumption and real GDP in G7 countries: new evidence from panel cointegration with structural breaks. Energy Econ. 30 (5), 2331–2341.

Nazlioglu, S., Kayhan, S., et al., 2014. Electricity consumption and economic growth in Turkey: cointegration, linear and nonlinear Granger causality. Energy Sources, Part B 9 (4), 315–324.

Nordhaus, W.D., Tobin, J., 1972. Is growth obsolete? NBER. Available from: http://www.nber.org/chapters/c7620.pdf.

Nourry, M., 2008. Measuring sustainable development: some empirical evidence for France from eight alternative indicators. Ecol. Econ. 67 (3), 441–456.

Ocal, O., Aslan, A., 2013. Renewable energy consumption-economic growth nexus in Turkey. Renew. Sustainable Energy Rev. 28, 494–499.

Ocal, O., Ozturk, I., et al., 2013. Coal consumption and economic growth in Turkey. Int. J. Energy Econ. Policy 3 (2), 193–198.

Odhiambo, N.M., 2009. Electricity consumption and economic growth in South Africa: a trivariate causality test. Energy Econ. 31 (5), 635–640.

Odhiambo, N.M., 2010. Energy consumption, prices and economic growth in three SSA countries: a comparative study. Energy Policy 38 (5), 2463–2469.

Odhiambo, N.M., 2012. Economic growth and carbon emissions in South Africa: an empirical investigation. J. Appl. Bus. Res. 28 (1), 37–46.

OECD, 2016. World Hunger and poverty facts and statistics. Available from: http://www.worldhunger.org/2015-world-hunger-and-poverty-facts-and-statistics/.

Oh, W., Lee, K., 2004. Energy consumption and economic growth in Korea: testing the causality relation. J. Policy Model. 26 (8–9), 973–981.

Omri, A., Ben Mabrouk, N., et al., 2015. Modeling the causal linkages between nuclear energy, renewable energy and economic growth in developed and developing countries. Renew. Sustainable Energy Rev. 42, 1012–1022.

Ouedraogo, N.S., 2013. Energy consumption and economic growth: evidence from the economic community of West African States (ECOWAS). Energy Econ. 36, 637–647.

Ozturk, I., 2010. A literature survey on energy-growth nexus. Energy Policy 38 (1), 340–349.

Ozturk, I., 2015. Sustainability in the food-energy-water nexus: evidence from BRICS (Brazil, the Russian Federation, India, China, and South Africa) countries. Energy 93 (Part 1), 999–1010.

Ozturk, I., Acaravci, A., 2010. The causal relationship between energy consumption and GDP in Albania, Bulgaria, Hungary and Romania: evidence from ARDL bound testing approach. Appl. Energy 87 (6), 1938–1943.

Ozturk, I., Aslan, A., et al., 2010. Energy consumption and economic growth relationship: evidence from panel data for low and middle income countries. Energy Policy 38 (8), 4422–4428.

Ozturk, I., Bilgili, F., 2015. Economic growth and biomass consumption nexus: dynamic panel analysis for Sub-Sahara African countries. Appl. Energy 137, 110–116.

Ozturk, I., Kaplan, M., et al., 2013. The causal relationship between energy consumption and GDP in Turkey. Energy Environ. 24 (5), 727–734.

Ozun, A., Cifter, A., 2007. Multi-scale causality between energy consumption and GNP in emerging markets: evidence from Turkey. Invest. Manage. Finan. Inn. 4 (2), 60–70.

Pao, H.T., Fu, H.C., 2013a. Renewable energy, non-renewable energy and economic growth in Brazil. Renew. Sustainable Energy Rev. 25, 381–392.

Pao, H.T., Fu, H.C., 2013b. The causal relationship between energy resources and economic growth in Brazil. Energy Policy 61, 793–801.

Pao, H.T., Li, Y.Y., et al., 2014. Clean energy, non-clean energy, and economic growth in the MIST countries. Energy Policy 67, 932–942.

Payne, J., Taylor, J., 2010. Nuclear energy consumption and economic growth in the US: an empirical note. Energy Source Part B 5 (3), 301–307.

Payne, J.E., 2009. On the dynamics of energy consumption and output in the US. Appl. Energy 86 (4), 575–577.

Payne, J.E., 2011. On biomass energy consumption and real Output in the U.S. Energy Sources B 6, 47–52.

Pielli, K., Dhungel, H., et al., 2016. Examining energy efficiency issues in Sub-Saharan Africa. Available from: www.usaid.gov/powerafrica.

Pulselli, F.M., Bravi, M., et al., 2012. Application and use of the ISEW for assessing the sustainability of a regional system: a case study in Italy. J. Econ. Behav. Organ. 81 (3), 766–778.

Rafindadi, A.A., Ozturk, I., 2015. Natural gas consumption and economic growth nexus: is the 10th Malaysian plan attainable within the limits of its resource? Renew. Sustainable Energy Rev. 49, 1221–1232.

Raheem, I.D., Yusuf, A.H., 2015. Energy consumption-economic growth nexus: evidence from linear and nonlinear models in selected African countries. Int. J. Energy Econ. Policy 5 (2), 558–564.

Rasul, G., 2016. Managing the food, water, and energy nexus for achieving the Sustainable Development Goals in South Asia. Environ. Dev. 18, 14–25.

Raza, S.A., Jawaid, S.T., et al., 2016. Electricity consumption and economic growth in South Asia. South Asia Econ. J. 17 (2), 200–215.

Razzaqi, S., Bilquees, F., et al., 2011. Dynamic relationship between energy and economic growth: evidence from D8 countries. Pak. Dev. Rev. 50 (4), 437–458.

Richard, O.O., 2012. Energy consumption and economic growth in sub-Saharan Africa: an asymmetric cointegration analysis. Econ. Int. 129 (1), 99–118.

Riti, J.S., Shu, Y., 2016. Renewable energy, energy efficiency, and eco-friendly environment (R-E5) in Nigeria. Energy Sustain. Soc. 6 (1).

Saboori, B., Sulaiman, J., 2013. Environmental degradation, economic growth and energy consumption: evidence of the environmental Kuznets curve in Malaysia. Energy Policy 60, 892–905.

Saidi, K., Hammami, S., 2015. The impact of CO_2 emissions and economic growth on energy consumption in 58 countries. Energy Rep. 1, 62–70.

Sari, R., Ewing, B.T., et al., 2008. The relationship between disaggregate energy consumption and industrial production in the United States: an ARDL approach. Energy Econ. 30, 2302–2313.

Saunoris, J.W., Sheridan, B.J., 2013. The dynamics of sectoral electricity demand for a panel of US states: new evidence on the consumption-growth nexus. Energy Policy 61, 327–336.

Shahbaz, M., Hye, Q.M.A., et al., 2013. Economic growth, energy consumption, financial development, international trade and CO_2 emissions in Indonesia. Renew. Sustainable Energy Rev. 25, 109–121.

Shahbaz, M., Loganathan, N., et al., 2015. Does renewable energy consumption add in economic growth? An application of auto-regressive distributed lag model in Pakistan. Renew. Sustainable Energy Rev. 44, 576–585.

Shahbaz, M., Zeshan, M., et al., 2012. Is energy consumption effective to spur economic growth in Pakistan? New evidence from bounds test to level relationships and Granger causality tests. Econ. Model. 29 (6), 2310–2319.

Shakeel, M., Iqbal, M.M., 2015. Energy consumption and GDP with the role of trade in South Asia. 2014 International Conference on Energy Systems and Policies, ICESP 2014.

Śmiech, S., Papież, M., 2014. Energy consumption and economic growth in the light of meeting the targets of energy policy in the EU: the bootstrap panel Granger causality approach. Energy Policy 71, 118–129.

Soares, J.A., Kim, Y.K., et al., 2014. Analysis of causality between energy consumption and economic growth in Indonesia. Geosystem Eng. 17 (1), 58–62.

Soytas, U., Sari, R., 2003. Energy consumption and GDP: causality relationship in G-7 countries and emerging markets. Energy Econ. 25, 33–37.

Soytas, U., Sari, R., 2009. Energy consumption, economic growth, and carbon emissions: challenges faced by an EU candidate member. Ecol. Econ. 68 (6), 1667–1675.

Spiegelberg, M., Baltazar, D.E., et al., 2017. Unfolding livelihood aspects of the water–energy–food nexus in the Dampalit Watershed, Philippines. J. Hydro 11, 53–68.

Squalli, J., 2007. Electricity consumption and economic growth: bounds and causality analyses of OPEC members. Energy Econ. 29 (6), 1192–1205.

Talbreth, J., Cobb, C., et al., 2007. The genuine progress indicator 2006, a tool for sustainable development. Available from http://rprogress.org/publications/2007/GPI%202006.pdf.

Tang, C.F., 2008. A re-examination of the relationship between electricity consumption and economic growth in Malaysia. Energy Policy 36 (8), 3067–3075.

Tang, C.F., Tan, B.W., 2014. The linkages among energy consumption, economic growth, relative price, foreign direct investment, and financial development in Malaysia. Qual. Quant. 48 (2), 781–797.

Tang, C.F., Tan, E.C., 2013. Exploring the nexus of electricity consumption, economic growth, energy prices and technology innovation in Malaysia. Appl. Energy 104, 297–305.

The Conference Board of Canada, 2016. Energy intensity. Available from: www.conferenceboard.ca/hcp/details/environment/energy-intensity.aspx.

Thoma, M., 2004. Electrical energy usage over the business cycle. Energy Econ. 26, 463–485.

Tiwari, A.K., Apergis, N., et al., 2015. Renewable and nonrenewable energy production and economic growth in sub-Saharan Africa: a hidden cointegration analysis. Appl. Econ. 47 (9), 861–882.

Tiwari, A.K., Mensi, W., et al., 2016. CO_2 emissions, energy consumption & economic growth: fresh evidence from Saudi Arabia using multi-horizon, time varying and Fourier based Granger causality approaches, Working Paper.

Torras, M., 2005. Ecological inequality in assessing well-being: some applications. Policy Sci. 38 (4), 205–224.

Tugcu, C.T., Ozturk, I., et al., 2012. Renewable and non-renewable energy consumption and economic growth relationship revisited: evidence from G7 countries. Energy Econ. 34, 1942–1950.

U.S. Department of Energy, 2016. Energy goals and standards for federal government. Available from: energy.gov/savings/energy-goals-and-standards-federalgovernment.

U.S. Government, 2008. National action plan for energy efficiency Vision Plan 2025: a framework for change. Available from: www.epa.gov/eeactionplan.

UN-ESCAP, 2013. The status of the Water-Food-Energy Nexus in Asia and the Pacific. Available from: http://www.unescap.org/sites/default/files/Water-Food-Nexus%20Report.pdf.

UN-Water, 2014. Water, food and energy nexus. Available from: http://www.unwater.org/topics/water-food-and-energy-nexus/en/.

United Nations, 1987. Report of the World Commission on Environment and Development: Our Common Future. Available from: http://www.un-documents.net/our-commonfuture.pdf.

United Nations, 2016. Kyoto Protocol. Available from: http://unfccc.int/kyoto_protocol/items/2830.php.

US Energy Information Administration, 2016. Energy efficiency and conservation. Available from: http://www.eia.gov/energyexplained/index.cfm?page=about_energy_efficiency.

Vidyarthi, H., 2015. Energy consumption and growth in South Asia: evidence from a panel error correction model. IJESM 9 (3), 295–310.

Wolde-Rufael, Y., 2010. Bounds test approach to cointegration and causality between nuclear energy consumption and economic growth in India. Energy Policy 38 (1), 52–58.

Wolde-Rufael, Y., 2014. Electricity consumption and economic growth in transition countries: a revisit using bootstrap panel Granger causality analysis. Energy Econ. 44, 325–330.

Wolde-Rufael, Y., Menyah, K., 2010. Nuclear energy consumption and economic growth in nine developed countries. Energy Econ. 32, 550–556.

World Energy Council, 2013. World Energy Perspective, Energy Efficiency policies: what works and what does not. Available from: www.worldenergy.org/wp-content.

World Energy Council, 2016. Energy intensity of agriculture (to value added). Available from: www.wec.-indicators.enerdata.eu/agriculture.

World Resources Institute, 2016. What is an INDC? Available from: http://www.wri.org/indc-definition.

Worldbank, 2010. Energy in MENA. Available from: web.worldbank.org.

Yang, Y.C.E., Wi, S., et al., 2016. The future nexus of the Brahmaputra River Basin: climate, water, energy and food trajectories. Glob. Environ. Chang. 37, 16–30.

Yildirim, E., Aslan, A., et al., 2014. Energy consumption and GDP in Asean countries: bootstrap-corrected panel and time series causality tests. Singapore Econ. Rev. 59 (2).

Yildirim, E., Sarac, S., et al., 2012. Energy consumption and economic growth in the USA: evidence from renewable energy. Renew. Sustainable Energy Rev. 16, 6770–6774.

Yoder, S., 2013. Energy intensity of global economy rises, reversing longtime trend. Available from: www.worldwatch.org.

Yoo, S.H., 2005. Electricity consumption and economic growth: evidence from Korea. Energy Policy 33 (12), 1627–1632.

Yoo, S.H., 2006. The causal relationship between electricity consumption and economic growth in the ASEAN countries. Energy Policy 34 (18), 3573–3582.

Yoo, S.H., Ku, S.J., 2009. Causal relationship between nuclear energy consumption and economic growth: a multi-country analysis. Energy Policy 37, 1905–1913.

Yoo, S.H., Kwak, S.Y., 2010. Electricity consumption and economic growth in seven South American countries. Energy Policy 38 (1), 181–188.

Yu, E.S.H., Choi, J.Y., 1985. The causal relationship between energy and GNP: an international comparison. J. Energy Dev. 10, 249–272.

Yu, E.S.H., Hwang, B.K., 1984. The relationship between energy and GNP: further results. Energy Econ. 6, 186–190.

Yu, E.S.H., Jin, J.C., 1992. Cointegration tests of energy consumption, income and employment. Resour. Energy 14, 259–266.

Yuan, J.H., Kang, J.G., et al., 2008. Energy consumption and economic growth: evidence from China at both aggregated and disaggregated levels. Energy Econ. 30 (6), 3077–3094.

Zaman, K., Abdullah, A b., et al., 2016. Dynamic linkages among energy consumption, environment, health and wealth in BRICS countries: green growth key to sustainable development. Renew. Sustainable Energy Rev. 56, 1263–1271.

Zeshan, M., Ahmed, V., 2013. Energy, environment and growth nexus in South Asia. Environ. Dev. Sustain. 15 (6), 1465–1475.

Ziramba, E., 2009. Disaggregate energy consumption and industrial production in South Africa. Energy Policy 37 (6), 2214–2220.

Further Readings

Menegaki, A.N., Tugcu, C.T., 2016a. The sensitivity of growth, conservation, feedback & neutrality hypotheses to sustainability accounting. Energy Sustainable Dev. 34, 77–87.

United Nations, 2016. Status of ratification of the Kyoto protocol. Available from: http://unfccc.int/kyoto_protocol/status_of_ratification/items/2613.php.

United Nations, 2016. Sustainable development goals. Available from: http://www.un.org/sustainabledevelopment/sustainable-development-goals.

The Econometrics of the Energy-Growth Nexus

Practical Issues on Energy-Growth Nexus Data and Variable Selection With Bayesian Analysis

Aviral K. Tiwari*, Anabel Forte, Gonzalo Garcia-Donato†,
Angeliki N. Menegaki‡,¶**

**Montpellier Business School, Montpellier Cedex 4, France; **University of Valencia, Valencia, Spain;
†University of Castilla-La Mancha, Albacete, Spain; ‡TEI STEREAS ELLADAS, University of Applied
Sciences, Lamia, Greece; ¶Hellenic Open University, Patras, Greece*

Chapter Outline

1 Introduction: The Virtues of Panel Data

Energy-growth nexus (EGN) researchers may become overwhelmed with the choice and origin of the variables used in their research field. During the years, the production function characterizing the economy or a particular economic sector is continually enriched with additional variables that are found to have a causal relationship with energy consumption and economic growth.

Unfortunately, sufficiently long time series do not exist for those variables and researchers can use only panel data. Therefore, the practical tips provided in the first part of this chapter will focus on panel data. Though there are several issues in the panel data transformation, however in this chapter, we will deal with the identification of relevant variables in the EGN, obtaining the data for the identified variable and dealing with missing values. The second part of this chapter will demonstrate the variable identification through Bayesian analysis. However, before directly addressing the issues related to panel data, we begin with the definition of the panel data, its advantages and shortcomings, and panel restructuring or transforming the data to form a panel.

Panel data are also referred to as longitudinal data or cross-sectional time series data, because these data are collected for cross-section units over a period of time. Just to be cautious, there is a small difference between panel data and pooled cross-sections, and this is that in the panel data, the same cross-section units (e.g., individuals, firms, or countries) are surveyed over a given time period, while in the pooled cross-sections, the units can be different.

There are two types of panel data models, macro panels with large N (number of cross-sections) and large T (length of the time series) and micro panels with large N and small T. However, in practice, sometimes we also face the issue of a large T and a small N. This occurs particularly when a researcher is interested in a very small group of cross-sections (e.g., when four or five cross-section units are studied with $T > 50$). There are several advantages of using panel data and are summarized in Fig. 6.1.

Based on Fig. 6.1, the advantages of using panel data settings are the following:

1) To increase the sample size. This is because observations under a time series data setup are practically limited. This leads to imprecise estimates of the parameters, while by pooling random samples drawn from the same population, even at different points in time, researchers can get more precise estimates and test statistics with more power.
2) It is helpful when the relationship between the explained variable and at least some of the explanatory variables remains constant over time.

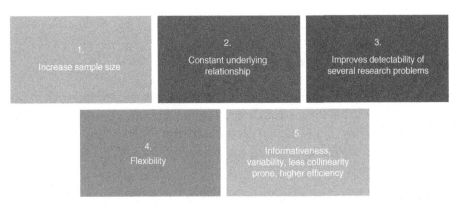

Figure 6.1: Advantages in Using Panel Data.

3) Panel data are helpful in providing answers to several research problems such as the identification and measurement effects that cannot be detected through either a time series or a cross-sectional data setup.

4) As documented in Green (2010, p. 284, 5ed) "The fundamental advantage of a panel data set over a cross section is that it will allow the researcher great flexibility in modeling differences in behavior across individuals."

5) Panel data, according to Baltagi (2005, p. 5, 3ed), provide more informative datasets, lend more variability, less collinearity among the variables, more degrees of freedom, and higher efficiency. Several other benefits attributed to panel data are documented in Green (2010, 5ed), Baltagi (2005, 3ed) and Gujarati (2015, 4ed) and the interested readers can refer to them.

One should note that the gains from the pooled data do not come without any complications. This is because the panel data are the collection of independently drawn samples from a different population (because total population size is likely to change over years) over a period of time. Thus, under a pooled data setup, it is quite likely that different populations have different distributions, and that their distribution also changes over time. However, this problem can be easily addressed by simply allowing the intercept to change each year by including dummy variables for each year (except one in order to avoid the so-called dummy variable trap).

In continuation with the above, there is also another possibility, in a pooled data framework that error variance changes over time, which may also create issues in their estimation. These issues are very well discussed in Green (2010, 5ed), Baltagi (2005, 3ed), and Gujarati (2015, 4ed), and the interested readers can refer to these textbooks. An additional problem may also arise [as documented in Baltagi (2005, 3ed)] related to the designing of panel surveys as well as their data collection and data management issues. There is also the possibility of measurement errors that may arise because of faulty responses due to unclear questions, memory errors, errors in data compilation, and so on. However, this can most likely occur when the EGN analysis involves micro data and discrete choice analysis. To the best of our knowledge, we have not seen such studies in the EGN, but we believe they would be very promising and are likely to appear in the near future.

Further, there is a common misunderstanding that for all panel data one should use either the fixed or the random effect models based on the Hausman's test. Given the data availability and the nature of the data, selecting the most appropriate modeling procedure is another big question to answer. Last but not least, the question remains on how to interpret the results.

Hence, in the first part of this chapter, we will focus on sources for the EGN data, the organization of that data in panel form, recognizing and handling ill-organized data, dealing with missing data, selecting the appropriate panel data model, correctly reading and reporting output generated from statistical software such as Stata, interpreting the result substantively, and presenting the results in a professional manner. Since this part of the chapter aims at providing practical and rule-of-thumb advice, we will not extend on any mathematical models since they are sufficiently documented in standard econometric textbooks for panel data. Also, to avoid unnecessary complication, this document mainly focuses on linear regression models rather than nonlinear ones or binary response and event count data models. Last, examples will be drawn from rather macroeconomic unit level data that are less likely to find for free and are more complicated to handle. The second part of the chapter will deal with variable selection through Bayesian analysis. It provides an introduction to Bayesian analysis, the essentials of Bayesian estimation, and prediction, as well as a worked example on the selection of the most influential variables in the EGN.

2 Part 1: Practical Issues on Data Use in the EGN

2.1 Data Choice in the EGN Research

Doing research demands trustworthy, accurate, and complete data. If they are open, namely they are provided for free of charge, that is even better, particularly for researchers who cannot afford buying access to expensive databases. Quite often, the research design in the EGN is dictated by the availability of data in these databases. The following databases are most common in the EGN research.

2.1.1 The Worldbank Databases (Available from: at://databank.worldbank.org/data.databases.aspx.)

Worldbank offers a wide variety of databases that the interested reader could reach at (//databank.worldbank.org/data/databases.aspx). The EGN uses the World Development Indicators (WDI) database quite often. Usually, the used data are selected to be in "per capita" terms, although they exist in many other versions and forms, for example the GDP is offered in the following eight versions.

The best strategy is to use constant data, but when current values are used instead, inserting inflation in the EGN equation can give an indication of whether the usage of current values is an issue (Fig. 6.2).

Sometimes data transformations are necessary in order to get the variable a researcher needs. For example, while OECD offers total labor force divided into total employment and total unemployment, WDI offers total labor force with total unemployment and one needs to make the subtraction of 1 minus total unemployment to get the employment. However, the necessary variable to use in the EGN is total labor, which includes both the employed population and those who are currently seeking a job. WDI data are typically provided in annual form.

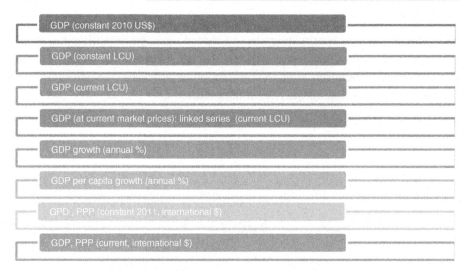

Figure 6.2: WDI Provide Eight Versions of GDP.

2.1.2 The OECD (at: stats.oecd.org)

Different databases provide different definitions of variables, for example GDP in the OECD database is offered as GDP (at current prices and PPPs), real GDP forecast (growth rates compared to previous years), nominal GDP forecast, GDP long-term forecast (which includes long-term baseline projections up to 2060) and quarterly GDP. The variety is huge and depends on what the researcher needs to investigate or prove.

2.1.3 Eurostat and other country statistical agencies (at: ec.europa.eu/eurostat/data/database)

Eurostat is the official statistical agency of the European Commission and offers a quite rich variety of sophisticated data only for countries of the European continent together with Turkey and some Eurasian countries. GDP is found under the title of "annual national accounts" and is offered at current prices (euro per inhabitant or in volumes as, e.g., percentage change on previous years and price indices as e.g. percentage change on previous period, based on 2005 = 100). Eurostat sometimes offers data in quarterly or monthly basis. Eurostat database is divided by themes and navigation trees. Each theme may be divided in further sub-themes. For example, the "Energy and Environment" theme is further divided into one sub-theme for environment and another for energy. Energy sub-theme further contains six broad sub-themes such as energy quantities in annual data and monthly data, short-term monthly data, prices of natural gas and electricity, market structure indicators, and heating degree days. These data categories contain many more variables and we could safely state that the Eurostat database is one of the richest databases.

2.1.4 International Energy Agency (at: www.eia.gov/tools)

A rich variety of data is also provided by the International Energy Agency. Useful databases for EGN researchers are, for example, short-term energy outlook (STEO) which contains the short-term forecasts of US energy supply, demand, prices, imports/exports for the next 18–24 months, the Alternative Fuel Vehicle Data Browser, which contains data about the fuel use and the number of vehicles for alternative fuel vehicle fleets, the US Electric system operating data which contains hourly actual and forecast electricity generation and demand by the US region and power authority. The IEA database offers a bulk download facility which contains the entire contents of each Energy Information Administration (EIA) and Application Programming Interface (API) data (time-series data for electricity, petroleum, and natural gas) set in a single zip file, the EIA Excel data add-in facility, which enables downloading the EIA energy data and economic data for the St. Louis Federal Reserve (FRED) directly to your spreadsheet, Google Sheets add-on facility which facilitates searching, filtering, and loading the US energy information administration free and open energy data sets into sheets and the electricity data browser which offers data on the US electricity generation, transmission, consumption, and prices as well as plentiful other facilities and outlets.

2.1.5 British Petroleum data (at: www.bp.com)

British Petroleum (BP) provides typical energy data, and also provides various historical energy data, which are not available in other databases, at least in an open form. For example, proved oil reserves and their history from 1980 to 2015 exists in this database. Furthermore, the BP Statistical Review of the World Energy is published every June and provides the highlights of what is provided by BP database.

2.1.6 Thomson Reuters Datastream (at: financial.thomsonreuters.com)

This database offers a wide collection of company data (data on stocks and stock indices), economic data (interest rates, currencies, and wide range of macroeconomic indices), market and pricing data (bonds, futures, and options). The data are both historical and real time. The data can be purchased and relevant inquiries can be addressed to the official site of the company. Datastream is both a financial and macroeconomic data platform, so it is useful for both energy economics and energy finance analysts. Since this is not an open access database, it is less quoted by independent researchers or EGN academics who are affiliated in departments with less abundant finances. For more information, the interested researcher could visit: http://financial.thomsonreuters.com/en/products/tools.

2.1.7 ODYSSE and MURE database (at: indicators.odyssee-mure.eu)

The ODYSSE-MURE databases are supported by the H2020 program of the European Commission and are coordinated by Agence de l' Environnement et de la Maitrise de l' Energie with the technical support of various companies.

ODYSSE database offers energy efficiency and energy consumption indicators, not only in their total aggregate form, but also in a disaggregated form (e.g., industry, transport, household, services, and agriculture). This database also provides five data facilities shown in Fig. 6.3. Furthermore, MURE database contains a description of all energy

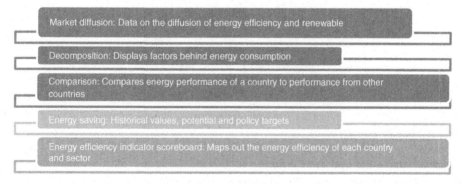

Figure 6.3: Data Facilities on ODYSSE Database.

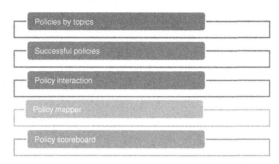

Figure 6.4: Data Facilities on MURE Database.

efficiency policy measures by sector and country together with their impact evaluation (Fig. 6.4).

2.1.8 Penn World Tables (PWT) (at: cid.econ.ucdavis.edu)

They were constructed by Robert Summers and Alan Heston (University of Pennsylvania) with the late Irving Kravis. They offer national accounts for each country in their own currencies and in adjusted form to be presented in US dollars across countries. The three sites that support the Penn World Tables are in the University of Groningen (Version 9.0.), University of Pennsylvania (Version 7.3.), and University of Toronto (Version 6.3.). The 8th version offers data on relative levels of income, output, input, and productivity covering 182 countries between 1950 and 2014 (Feenstra et al., 2015).

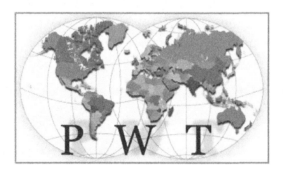

2.1.9 "Data in Brief" Journal by Elsevier

EGN researchers have the option of working on other colleagues' datasets after they receive their permission and agreement on that. This type of reusage of data contributes to the reproducibility of research in the EGN field, which is very important and longed for in this research field. The "Data in Brief" is a new journal in ELSEVIER, which constitutes an outlet for publishing datasets.

2.1.10 *Database of Political Institutions (DPI) (at: econ.worldbank.org)*

This is a Worldbank database which is open. It covers 180 countries for 40 years, from 1975 to 2015. It hosts political economy nature data such as institutional and electoral results, tenure and stability of government, identification of parties' ideology, opposition and government parties in the legislature, and so on. For EGN researchers who use political economy variables as control variables in their production function for the economy (Menegaki and Özturk, 2013), this is a very useful database.

2.1.11 Various micro data

At the time of writing this chapter, the EGN has taken place at a country or sector level. As the field develops from up-bottom direction and more disaggregate levels of analysis appear, it may not be far in future, that EGN researchers will move to household or businesses as a level of analysis. Finding data resources from socioeconomic panels such as the German SOEP (Socioeconomic Panel) would be one of the options. However, these data cannot always be found for free and are also protected from specific regulations. Usually researchers can have access to them, only after they submit a detailed research proposal to the organization that has compiled and updates the data. However, sometimes research ideas appear and develop as one is confronted with data availability. Therefore, it would facilitate future research if these databases offered, in public, at least a list of contents of the data they have. An indicative list of the data in the SOEP are: demographic and parental characteristics, labor market, health, personality, preferences and subjective orientations, subjective well-being, political involvement and participation, and so on.

Other very important databases that belong to this group are: LIS—Cross Sectional Data center in Luxembourg (available at: http://www.lisdatacenter.org), IECM-CED—Integrated European Census Microdata (available at: www.iecm-project.org), UNIPI-DEM University of Pisa (available at: www.ec.unipi.it), TARKI-POLC (available at: www.tarki.hu), UNI-TRIER (available at: www.statistik.uni-trier.de), S3RI-SOTON (available at: www.southampton.ac.uk.s3ri), SOFI-SU (available at: www.sofi.su.se), HIVA-KU Leuven (available at: www.hiva.kuleuven.be). To save space, we have not devoted a section for each one of them but the interested reader can use the link provided for each one of them.

2.2 Data Type and Frequency in the Energy-Growth Nexus

Cross-sectional, time series, pooled cross-sectional, and panel data are the most common types of data structures that are used in applied econometrics and consequently the EGN field. Data sets involving a time dimension, such as time series and panel data, require special treatment because of the correlation across time of most economic time series. Other issues, such as trends and seasonality, arise in the analysis of time-series data, but not in the cross-sectional data.

The selection of one of the above databases depends both on the topic and the hypotheses we aim to investigate, as well as whether we can afford buying the data or not. Researchers should always have in mind the economy characteristics one has to study and try to represent the economy as accurately as possible. The interested reader should go back to Chapter 4 for an overview on the aspects that have been examined so far and the new options opening for them in the near future. We present later four case studies with different data frequencies.

Case study 1: daily data

Daily data are the least frequent type of data used in the EGN papers. Salisu and Fasanya (2013) utilized daily data from Thomson Reuters over the period from April 1, 2000 to February 3, 2012. The data are oil prices and their returns. The authors examine the frequency domain Granger causality between the two series. Particularly they analyzed oil prices of West Texas Intermediate (WTI) and Brent using the tests by Narayan and Popp (2010) and Liu and Narayan (2010), both of which allow for two structural breaks in the data series, while modifying the method to include both symmetric and asymmetric volatility models. They identify two structural breaks that occurred in 1990 and 2008, which coincidentally corresponded to the Iraqi/Kuwait conflict and the global financial crisis, respectively. Also, they find evidence of persistence and leverage effects in the oil price volatility.

Case study 2: weekly data

Weekly data are also a not-so-frequent type of data in the EGN studies. Maslyuk and Smyth (2008) used weekly spot and future prices at 1, 3, and 6 months to maturity for the US WTI and the UK Brent for the period from January 1991 to December 2004. Spot prices were sourced from the Energy Information Agency (EIA), while future prices were taken from the New York Mercantile Exchange (NYMEX) or the Intercontinental Exchange (ICE). They examined the existence of a unit root with one and two structural breaks with Lagrange multiplier (LM) unit root tests with one and two endogenous structural breaks and they find that each of the oil price series can be characterized as a random walk process and that the endogenous structural breaks are significant and meaningful in terms of events that have impacted on world oil markets.

Case study 3: monthly data

Monthly data are sometimes used in the EGN studies. Tiwari (2014) used monthly data on primary energy consumption and electricity consumption of the United States for the period from January 1973 to December 2008. He sourced energy data from the IEA and the economic data from the Bureau of Economic Analysis (BEA). He investigated Granger causality by employing the approach of Lemmens et al. (2008). He found that causal and reverse causal relations between primary energy consumption and GDP and electricity consumption and GDP vary across frequencies. Foremost he decomposes the causality on the basis of time horizons and finds evidence for the feedback hypothesis.

Case study 4: quarterly data

Quarterly data are often used in the EGN studies. Lee and Lee (2009) use real energy prices with base year 2000 = 100 from 1978–2006 for 21 OECD countries for natural gas price and 15 OECD countries for coal price. Data have been sourced from the IEA. They investigate the efficient market hypothesis using total energy price and four kinds of various disaggregated energy prices—coal, oil, gas, and electricity. They employ a highly flexible panel data stationarity test of Carrion-i-Silvestre et al. (2005), which incorporates multiple shifts in level and slope, thereby controlling for cross-sectional dependence through bootstrap methods. Overwhelming evidence in favor of the broken stationarity hypothesis is found, implying that energy prices are not characterized by an efficient market and thus profitable arbitrage opportunities among energy prices were present. The estimated breaks were shown to be meaningful and coincided with the most critical events, which affected the energy prices.

Case study 5: annual data

The annual data frequency is the most often used type of data in the EGN. Tiwari and Albulescu (2016) used annual data on total electricity and renewable energy electricity net consumption from the EIA database. The data span ranged from 1980 to 2011 for 90 countries. They examined the stationarity properties of the renewable-to-total electricity consumption ratio using Becker et al. (2006) flexible Fourier stationarity test as a benchmark, and the recent advanced Fourier ADF unit root test. The results of the first test document the stationarity of the renewable-to-total electricity consumption ratio for 65 countries from different geographic areas, namely Africa, America, Asia-Pacific, and Europe. However, when validated, the Fourier ADF test shows that the null hypothesis of the unit root is rejected in all the cases, except for the United Kingdom. The fact that the stationarity in the share of renewable-to-total electricity consumption is documented by the newly proposed, more powerful tests does not necessarily mean that renewable energy policies are ineffective, as the previous literature states.

2.3 Missing Data, Solutions, and Transformations

Quite often, data are missing for some series. Countries fail to report them due to structural or political hindrances, or they may be missing at random, namely due to a technical problem faced while reporting. Random missing data usually are not a problem in the analysis, although some software might not be operating well under the missing values situation. Missing data cause the occurrence of unbalanced panels. One way to face this is either to exclude some countries from a panel data set, namely those for which the data are not available for some variables, or curtail the data span if data gaps exist at the two ends of the series. Also, despite the usefulness of a certain variable, sometimes it has to be dropped because the series is highly discontinuous. Furthermore, there are various statistical methods, which deal with missing data.

In the EGN framework, basic variables such as income are rarely missing; on the contrary, they exist for quite long data-spans. As the data become more sophisticated, for example, financial development data, they have increasingly more gaps.

To deal with random missing data, one can use various remedies. For example, if a value is missing from the beginning of the series, one can transfer the next available value and copy that in the place of the missing value. Other solutions adhere to certain convenience assumptions, which apply mostly to cases with qualitative data. It is easier to assume that the missing data take one or the other value in a binary context (e.g., yes or no variables), but this is not the case usually in the EGN modeling and estimation. In that case, replacement should be made explicit and sensitivity analysis should be applied to investigate the effect of replacement. Another mode of replacement for missing values can take place with the average of a series or predict those values through a regression.

While analyzing the time series data, particularly if we are using a single variable, single missing values or longer periods for which observations are not available is not an unsolved problem, as we can remove those periods from the sample and go ahead with the analysis. Removal of missing data in this case can be a more serious problem when the data span is already short and the removal of missing values causes it to become even shorter.

However, when we have time series studies that involve many variables, and in several of those series, there are no observations for few periods, there are two ways to solve this problem. One solution is to remove those periods for which there are no data available. However, this approach will sometimes lead us to a very small sample size and if we do not have a sufficiently large number of observations, we may face serious issues in the estimation process due to small sample bias. An alternative way is to interpolate the data if missing values are close to the start and the end date of a data series. However, if there are no data toward the end date, one may use extrapolation methods. There are so many interpolation and extrapolation methods available in different pieces of software that one has to use them with great care after looking into the properties and assumptions of the methods.

A possible option is to use the mean value of the data to replace the missing values. In practice, sometimes researchers also use four years average to replace the missing values. For example, suppose we had data from 1991 to 1998 and there are no values for 1995 in a particular series. Thus, sometimes researchers use the average values generated from preceding two years of 1995 and succeeding two years of 1995 (i.e., (1993 + 1994 + 1996 + 1997)/4). Sometimes, this approach gives closer approximations to the actual values than the values generated through some interpolation method. Some researchers may prefer to use the Expectation Maximization (EM) algorithm proposed by Stock and Watson (2002) for the extrapolation and interpolation processes. However, one should be aware that each interpolation or extrapolation method has its advantages or disadvantages, and sometimes, if there are many missing values, the use of an interpolation method may lead to misleading results.

Besides, data compilation, which may be tough due to coding and formatting requirements to which the uninitiated researcher may not be accustomed to the modeling procedure in the panel data framework, is not as easy as it may appear. For that purpose, we summarize below the most frequent Stata commands that are helpful in preparing the panel data for Stata. Note that xt commands require data in long form; therefore, use reshape long command to convert from wide to long form as shown in Figs. 6.5 and 6.6.

There are some other useful Stata commands to understand the nature of data in the Stata. Note however, that these commands work after data set *xtset*.

Note that if one gets the following error after using xtset: *varlist: country: string variable not allowed,* one need to convert "panel_id" to numeric, type: encode country, gen(panel_id1). Then use "panel_id1" instead of "country" in the xtset command.

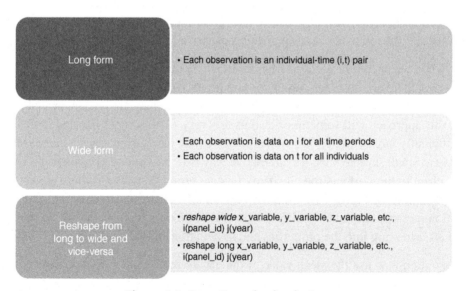

Figure 6.5: Data Organization in Stata.
Authors' selection and compilation based on STATA Manual.

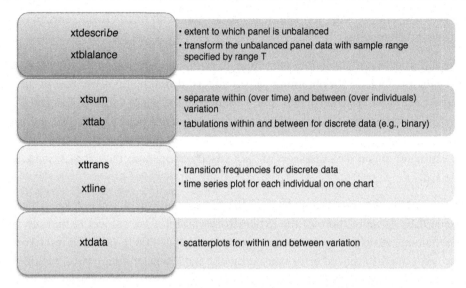

Figure 6.6: Understanding the Nature of Data with Stata.
Authors' selection and compilation based on STATA Manual.

2.3.1 Variable transformation into logs and first differences

To model the data appropriately, it is important to use the best possible functional form. There are several functional forms available, based on the nature of the available data and the research question one has in mind. However, most frequently, the log-log model is used, either in time series or cross-section or panel data. There are two main advantages in using the log-log model: (1) It reduces data variation, (2) the estimated parameters can directly be interpreted as elasticities. This is very helpful for policy makers to understand the behavior of the dependent variable, if there is one percent change in the explanatory variable. However, in some cases, data may also contain negative values. In that case, using the log transformation is not possible. A solution to such problems may be to use the negative variables expressed as a percentage of some other variables that are not already added in the estimated function. Otherwise, the problem of endogeneity may be injected among the variables.

Furthermore, if T is very large, one may face the issue of nonstationarity of the variable in the panel structure. Just like in a time series, if a panel series is non-stationary, and it is used for estimation, the obtained estimates may not be reliable, because they will produce a spurious regression. For example, for $T \rightarrow$ infinity and $N \rightarrow$ finite, Entorf (1997) found that the nonsense regression phenomenon holds for spurious fixed effects models. Also, inference based on t-values can be highly misleading, when studying the spurious fixed effects' regressions, whereas the true model involves independent random walks with and without drifts. Thus, when dealing with non-stationary panel data with fixed effects, one may go ahead with first differencing.

Even if the differencing of panel data over time, to eliminate a time-constant unobserved effect, is a valuable method for obtaining causal effects, differencing is not free of difficulties. For example, even if we have sufficient time variation in the panel variable, the first-differenced (FD) estimation can be subject to serious biases. Wooldridge (2002, section 11.1) document despite having more time periods, generally does not reduce the inconsistency in the FD estimator, when the regressors are not strictly exogenous (say, if y_i, t_1 is included among the x_{itj}).

Furthermore, another important drawback to the FD estimator is that it can be worse than pooled OLS, if one or more of the explanatory variables are subject to measurement error. Differencing some poorly measured regressors reduces their variation relative to its correlation with the differenced error caused by classical measurement error, resulting in a potentially sizable bias. Solving such problems can be very difficult [see section 15.8 and Wooldridge (2002, chapter 11)]. Furthermore, as we know that panel data sets are most useful when controlling for time-constant unobserved features—of people, firms, cities, and so on—which we think might be correlated with the explanatory variables in our model. One way to remove the unobserved effect is to take differences of data in adjacent time periods. Then, a standard OLS analysis on the differences can be used. Using two periods of data results in a cross-sectional regression of the differenced data. The usual inference procedures are asymptotically valid under homoskedasticity; exact inference is available under normality. For more than two time periods, we can use pooled OLS on the differenced data; we lose the first time period because of the differencing. In addition to homoskedasticity, we must assume that the differenced errors are serially uncorrelated to apply the usual t and F statistics.

2.3.2 Panel data estimations

In this section we mention few other useful Stata commands that are commonly used in panel data estimations. In Figs. 6.7–6.9, commands are presented with a brief introduction, so that the reader/researcher can follow. However, additional details are given in the Stata books/ material.

2.3.3 Interpretation of the cointegration equation

It seems worthwhile devoting a sub-section on the interpretation of the cointegration equation since this book is addressed to EGN researchers with various econometrics backgrounds. As far as causality equations are concerned, we will not devote any space in this chapter, because there is a whole chapter in this book devoted to causality (the interested reader should refer to Chapter 9). Data in the EGN are typically transformed in logarithm values in order to represent elasticities. We provide two case studies for cointegration equations forms.

1. • **xtreg** Fixed-, between-, and random-effects, and population-averaged linear models
2. • **xtregar** Fixed-and random-effects linear models with an AR(1) disturbance xtgls Panel-data models by using GLS
3. • **xtpcse** Linear regression with panel-corrected standard errors
4. • **xthtaylor** Hausman–Taylor estimator for error-components models
5. • **xtfrontier** Stochastic frontier models for panel data
6. • **xtrc** Random-coefficients regression
7. • **xtivreg** Instrumental variables and two-stage least squares for panel-data models

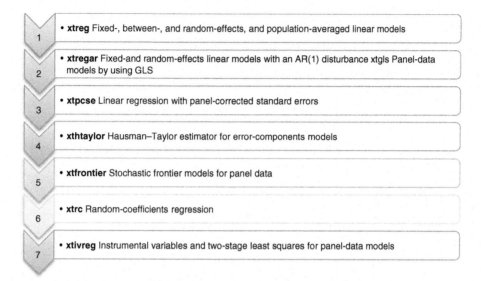

Figure 6.7: Linear Regression Estimators.
Authors' selection and compilation based on STATA Manual.

1. • **xtunitroot:** Panel-data unit-root tests
2. • **xtunitroot breitung:** TheBreitung (2000) panel unit root/stationarity test
3. • **xtunitroot ht:** The Harris & Tzavalis (1999) panel unit root/stationarity test
4. • **ipshin:** The Im, Pesaran & Shin (1997, 2003) panel unit root/stationarity test
5. • **levinlin:** The Levin, Lin & Chu (1992, 2002) panel unit root/stationarity test
6. • **xtfisher:** TheMaddala & Wu (1999) panel unit root/stationarity test
   pescadf: ThePesaran (2006) CIPS panel unit root/stationarity test
7. • **xtpanicca:** The Bai and Ng (2004) PANIC PURT along with Westerlund and Reese (2016) PANICCA PURT

Figure 6.8: Panel Unit-root Tests Estimators.
Authors' selection and compilation based on STATA Manual.

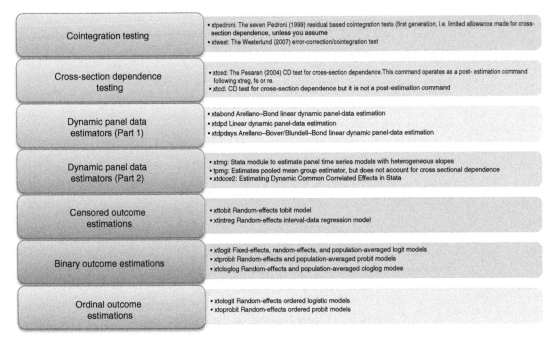

Figure 6.9: A Comprehensive Selection of Stata Commands for Cointegration, Cross Section Dependence, Dynamic Panel, Censored Outcome, Binary and Ordinal Outcome.
Authors' selection and compilation based on STATA Manual.

Case study 6: A Log Cointegration Equation

Apergis and Payne (2009) have studied energy consumption and economic growth for eleven Commonwealth independent countries for the years 1991–2005. The FMLS (Fully modified OLS) cointegration equation is as follows (numbers in parentheses are t-stats):

$$Y = \underset{(4.62)}{0.761} + \underset{(4.73)}{0.42E} + \underset{(5.89)}{012K} + \underset{(5.17)}{0.67L}, \text{Adj } R^2 = 0.68$$

Based on the above results, 1% increase in energy use increases real GDP by 0.42%. 1% increase in capital, increases real GDP by 0.12%, 1% increase in labor force increases real GDP by 0.67%.

Source: Apergis and Payne (2009).

Sometimes, variables contain negative values, which cannot be transformed into logarithms. In that case, we adopt semilogarithmic transformations.

Case study 7: A Semi-Log Cointegration Equation

Menegaki and Tiwari (2017) have estimated the following long-run equation for the Index of Sustainable Economic Growth-ISEW (as a proxy for GDP) and renewable energy use for a sample of twenty American countries in the period 1990–2013.

$$ISEW\ per\ capita = \frac{-126,935.60}{(0.000)*} + \frac{1959.18K}{(0.000)*} + \frac{49.78CO_2}{(0.931)} - \frac{8807.23L}{(0.000)*}$$

$$- \frac{2303.44E}{(0.000)*} + \frac{3984.43RES}{(0.000)*} + \frac{371.67R}{(0.032)*} - \frac{1395.26T}{(0.000)*}$$

where E: energy use per capita, RES: renewable energy use as % of total energy, K: gross capital formation, L: labor, CO_2: carbon dioxide emissions, R: rents, T: trade

Since all the variables, except for the ISEW per capita, had been expressed in logarithms, the coefficients cannot represent elasticities. The reason, they have not used logarithms in the ISEW per capita, is that for some countries and for some periods, the ISEW turned negative. If they insisted on taking logarithms in the ISEW per capita, that would mean they would have to drop observations and hence the data span would be much shorter.

Given the semilog formulation of the equation, dividing each coefficient with the mean ISEW produces elasticities. Thus, given that our main concern is energy-related variables, 1% increase in E the ISEW will cause a 0.72% decrease in the ISEW, and 1% increase in RES will cause 1.24% increase in the ISEW.

Source: Menegaki and Tiwari (2017).

2.4 Principal Component Analysis as a Method of the Variable Number Reduction

It may be that a multitude of variables could be used in the EGN equation and all of them are important. However, not all of them can be included in the analysis due to their multitude. In this case, principal component analysis can narrow their number by keeping the ones that are most eloquent. Principal component analysis is a statistical method, which "compresses" data and reduces dimensionality. The eigenvectors from the covariance matrix are ordered from the highest to the lowest and their transpose is multiplied on the left with the original dataset.

Case study 8: A Principal Component Analysis in the Tourism-Growth Nexus

Shahzad et al. (2017) examined a bivariate tourism—growth nexus for ten countries from 1990Q1 to 2015Q2. Given that the volume of tourist flows is measured by the total number of tourist arrivals, international tourist receipts, and international tourist expenditure, the simultaneous use of the three of them in the tourism-growth model would inject multicollinearity. Through the principal component analysis, a weighted index of the three variables is calculated.

Based on the information from Table 6.1, the eigenvalue of the first principal component exceeds unity, which is evidence for the relevance of the first factor. The eigenvalues of the 2nd and 3rd principal components are not reported by authors, because they are below 1. With the exception of Mexico, the first principal component explains more than 80% of total variability for all countries and it can be considered as a good summary indicator of tourism activity. Part 2 of the table contains the factor loadings of the first principal component. The three standard tourism variables enter the 1st principal component with the same weight for all countries.

Part 3 of the table contains the correlations between the weighted index and each of the three variables individually. There is a high correlation (above 75%) between the weighted index and the three common tourist indicators for all countries.

Source: Shahzad, S.J.H., Shahbaz, M., Ferrer, R., Kumar, R.R., 2017. Tourism-led growth hypothesis in the top ten tourist destinations: new evidence using the quantile-on-quantile approach. Tourism Manage. 60 (June), 223–232.

Table 6.1: PCA results for the weighted tourism activity index.

Country	Eigenvalue	Part 1 Proportion Explained	Part 2 Factor Loadings Arrivals	Receipts	Expenditure	Part 3 Correlation with Weighted Tourism Index Arrivals	Receipts	Expenditure
China	2.678	0.893	0.576	0.608	0.546	0.777	0.959	0.685
France	2.436	0.812	0.520	0.634	0.572	0.766	0.996	0.895
Germany	2.843	0.948	0.572	0.590	0.570	0.909	0.976	0.991
Italy	2.524	0.841	0.545	0.569	0.616	0.680	0.946	0.920
Mexico	2.277	0.759	0.444	0.618	0.648	0.521	0.962	0.980
Russia	2.627	0.876	0.550	0.603	0.577	0.680	0.925	0.986
Spain	2.802	0.934	0.560	0.588	0.583	0.820	0.981	0.992
Turkey	2.929	0.976	0.580	0.580	0.572	0.966	0.967	0.981
UK	2.796	0.932	0.591	0.574	0.567	0.881	0.773	0.989
USA	2.861	0.954	0.577	0.586	0.569	0.850	0.921	0.980

In essence, the principal component analysis solves the problem of multicollinearity among the explanatory variables. In time series analysis, one might use the dynamic principal component method to obtain a particular index form from a set of collinear variables. One should be cautious with the normal principal component method, which does not consider the properties of the time series data. Also, one has to check whether the series are stationary or non-stationary. For example, one may use the normal principal component method available in various econometrics software, if the series are stationary. However, if series is non-stationary, one has to consider the dynamic factor models (deJong's diffuse Kalman filter to provide for the transition model in the state space form).

Now the question remains how to test the multicollinearity in the time series and panel data. In the time series framework, the Variance Inflation Factor (VIF) is the mostly used test and multicollinearity is regarded as present, if the VIF exceeds the value 10 (as documented in the Gujarati (textbook)).

As far as the question of how to test multicollinearity in the panel data is concerned, as a rule of thumb, if correlation is close to 0.4, this is acknowledged to be high and there may be evidence of multicollinearity between the variables. One suggestion could be to use the simple VIF (without accounting for country specific effects) to test if there is evidence of multicollinearity. And if there is evidence of multicollinearity, one may use the three way

PCA (e.g, parafac, Tucker 3, etc.) to construct the index. However, one needs to be cautious for the implementation of appropriate analysis. In order to perform the PCA, there are certain requirements that the data series must be stationary (i.e., without deterministic or stochastic trends), they should be of a comparable range of variation (i.e., have similar means and volatilities), and they should be defined over a common range of dates (i.e., either annual or quarterly or weekly etc. but not with mixed frequencies).

We reproduce the steps for principal component analysis on an unbalanced panel with mixed frequencies (Extract from page 40 of *BIS Quarterly Review*, March 2013):

1) The first step is to test the stationarity of the series using normal or advanced unit root tests.
2) If series are non-stationary, we apply first differences to make series stationary.
3) Then, we normalize all the series by dividing them by their standard deviation.
4) The missing observations may be filled by applying the EM algorithm (also known as maximum likelihood approach) proposed by Stock and Watson (2002). Note that in the case of mixed frequencies, missing values are generated. Since some series are annual and some quarterly, one can extrapolate quarterly series beyond their observed range to obtain the required data format.
5) The algorithm is embedded in the process of the estimation of the Principal Components (PCs), which comprises two steps. The first step involves the linear projection (regression) of those variables with missing observations, on a balanced panel of PCs estimated on the basis of the quarterly series observed over the entire sample period. This projection is used in the second step to fill in the missing observations before a new set of PCs is estimated on the basis of the complete and projected series. The procedure is repeated until the process converges, namely the subsequent estimates of PCs are sufficiently close between iterations. As prescribed in Stock and Watson (2002), the details of the algorithm are slightly different depending on whether the interpolated series refers to a time constant or time varying variable, and whether it is in levels or first differences.
6) In the final step, the balanced panel of variables at a high frequency together with a one-period lag of high frequency can be used to calculate a final set of factors that can be used in the forecasting exercise for the real variables. Stock and Watson (2002) argue that the inclusion of a one-period lag, can go some way towards capturing the time dynamics of the financial variables in the estimated factors.

3 Part 2: Variable Selection With Bayesian Analysis

3.1 Introduction to Bayesian Inference

The use of statistical models to explain the random behavior of samples is an essential ingredient of many inferential procedures. For instance, a sensible model to explain energy consumption in different countries (the sample), could be a Gaussian model with a mean given by a linear combination of exogenous variables and an unknown variance. These models depend on unknown parameters (e.g., regression parameters) that contain the information about the population and constitute the aim of the inference. Mathematically, a model defines the likelihood function, which is a central piece common to all predominant statistical paradigms.

Within the Bayesian paradigm, the parameters are treated as random and inferences are obtained conditioned on what has been observed: the sample. This way, the uncertainty is explicitly acknowledged. It stems from the unknown value of the parameters, which are updated once the data is collected. This learning process is implemented by means of the very well-known Bayes' Theorem, resulting in the posterior distribution, which is a density function encapsulating all the available information about the parameters. In this setting, performing inferences is just a matter of conveniently summarizing the posterior distribution. For instance, one can analyze the impact of private or public investment over energy consumption, through the posterior distribution of the difference between the corresponding regression parameters.

3.1.1 Prior information, data, and posterior inference

Combining the "I know something" with "what I observe" to update our knowledge is done in the Bayesian framework, by means of the Bayes' Theorem, which has a naive interpretation in terms of the conditional probability of an event A given any other event B in Eq. (6.1):

$$P(A|B) = \frac{P(B|A)P(A)}{P(B)} \tag{6.1}$$

It also serves as a tool to update our knowledge about a set of unknown parameters $\theta\,(\in \Theta)$ after a data set D is observed.

$$\pi(\theta|D) = \frac{L_D(\theta)\pi(\theta)}{P(B)} \tag{6.2}$$

The elements of Eq. (6.2) are the following: The likelihood of θ given the observed data set D: $L_D(\theta)$. This function summarizes the available information in the data about the unknown parameters. The prior distribution for θ: $\pi(\theta)$. A probability or density function, which represents the previous knowledge about θ. The marginal likelihood of the data P(D), which is calculated by integrating out parameters θ as in Eq. (6.3):

$$P(D) = \int_\Theta L_\Theta(\theta)\pi(\theta)d\theta \tag{6.3}$$

Notice that, for a given data set, this represents a constant, which allows the posterior distribution $\pi(\theta|D)$ to integrate to 1. Finally, the posterior distribution $\pi(\theta|D)$ contains all the available knowledge about the parameters after data was observed. The process is identical for any scenario, irrespective of the complexity of the model or the number of the unknown parameters. However, there are two big challenges to complete this process: (1) the definition of the prior distribution and (2) the calculation of the marginal likelihood, $P(D)$.

With respect to the prior distribution, it is not simple, not even for an expert, to translate personal or experimental thoughts into a probability or density function, particularly when a complete lack of prior knowledge is the case. Many papers and books have been devoted to the elicitation of priors: Berger (1985), Berger et al. (2009), Garthwaite et al. (2005), Ibrahim and Chen (2000), Jeffreys (1961), Kadane and Wolfson (1998), and Moala and O'Hagan (2010).

The other big challenge which made Bayesian statistics Unaffordable for many years is the calculation of the marginal likelihood P(D). Notice that the numerator in the Bayes' Theorem is computationally easy to obtain. However, if it does not represent any known density for θ, integrating it out to obtain P(D) can be quite difficult or even unfeasible. This undesirable situation was overcome thanks to the development of stochastic integration, in particular to Markov chain Monte Carlo (MCMC) methods such as Gibbs sampling or Metropolis Hasting algorithms (Carlin and Chib, 1995; Gamerman and Lopes, 2006; Gelfand, 2000; Gelfand and Smith, 1990; Geman and Geman, 1984). Moreover, the evolution of technology and the rise of the processing capacity of computers placed Bayesian statistics again in the spotlight of research (Robert and Casella, 2011).

3.1.2 Literature review: Bayesian statistics in the EGN context

Despite the rising relevance of the Bayesian methodology, it seems that it has not yet been introduced in the context of the EGN. In fact a literature search shows that the appearance of the word "Bayes" in this area is reduced to a number of papers using the Schwarz criterion (also known as Bayesian Information Criteria or BIC, Schwarz, 1978) to assess the performance of models estimated following a frequency perspective (Wolde-Rufael, 2006). Only the paper by Camarero et al. (2015) adopts a fully Bayesian point of view for selecting energy variables related to growth.

It is worth mentioning that, in general, Variable Selection is an important but difficult to solve problem, as we will see in Section 3.2.3 of this chapter, for which great attention has been placed from a Bayesian perspective as shown (in an economic framework) in the papers by Fernández et al. (2001a,b), Sala-I-Martin et al. (2004), and Moral-Benito (2010).

In the following sections we will review the ideas of estimation, prediction, and hypothesis testing under the Bayesian paradigm paying particular attention to the problem of variable selection in the EGN context.

3.2 Bayesian Estimation and Prediction

3.2.1 Bayesian estimation, credible intervals, and its interpretation

Whenever we adopt a Bayesian approach, our final knowledge about any parameter in the model is summarized by a complete probability distribution, the posterior distribution. In this sense, we can use all the posterior information to estimate θ. In particular, we can obtain point estimates using the expected value or the median of the posterior distribution or interval estimates from the variance, or directly as $P(a < \theta < b \mid D) = \alpha$.

This can be done even if the posterior distribution cannot be analytically derived. We can obtain a random sample of it using Markov Chain Monte Carlo (MCMC) methods. Indeed if $\pi(\theta|D)$ is known then the expected value is shown in Eq. (6.4).

$$E(\theta \mid D) = \int \theta \pi (\theta \mid D) d\theta \tag{6.4}$$

while, if we have a posterior sample we can approximate it as shown in Eq. (6.5):

$$E(\theta \mid D) = \approx \frac{1}{M} \sum_{m-1}^{M} \theta^{(m)} \tag{6.5}$$

Similarly we can obtain the posterior variance as shown in Eq. (6.6):

$$Var(\theta \mid D) = E(\theta^2 \mid D) - E(\theta \mid D)^2 \tag{6.6}$$

Or, from a posterior sample as in Eq. (6.7)

$$Var(\theta \mid D) \approx \frac{1}{M} \sum_{m-1}^{M} \left[\theta^{(m)} \right]^2 - \left(\frac{1}{M} \sum_{m-1}^{M} \theta^{(m)} \right)^2 \tag{6.7}$$

We can also approximate the posterior quantiles from a sample just by obtaining the empirical quantiles of that sample and use them to give an interval over the true value of θ. This type of intervals under the Bayesian reasoning are called credible intervals and differ from the classical confidence intervals in their interpretation. When we specify a confidence interval with a given probability, that value is the probability of the interval containing the true value of the parameters, while for a credible interval, it refers to the probability of the parameter being in that interval. This subtle but important difference in the interpretation can lead to incorrect decisions as exposed in the editorial by Trafimow and Marks (2015).

Example 1. Suppose that we are interested in the mean μ of a normally distributed variable $Y \sim N(\mu, \sigma^2)$ with a known σ^2. The classical approach gives the mean of the data $\bar{y}$ as a point estimate, while the confidence interval for a significance level of α can be obtained using the $1 - \alpha/2$ quantiles of a standard normal as in Eq. (6.8):

$$\left(\bar{y} \pm z_{(1-a/2)} \, \sigma/\sqrt{n}\right) \tag{6.8}$$

Exactly the same results can be obtained from the Bayesian approach, using a noninformative prior over μ, which admits that any real value is equally plausible. This is: Prior distribution: $\pi(\mu) = $ Constant, for $-\infty < \mu < \infty$, Likelihood function for μ using data $D = \{y_1,...,y_n\}$

$$L_D(\mu) = \prod_{i=1}^{n} N\left(y_i \mid \mu, \sigma^2\right) \tag{6.9}$$

Posterior distribution for μ is given by Eq. (6.10):

$$\pi\left(\mu \mid D = N\left(\mu \mid \bar{y}, \sigma^2/n\right)\right) \tag{6.10}$$

The difference relies, as mentioned above in the interpretation of the results as shown in Table 6.2.

Using the posterior distribution of the parameters, we can also estimate any other quantity of interest, which can be expressed as a function of them. In fact, the expectation of a function of θ, $h(\theta)$ is shown in Eq. (6.11):

$$E\left(h(\theta) \mid D\right) = \int h(\theta)\pi\left(\theta \mid D\right)d\theta \tag{6.11}$$

or, if using a sample of the posterior shown in Eq. (6.12):

$$E\left(h(\theta) \mid D\right) = \int h(\theta)\pi\left(\theta \mid D\right)d\theta \approx \frac{1}{M}\sum_{m=1}^{M} h\left(\theta^{(m)}\right) \tag{6.12}$$

3.2.2 Posterior predictive inference

Again, when it comes to the prediction of a new observation in a setting of unknown parameters, the Bayesian approach considers the uncertainty of the parameters through the

Table 6.2: Difference in the interpretation of results.

	Classical Inference	Bayesian Inference
Point estimates	$E\left(\bar{Y}\right) = \mu$	$E\left(\mu \mid D\right) = \bar{y}$
Intervals	$P\left(\mu \ni I\left(\bar{Y}\right)\right) = \alpha$	$P9\left(\mu \in I(\bar{y}) \mid D\right) = \alpha$

posterior distribution by using a weighted average of the distribution of a new observation under all possible values of the parameters, as shown in Eq. (6.13):

$$\pi(y_{new} \mid D) = \int f(y_{new} \mid \theta)\pi(\theta \mid D)d\theta \qquad (6.13)$$

When the posterior distribution of θ is available only through a MCMC sample, the posterior predictive density can be sampled as prediction of a new observation and can be done by sampling new observations y_{new} from $f(y_{new} \mid \theta^{(m)})$ for $m = 1, \cdots, M$.

3.2.3 Hypothesis testing

As Hilborn and Mangel (1997) stated in their book entitled "The Ecological Detective": "Science is a process for learning about nature in which competing ideas about how the world works are evaluated against observations. Because our description of the world is almost always incomplete and our measurements involve uncertainty and inaccuracy, we require methods for assessing the concordance of the competing ideas and the observations."

Each idea or hypothesis mentioned above can be translated into a mathematical expression or model, which will never be exactly true and so, uncertainty should be considered. The idea of uncertainty and inaccuracy reflected by Hilborn and Mangel (1997) accords with the fact that, in most situations and even if we have a large set of potential hypotheses/models, all of them are possibly wrong. Hence, the question is: Does it make sense to compare among models that we do not trust? In this sense Wasserman (2000) gave a very nice example: "Newtonian physics and general relativity are both wrong. Yet, it makes sense to compare the relative evidence in favor of one or the other. Our conclusion would be that, under the tentative working hypothesis that one of these two theories is correct, we find that the evidence strongly favors general relativity."

So, selecting among hypotheses/models is important and we should do it correctly. From a Classical perspective, the usual methodology involves the selection between two competing hypotheses (the null and the alternative) based on the well-known p-values. However this methodology has some problems. One of them is directly related to the use of p-values whose interpretation is not as direct as it looks like (Hubbard and Bayarri, 2003). A recent discussion about this topic can be found in Wasserstein and Lazar (2016). The other issue has to do with the number of the competing hypotheses. Whenever we consider more than two hypotheses, classical statistics need an error control better known as multiplicity control (Bonferroni, 1936; Dunnett, 1955; Holm, 1979; Tukey, 1949).

Moreover, every type of hypothesis test needs a specific statistic (a function of data) and requires some applicability conditions for the sample distribution of this statistic to be known. All these issues are overcome by the use of Bayesian methodology, where p-values and ad hoc tests are no longer used, and multiplicity control can be automatically introduced (Scott and Berger, 2010). Let us introduce now the procedure for hypothesis testing from a Bayesian point of view, which is based on the use of Bayesian factors.

3.2.4 Bayesian Hypothesis testing and Bayes factors

Consider a classical hypothesis test setting with two competing theories H_0 and H_1 about a given scientific problem. Following the Bayesian setup, we can give prior probabilities to $H_0(H_1)$ and then actualize them following the Bayes' Theorem to obtain posterior probabilities of the hypothesis, given the data as shown in Eq. (6.14):

$$P\left(H_j \mid D\right) \propto L_D(H_j)P(H_j) \tag{6.14}$$

where $L_D(H_j) = p\left(D \mid H_j\right)$ for $j = 0,1$

Then, we can summarize this information as the posterior odds indicating how much more probable one hypothesis than the other is. In this calculation, we observe that prior odds are multiplied by a quantity depending on the information available in the data which is usually known as Bayes factors shown in Eq. (6.15) (for a review see Kass and Raftery, 1995).

$$\underbrace{\frac{P\left(H_1 \mid D\right)}{P\left(H_0 \mid D\right)}}_{\text{Posterior odds}} = \underbrace{\frac{L_D(H_1)}{L_D(H_0)}}_{\text{Factor Bayes, } B_{10}} \times \underbrace{\frac{P(H_1)}{P(H_0)}}_{\text{Prior odds}} \tag{6.15}$$

Hence B_{10} indicates the evidence available in data in favor of H_1 and against H_0. Following Jeffreys (1961), this measure can be quantified as shown in Table 6.3.

Notice that using Bayes factors in this context, it is equivalent to selecting between the two hypotheses based on their posterior probabilities. This can be generalized to the case where more than two hypotheses are considered. This is a framework better known as Model selection. In this sense, and from now, on we will use the term "model" instead of "hypothesis" and the notation M instead of H.

Table 6.3: Quantification of the evidence in favor H_1.

B_{10}	Evidence against H_0
1–3	Small
3–20	Positive
20–150	Strong
>150	Very strong

Source: Authors.

3.2.5 An extension of hypothesis testing
3.2.5.1 Model selection

When our interest is to select among more than two models Bayes factors can still be defined in a 2 by 2 basis. That is, we will define the Bayes Factor in favor of model i and against model j as shown in Eq. (6.16):

$$B_{ij} = \frac{L_D(M_i)}{L_D(M_j)} \qquad (6.16)$$

If one of the models in the set of all possible ones, M, is the true model, by the Bayes' Theorem, the posterior probabilities can easily be expressed in terms of Bayes factors as shown in Eq. (6.17):

$$P\left(M_j \mid y\right) = \left(\sum_{i=1}^{m} B_{ij} \frac{P(M_i)}{P(M_j)}\right)^{-1} \qquad (6.17)$$

Since Bayes factors are transitive in the sense that $B_{ij} = B_{il}B_{lj}$ for any model M_l, the usual approach to compare $m > 2$ models consists in comparing each model with a fixed one, M_d, which we call the base model. Hence, since $B_{ij} = B_{ji}^{-1}$ we can write what is shown in Eq. (6.18):

$$P\left(M_j \mid y\right) = \frac{B_j P(M_j)}{\sum_{l=1}^{m} B_{ld} P(M_l)} \qquad (6.18)$$

Notice that for this equation to be true, we implicitly assume that one of the models is true. This requirement is often used as an argument against Bayes factors, because this is not always the case. However García-Donato (2003) showed that, even when the true model is not within M, the evidence in favor of a model, given some data, is always proportional to its posterior probability (for which Bayes factors are essential ingredients). Also, Dmochowski (1996) showed that, in this same scenario, the Bayes factors select the closest model to the true one in Kullback-Leibler sense (Kullback, 1999). Wasserman (2000) also advocated the use of posterior probabilities for comparing the relative evidence of models, even when they cannot be considered "true".

Although this approach seems theoretically simple, the calculation of Bayes factors is challenging. One of the main issues is the calculation of the marginal likelihood under each model i, $L_D(M_i)$. This calculation can be simple if M_i does not involve unknown parameters, but it is not so when it does. Indeed if model M_i is related to a density function $f(\cdot \mid \theta_i)$, the marginal likelihood requires for the elicitation of a prior distribution for θ_i, under model M_i, the corresponding (possibly multivariate and not analytical) integral shown as Eq. (6.19).

$$L_D(M_i) = \int_{\Theta_i} f(D \mid \theta) \pi_i(\theta_i) d\theta_i \qquad (6.19)$$

To make it more complicated, the corresponding Bayes factor is very sensible to the elicitation of $\pi_i(\theta_i)$, which cannot be simply set as a default noninformative prior (as we usually do when doing estimation and/or prediction) (Kass, 1993; Kass and Greenhouse, 1989; Kass and Raftery, 1995).

A lot of effort has been put in the recent years to throw some light to this issue with many resulting papers covering the topic (Bayarri et al., 2012; Berger and Pericchi, 2001; Liang et al., 2008; O'Hagan, 1995). In the next section, we explore the solution given by Bayarri et al. (2012) for a particular case of model selection and variable selection. The other challenge in the calculation of posterior probabilities over the model space has to do with the elicitation of priors $P(M_i)$ for $i = 1,...,m$. A smart choice can help to avoid multiplicity issues as mentioned by Scott and Berger (2010). Again the next section illustrates this choice in the specific context of variable selection. Once that $P(M_i|D)$ for $i = 1,...,m$, have been obtained, we can proceed by selecting the model with larger posterior probability or by predicting quantities of interest based on a weighted average of all the potential models. This last methodology is known as Bayesian Model Averaging (Hoeting et al., 1999).

3.2.6 A particular problem: variable selection

A particular and very relevant problem in model selection is variable selection, where the main goal is to understand which of p potential variables are related to the process of interest. This is translated into a model selection problem by considering a model for each possible subset of the covariates initially considered. Therefore, the cardinality of the model space is 2^p. In the particular case of a normal response variable, each competing model M_i for $i = 0,..., 2^p - 1$ relates the response variable to a subset of k_i covariates, such as shown in Eq. (6.20):

$$y = \alpha_0 X_0 + X_i \beta_i + \varepsilon, \quad \varepsilon \sim N_n\left(0, \sigma^2 I\right) \tag{6.20}$$

where y is the n dimensional vector of observations for the response variable; X_0 is a $n \times k_i$ design matrix of fixed covariates (present in all models) with associated vector of linear regressors α_0; X_i is the $X \times K_i$ design matrix containing potential covariates with β_i the k_i vector of linear regressors. Notice that X_0 can be just the intercept or may contain a lagged effect of a previous observation y_{-1}. This last option is particularly useful in economic contexts where any observation may depend on the previous scenario. Finally, ε is a white noise error. For computing posterior probabilities, we set the base model in Eq. (6.18) as the simplest one containing only the fixed part. We will abuse notation slightly and we denote this model by M_0. Hence the posterior probability of each model can be computed as shown in Eq. (6.21):

$$P\left(M_i \mid y\right) = \frac{B_{i0} P(M_i)}{\sum_{l=1}^m B_{l0} P(M_l)} \tag{6.21}$$

where the corresponding B_{i0} involves the calculation of the marginal likelihoods as in Eq. (6.19), namely $L_D(M_i)$ and $L_D(M_0)$. As M_i and M_0 involve parameters it is mandatory to define priors distributions $\pi_i(\theta_i)$ (for $i = 1,...,m$) and $\pi_0(\theta_0)$ with $\theta_0 = (\sigma, \alpha_0)$ and $\theta_i = (\theta_0, \beta_i)$. Following this notation, we refer to θ_0 as the common parameters and to β_i as the extra parameters. These prior distributions for the model-specific parameters are the most problematic element in the whole setting and many papers have been written on this topic (Liang et al., 2008; Zellner, 1986; Zellner and Siow, 1980, 1984).

3.2.7 The robust prior

Bayarri et al. (2012) adopted a new perspective to assign the prior density based on a list of criteria that should be fulfilled to solve a variable selection problem. The authors then use these criteria to propose a specific prior distribution over the parametric space, which has been proven to provide a reliable theoretical result at a relatively low computational cost. This prior, known as the Robust prior, is shown in Eq. (6.22):

$$\pi_j^R(\beta_0, \beta_i, \sigma) = \pi(\beta_0, \sigma) \times \pi_i^R(\beta_i \mid \beta_0, \sigma) = \sigma^{-1} \times \int_0^\infty N_{k_i}\left(\beta_i \mid 0, g \sum i\right) p_j^R(g) dg \quad (6.22)$$

where $\sum_i = \text{Cov}\left(\hat{\beta}_i\right) = \sigma^2 \left(V_i^t V_i\right)^{-1}$ is the covariance of the maximum likelihood estimator of β_i with $V_i = \left(I_n - X_0\left(X_0^t X_0\right)^{-1} X_0^t\right) X_i$ and $p_i^R(g)$ is given by Eq. (6.23):

$$P_i^R(g) = \frac{1}{2}\sqrt{\frac{1+n}{k_i + k_0}}(g+1)^{-3/2}, g \in \left(\frac{1+n}{k_i + k_0} - 1, \infty\right) \quad (6.23)$$

and zero otherwise. In Eq. (6.23), k_0 denotes the number of fixed covariates. Despite its complicated appearance, the main advantage of this prior, apart from its reliable theoretical properties, is that it provides marginal densities in an analytic way (i.e., integral which can be solved algebraically), which is an important computational advantage.

3.2.8 Prior over the model space

The other element in Eq. (6.21) is the prior probability of models. As mentioned earlier, Scott and Berger (2010) showed how these probabilities can serve as a multiplicity control tool, which is particularly needed in scenarios, like variable selection, with huge model spaces. Indeed, Scott and Berger (2010) argued that it is important to account for the increasing number of models and hence for the difficulty of detecting influential covariates when p and, presumably the background noise, grow larger. As Scott and Berger (2010) pointed out, a standard practice in variable selection is to assign a probability q to each variable being in the model, and consider their inclusion in a model as exchangeable Bernoulli trials. That is shown in Eq. (6.24).

$$P(M_i \mid q) = q^{k_i}(1-q)^{q-k_i}]$$ (6.24)

A fixed value of q (independent of p) does not control for multiplicity. For instance, selecting $q = 1/2$ gives the same results as giving an equal prior probability to each model. Scott and Berger (2010) showed that treating q as an unknown parameter and allowing learning from data, results in an automatic penalty for multiplicity. Choosing a uniform prior for q in (11), and integrating it out resulted into Eq. (6.25) (Scott and Berger, 2010).

$$P(M_i) = \frac{1}{p+1}\binom{p}{k_i}^{-1}$$ (6.25)

Note that the prior in Eq. (6.25) is equivalent to assessing a uniform prior to each dimension k, that is $P(k) = 1/(p+1)$ for $k = 0,...,p$, and then dividing this probability equally among the $\binom{p}{k}$ models of dimension k. It is interesting to remark that Scott and Berger (2010) proposal resulted in marginal prior inclusion probability of 1/2 for each variable, the same as the one for the constant prior $P(M_i)$, but the behavior is very different due to the way of apportioning the probability among models.

3.2.9 Posterior inferences in variable selection

Once that all needed elements have been defined and posterior probabilities for each model are available, it is important to summarize this information in the proper way. Notice that giving all these posterior probabilities is not always an option, for instance, if the number of competing models is large. As described in Section 2.2, when we want to do estimation or prediction, we simply report posterior means or medians, posterior standard deviations, and credible intervals. But our scenario is different. In model selection, the explored space is discrete, without any possible ordering, and so these summaries are neither appropriate nor well-defined. Then, one can report the highest posterior probability model (HMM) and its posterior probability but, again, if p is moderate to large, this probability is small and, maybe, very close to the probability of other models.

An interesting solution comes from the intrinsic nature of the problem. Recall that we are trying to understand which are the relevant variables in the study. Hence, reporting a probability of each variable to be in the true model, is a good summary of our posterior information. These probabilities are known as posterior inclusion probabilities (PIP) and can be obtained by summing over the posterior probabilities of all the models containing that covariate. This is shown in Eq. (6.26):

$$p(x_i \mid y) = \sum_{\{M_l : x_j \in M_l\}} P\pi(M_l \mid y), i = 1, 2, \ldots p \qquad (6.26)$$

Inclusion probabilities have a number of theoretical properties studied in Barbieri and Berger (2004). The authors showed that the model containing all the variables with a PIP over 0.5 (named after Median Probability Model, MPM) can be regarded, under general conditions, as the better predictive model. It is worth mentioning that when p is very large, assuming higher than 30, it becomes almost impossible to compute each of the 2^{30} in number, posterior probabilities. In those cases, a method for sampling from the model space is required and the approximated probabilities reported. A good review with a discussion about those sampling methods can be found in García-Donato and Martínez-Beneito (2013). In this paper, the authors also showed that the most stable posterior reported in those cases are the PIPs.

3.2.10 BayesVarSel an R package for Bayesian Variable Selection

To implement the described variable selection approach, we use the R package BayesVarSel. In particular, we use the function GibbsBvs to obtain approximations of the PIPs of covariates, based on the methodology proposed in García-Donato and Martínez-Beneito (2013). Other R packages for doing Bayesian Variable Selection can be found as is shown in Forte et al. (2017) but BayesVarSel is selected here since it is the only one implementing the Robust prior in Eq. (6.22).

3.3 Selecting Influential Variables in the EGN

The problem of variable selection becomes of great relevance in the EGN. In this context, different economic theories arise stating whether energy and growth are related variables or not. The adoption of one theory or the other has direct policy implications as pointed out in Camarero et al. (2015). Departing from the classical discussion about the direction of the relationship (which variable is the cause and which one the consequence), our concern is about if the relationship exists or not. An abundance of empirical literature has attempted to address this issue over the last 30 years, beginning with the seminal paper by Kraft and Kraft (1978) (Belke et al., 2011; Johansen, 1991; Sims, 1972).

But the evidence about the EGN is mixed as pointed out in Özturk (2010), Payne (2010), and Coers and Sanders (2013). The main reasons given in the literature for these discrepancies are the application of a variety of econometric approaches, the heterogeneity of the countries analyzed, and the differences in the time span of the samples. Additionally, certain authors argue that the main factors explaining the mixed evidence are the limitations of considering a bivariate approach just relating energy and growth followed by a complete overlooking of other variables, which may play a role in this complex scenario. These omitted variables, which should act as control variables for understanding the real EGN are, in the best case, selected ad-hoc, that is without any statistical motivation. Here is where variable selection can contribute allowing for the selection of control variables, as well as for establishing the incorporation of energy as a covariate when considering growth as an exogenous variable. The following sections are devoted to explore an example in this context. Specifically we consider a part of the study realized in Camarero et al. (2015). Next, we describe the scenario considered and present the results.

3.3.1 Data and sources

We consider panel data covering the period from 1949 to 2010 in the United States and aggregated among all economic sectors [for a separate study by sector we refer the reader to the original paper by Camarero et al. (2015)]. United States was chosen for two reasons: first, for the availability of data for both a longer time span and for a large set of related variables and due to sector disaggregation; second, it is responsible for one of the largest world shares of pollutant emissions.

The main covariates considered are energy consumption EC and GDP, where GDP is used as a proxy for economic growth and is taken to be the exogenous dependent variable. A lag of this dependent variable is considered as a fixed covariate in the model to account for its dynamic nature and for avoiding serial correlation in the results as pointed out by Keele and Kelly (2006). Finally, a total of 32 variables have been considered to address the omitted variable bias. These are variables previously used in the literature and which are available in the case of the United States at an aggregated level, as well as additional variables that we consider suitable for capturing the aforementioned multiple transmission channels. The data and their sources are described in Table 6.4.

3.3.2 Model and variable selection

For this study we consider the model presented in Eq. (6.20) with X_0 containing the lag of the GDP and the intercept and X_i being a subset of the covariates presented in Table 6.3. Of course, departures from the normality of the residuals can be an issue but, with the data in this chapter, we did not observe severe violations of such assumptions. Generally, when normality is the main concern, the recent study by Maruyama and Strawderman (2012) was quite revealing, since it theoretically demonstrated that, in a framework similar to ours, the Bayes factors are independent of the assumed distribution of ε, as long as they are spherically symmetric (a large family of distributions). This intuitively points to the conclusion that the results presented in this chapter are quite robust to the Gaussian hypothesis. Within this model, our goal is to give posterior inclusion probabilities of all these 32 covariates, together with the inclusion probability of EC. Notice that, as mentioned above, the large number of potential covariates makes it mandatory to sample in the model space and thus, the posterior probabilities provided would be approximated. Results are shown in the following section.

Table 6.4: Variables, measurement, and data source.

Variables	Measurement	Data Source
Growth	Real = VA/VAPI (millions $US)	US Bureau of Economic Analysis (https://www.bea.gov/)
Employment (EMP)	Full time and part time employees (millions)	US Bureau of Economic Analysis (https://www.bea.gov/)
Energy Consumption (EC)	Billion BTU	US Energy Information Administration (https://www.eia.gov/)
Consumption of: Total energy non-renewable (TNR), of Total Renewable Energy (TR)	Billion BTU	US Energy Information Administration (https://www.eia.gov/)
Energy Prices: Natural gas price (NG_P), Coal Price (C_P)	NG_P: Natural Gas Wellhead, Price C_P: $US per short ton. All the prices are in constant (2005) $US, calculated by using GDP implicit price deflators.	US Energy Information Administration (https://www.eia.gov/)
Oil Price (O_P)	Real oil price (in $US/bbl.) Prices are based on historical free-market (stripper) prices of Illinois Crude as presented by Illinois Oil and Gas Association (IOGA). Prices are adjusted for inflation to December 2012 prices, using the Consumer Price Index (CPI-U) as presented by the Bureau of Labor Statistics.	https://inflationdata.com/Inflation_Rate/Historical_Oil_Prices_Table.asp
US Bureau of Economic Analysis (https://www.bea.gov/)	Government Spending (Real). Total spending-total ($US/bbl.) 2005	https://www.spending.com/usgovernmentspending_chart_1940_2017USk_13s11i011men_F0t
US Bureau of Economic Analysis (https://www.bea.gov/) Fixed Investment (FI), No Residential Investment (NR), Structure Investment, Equipment & Software Investment (ESI), Residential Investment (R), Public Investment (IPU), Private Investment (PI), Structure Investment (SI), Total Investment (IT)	Investment in Fixed Assets and Consumer Durable Goods ($US/bbl.)	US Bureau of Economic Analysis (https://www.bea.gov/)
Money Supply (RMO) Energy Intensity (EIN)	Real money. Reserve Assets, SDR millions. Primary energy (billion BTU)/GDP in billions of constant 2005 $US.	OCDE Primary Energy Consumption: EIA US Energy Information Administration (https://www.eia.gov/). DP: US Bureau of Economic Analysis (https://www.bea.gov/)

Variable	Definition	Source
Energy Efficiency (EEF)	GDP in billions of constant 2005 $US/primary energy consumption (billion BTU)	Primary Energy Consumption: EIA US Energy Information Administration https://www.eia.gov/ GDP: US Bureau of Economic Analysis (https://www.bea.gov/)
Source of energy production: (COAL), Natural Gas (GAS), Crude oil (OIL), Natural Gas Plant Liquids (NGPL), Nuclear (NUC)	Total energy production (Billion BTU)	https://www.eia.gov/
Consumer Price Index (CPI)	All urban consumers (CPI-U), US city average 1982–84 = 100	US Department of Labor Bureau of Labor Statistics https://www.bls.gov/data/
Business Sector Productivity (B_P), Nonfarm business sector productivity (NF_P), Nonfinancial corporate sector productivity (NFI_P)	Output per hour. Type of measure: Index, base year 2005 = 100	
Exports: Goods exports (X_G), Services Exports (X_S), Imports: Goods Imports (M_G), Services Imports (M_S)	Output per hour. Type of measure: Index, base year 2005 = 100. Millions of $US, seasonally adjusted.	US Bureau of Economic Analysis (https://www.bea.gov/)

bbl, Barrels of oil; *SDR*, special drawing rights.
Source: Authors' compilation.

3.3.3 Results

Based on the paper of Barbieri and Berger (2004), we believe that researchers who want to model growth should take into account all the variables with an associated probability greater than 0.5. We should note that the main objective of this chapter is not to interpret all the critical variables, as that would require further study, but rather to help researchers evaluate which variables are fundamental in explaining growth, and provide a guide to selecting the most relevant variables. The reader may notice that the results presented in this chapter slightly differ from the ones in Camarero et al. (2015). This is due to a change in the prior distribution over the model space, which is now taken to be the prior in Scott and Berger (2010). We decided to change priors after we gain a better knowledge of the behavior of these priors and the good performance of the Scott and Berger prior to controlling for multiplicity.

3.3.3.1 Aggregate growth results

Concerning the aggregate growth, our results confirm the importance of energy consumption (EC) in explaining US aggregate growth, given that it has a posterior inclusion probability of 0.83. Therefore, the application of our probabilistic model shows EC and growth to be highly correlated, highlighting the energy dependence, which is the main issue raised in the literature. The fact that EC is a significant explanatory variable of growth, can be interpreted in favor of the growth hypothesis. However, EC is not included as an endogenous variable in our model, and thus it is not possible to test or to reject the feedback hypothesis.

Concerning the role of the potential control variables, our study demonstrates that only certain candidate variables explain aggregate growth. Table 6.5 provides strong evidence for the inclusion of energy intensity (with probability 1), nuclear power (with probability 0.89), public spending (with probability 0.89) and energy efficiency (with probability 0.79).

According to our probabilistic model, the variable with the highest probability of explaining growth is the energy intensity (EIN). Historically, total the US primary energy consumption has been growing at a similar rate as economic activity. Present day energy consumption continues to increase (with this trend set to continue according to AEO, 2010), but at a slower rate than economic activity. This implies that there has been a progressive improvement in the US energy intensity ratio. Two factors may be responsible: first, the larger share of services

Table 6.5: Aggregate analysis-posterior inclusion probabilities larger than 0.5.

	Incl. prob.
Ln (EC)	0.8349
Ln (EIN)	1.0000
Ln (NUC)	0.8935
Ln (SPE)	0.8892
EEF	0.7908

Source: Authors

in growth and, second, the increase in efficiency in other, more energy intense sectors. Our methodology has been able to capture the direct link that exists between energy intensity and growth. An alternative interpretation of energy intensity is the rate of output return achieved by energy consumption (i.e. energy efficiency [EEF]). As economies develop, they tend to improve the energy efficiency of their industrial sectors; however, higher living standards imply more energy-consuming human activities, as shown in the study by Corless (2005) that analyzed the top 40 largest national economies (GDP) by plotting GDP per capita against energy efficiency.

We found that nuclear power (NUC) has the highest probability of inclusion after energy intensity (EIN). This is not surprising considering that the United States is the country with the largest installed nuclear power capacity: approximately 20% of the total amount of electricity generated comes from nuclear reactors. Since 1951, when the first reactors were installed, nuclear power has had a predominant role in the US energy mix1. The uncertainty with respect to oil and gas reserves, together with the scarcity of renewable energy has increased the relative importance of nuclear power. According to the IEA, a nuclear energy contribution of approximately 3.8 trillion kilowatt hours is expected in 2030, in contrast to a contribution of 2.7 trillion kilowatt hours in 2006. Apergis and Payne (2009) have argued that nuclear energy plays a crucial role in the design of environmental strategies. This energy source can address the needs of countries with a rapidly growing energy demand.

The next explanatory variable with a high probability, as shown in Table 6.4, is public spending (or SPE). There is no discussion in the literature regarding the crucial role that fiscal policies play in the output growth of a country. The debate only concerns the cyclical or counter-cyclical nature of public spending. We find that this is one of the variables with a higher probability (0.8862) in explaining aggregate growth.

4 Concluding Remarks

This chapter hosted two parts related with data analysis in the EGN. The first part had a more practical advice orientation, while the second focused on Bayesian data analysis and it was more technical. Next, we provide conclusions separately for each part.

4.1 Part 1

The first part of the chapter gives condensed knowledge of the most important caveats in data analysis in the EGN. Without following the flow of a typical econometrics textbook, nor aiming to replace it either, this part gives valuable key knowledge for data sources, data organization, and data analysis in the EGN. It summarizes a comprehensive list of STATA commands and provides various case studies that equip readers and researchers with a "hands on" experience. This part is ideal both for new and experienced researchers, because they have the opportunity to acquire fundamental data analysis knowledge, gathered in a single outlet—this chapter part.

4.2 Part 2

A robust statistical approach is of great importance for selecting the variables that explain growth. Moreover, it is mandatory to perform this variable selection process if we want to understand the relationship between energy and growth without ignoring the effect of other related covariates. Inclusion probabilities of variables are a good summary of their role in the process and should be obtained prior to the performance of cointegration or causality tests. Still, this has been neglected in the empirical literature.

A limitation in the methodology used here, and as with any model selection technique, is that no model-specific parameters are estimated. Hence, we can say for instance that residential investment affects growth, but we cannot specify the magnitude of that effect. To the best of our knowledge, this is still an open question in the field of model selection with only partial answers [an interesting exception being the study in Scott and Berger (2006) within a context much simpler than ours]. Nevertheless, this limitation is not a drawback in this study, since our main motivation is the identification of variables that affect growth. It could, however, prove problematic for other researchers intending to apply this methodology.

Our results are twofold. First, the empirical evidence confirms the prior expectation that energy consumption is a critical variable to understanding the path of growth. Second, the results highlight the importance of the disaggregate analysis of economic activity, because the relevant explanatory variables are not the same for the different sectors under study, namely, the commercial sector, and transport and industry. In fact, nuclear energy production and employment are fairly relevant for only two sector outputs, but for these sectors are quite critical variables. Finally, the results reveal the complexity of policy making: the interaction found among the group of variables considered in this chapter indicates that policy makers not only have to design policies that focus on reducing energy consumption, but must also take into account other important macro variables. This complexity is further compounded by the sector differences that prevent the design of an overall policy.

References

Annual Energy Outlook (AEO), 2010. US Energy Information Administration, USA. Available from: www.eia.gov/aeo.
Apergis, N., Payne, J.E., 2009. Energy consumption and economic growth: evidence from the Commonwealth of Independent States. Energy Econ. 31 (5), 641–647.
Baltagi, 2005. Econometric Analysis of Panel Data, third ed., USA.
Barbieri, M.M., Berger, J.O., 2004. Optimal predictive model selection. Ann. Stat. 32 (3), 870–897.

Bayarri, J.O., Berger, A.F., García-Donato, G., 2012. Criteria for Bayesian model choice with application to variable selection. Ann. Stat. 40 (3), 1550–1577.

Becker, R., Enders, W., Lee, J., 2006. A stationarity test in the presence of an unknown number of smooth breaks. J. Time Ser. Anal. 27 (3), 381–409.

Belke, A., Dobnik, F., Dreger, C., 2011. Energy consumption and economic growth: new insights into the cointegration relationship. Energy Econ. 33, 782–789.

Berger, J.O., 1985. Statistical Decision Theory and Bayesian Analysis, 2nd ed. Springer, New York, NY, USA.

Berger, J.O., Bernardo, J.M., Sun, D., 2009. The formal definition of reference priors. Ann. Stat. 37 (2), 905–938.

Berger, J.O., Pericchi, L.R., 2001. Objective Bayesian methods for model selection: introduction and comparison. Lect. Notes Monogr. Ser. 38 (3), 135–207.

Bonferroni, C.E., 1936. Teoria statistica delle classi e calcolo delle probabilità. R Istituto Superiore di Scienze Economiche e Commerciali di Firenze.

Camarero, M., Forte, A., García-Donato, G., Mendoza, Y., Javier Ordoñez, J., 2015. Variable selection in the analysis of energy consumption-growth nexus. Energy Econ. 52 (Part A), 207–216.

Carlin, B.P., Chib, S., 1995. Bayesian model choice via Markov chain Monte Carlo methods. J. R. Stat. Soc. Series B Methodol. 57 (3), 473–484.

Carrion-i-Silvestre, J.L., Del Barrio-Castro, T., Lopez-Bazo, E., 2005. Breaking the panels: an application to GDP per capita. J. Econom. 8, 159–175.

Coers, R., Sanders, M., 2013. The energy-GDP nexus: addressing an old question with new methods. Energy Econ. 36, 708–715.

Corless, P., 2005. Analysis of top 40 largest national economies (gdp) by plotting GDP per capita vs. energy efficiency (GDP per million btus consumed); an inverse examination of energy intensity. Available from: http://en.wikipedia.org/wiki/File:Gdp-energy-efficiency.jpg.

Dmochowski, J., 1996. Intrinsic priors via Kullbakc-Leibler geometry. In: Bernardo, J.M., DeGroot, M.H., Lindley, D.V. (Eds.), Bayesian Statistics 5. Oxford University Press, London, pp. 543–549.

Dunnett, C.W., 1955. A multiple comparison procedure for comparing several treatments with a control. J. Am. Stat. Assoc. 50 (272), 1096–1121.

Entorf, H., 1997. Random walks with drifts: nonsense regression and spurious fixed-effect estimation. J. Econom. 80 (2), 287–296.

Feenstra, R.C., Inklaar, R., Timmer, M.P., 2015. The next generation of the Penn World Table. Am. Econ. Rev. 105 (10), 3150–3182, (available at: www.ggdc.net/pwt).

Fernández, C., Ley, E., Steel, M.F., 2001a. Benchmark priors for Bayesian model averaging. J. Econom. 100, 381–427.

Fernández, C., Ley, E., Steel, M.F., 2001b. Model uncertainty in cross-country growth regressions. J. Appl. Econom. 16, 563–576.

Forte, A., García-Donato, G., Steel, M.F., 2017. Methods and tools for Bayesian variable selection and model averaging in univariate linear regression. Technical report, arXiv:1612.02357v1 [stat.CO].

Gamerman, D., Lopes, H.F., 2006. Markov Chain Monte Carlo: Stochastic Simulation for Bayesian Inference, 2nd ed. Chapman & Hall/CRC, London, UK, (ISBN 9781584885870).

García-Donato, G., 2003. Factores Bayes y Factores Bayes Convencionales: Algunos Aspectos Relevantes. PhD thesis, Universidad de Valencia.

García-Donato, G., Martínez-Beneito, M.A., 2013. On sampling strategies in Bayesian variable selection problems with large model spaces. J. Am. Stat. Assoc. 108 (501), 340–352.

Garthwaite, P.H., Kadane, J.B., O'Hagan, A., 2005. Statistical methods for eliciting probability distributions. J. Am. Stat. Assoc. 100 (470), 680–701.

Gelfand, A.E., Smith, A.F.M., 1990. Sampling-based approaches to calculating marginal densities. J. Am. Stat. Assoc. 85 (410), 398–409.

Gelfand, E., 2000. Gibbs sampling. J. Am. Stat. Assoc. 95 (452), 1300–1304.

Geman, S., Geman, D., 1984. Stochastic relaxation, Gibbs distributions, and the Bayesian restoration of images. IEEE Trans. Pattern Anal. Mach. Intell., PAMI 6 (6), 721–741.

Green, 2010. Econometric Analysis, fifth ed., USA.

Gujarati, 2015. Basic Econometrics, fourth ed., USA.

Hilborn, R., Mangel, M., 1997. The Ecological Detective. Princeton University Press, Princeton, New Jersey, pp. 336.

Hoeting, J.A., Madigan, D., Raftery, A.E., Volinsky, C.T., 1999. Bayesian model averaging: a tutorial. Stat. Sci. 14 (4), 382–401.

Holm, S., 1979. A simple sequentially rejective multiple test procedure. Scand. J. Stat. 6 (2), 65–70.

Hubbard, R., Bayarri, M.J., 2003. Confusion over measures of evidence (p's) versus errors (α's) in classical statistical testing. Am. Stat. 57 (3), 171–178.

Ibrahim, J.G., Chen, M.-H., 2000. Power prior distributions for regression models. Stat. Sci. 15 (1), 46–60.

Jeffreys, H., 1961. Theory of Probability, 3rd ed. Oxford University Press, Oxford.

Johansen, S., 1991. Estimation and hypothesis testing of cointegration vectors in Gaussian vector autoregressive models. Econometrica 59, 1551–1580.

Kadane, J., Wolfson, L.J., 1998. Experiences in elicitation. J. R. Stat. Soc. Series B Stat. Methodol. 47 (1), 3–19.

Kass, R.E., 1993. Bayes factors in practice. J. Royal Stat. Soc. Ser. D 42, 551–560.

Kass, R.E., Greenhouse, J.B., 1989. Comment on investigating therapies of potentially great benefit: Ecmo by ware. Stat. Sci. 4, 310–317.

Kass, R.E., Raftery, A.E., 1995. Bayes factors. J. Am. Stat. Assoc. 90 (430), 773–795.

Keele, L.J., Kelly, N.J., 2006. Dynamic models for dynamic theories: the ins and outs of LDVs. Polit. Anal. 14 (2), 186–205.

Kraft, J., Kraft, A., 1978. On the relationship between energy and GNP. Energy Dev. 3, 401–403.

Kullback, S., 1999. Information Theory and Statistics. Dover, New York.

Lee, C.-C., Lee, J.-D., 2009. Energy prices, multiple structural breaks, and efficient market hypothesis. Appl. Energy 86, 466–479.

Lemmens, A., Croux, C., Dekimpe, M.G., 2008. Measuring and testing Granger causality over the spectrum: an application to European production expectation surveys. Int. J. Forecast. 24 (3), 414–431.

Liang, F., Paulo, R., Molina, G., Clyde, M.A., Berger, J.O., 2008. Mixtures of g- priors for Bayesian variable selection. J. Am. Stat. Assoc. 103 (481), 410–423.

Liu, R., Narayan, P.K., 2010. Are shocks to commodity prices persistent? Appl. Energy 88, 409–416.

Maruyama, Y., Strawderman, W.E., 2012. Bayesian predictive densities for linear regression models under α-divergence loss: some results and open problems. IMS Collections 8, 42–56.

Maslyuk, S., Smyth, R., 2008. Unit root properties of crude oil spot and futures prices. Energy Policy 36 (7), 2591–2600.

Menegaki, A.N., Özturk, I., 2013. Growth and energy consumption in Europe revisited: evidence from a fixed effects political economy model. Energy Policy 61, 881–887.

Menegaki, A.N., Tiwari, A.K., 2017. The index of sustainable economic welfare in the energy-growth nexus for American countries. Ecol. Indic. 72, 494–509.

Moala, F.A., O'Hagan, A., 2010. Elicitation of multivariate prior distributions: a nonparametric Bayesian approach. J. Stat. Plan. Inference 140 (7), 1635–1655.

Moral-Benito, E., 2010. Determinants of economic growth: a Bayesian panel data approach. Working Papers 1031, Banco de España;Working Papers Homepage. Available from: https://ideas.repec.org/p/bde/wpaper/1031.html.

Narayan, P.K., Popp, S., 2010. A new unit root test with two structural breaks in level and slope at unknown time. J. Appl. Stat. 37 (9), 1425–1438.

O'Hagan, A., 1995. Fractional Bayes factors for model comparison (with discussion). J. R. Stat. Soc. Series B 57. 99–138.

Özturk, S., 2010. A literature survey on energy-growth nexus. Energy Policy 38, 340–349.

Payne, J.E., 2010. Survey of the international evidence on the causal relationship between energy consumption and growth. J. Econ. Stud. 37, 53–95.

Robert, C., Casella, G., 2011. A short history of Markov chain Monte Carlo: subjective recollections from incomplete data. Stat. Sci. 26 (1), 102–115.

Sala-I-Martin, X., Doppelhofer, G., Miller, R.I., 2004. Determinants of long-term growth: a Bayesian averaging of classical estimates (BACE) approach. Am. Econ. Rev. 94 (4), 813–835.

Salisu, A.A., Fasanya, I.O., 2013. Modeling oil price volatility with structural breaks. Energy Policy 52, 554–562.

Schwarz, G., 1978. Estimating the dimension of a model. Ann. Stat. 6 (2), 461–464.

Scott, J.G., Berger, J.O., 2006. An exploration of aspects of Bayesian multiple testing. J. Stat. Plan. Inference 136, 2144–2162, (MR2235051).

Scott, J.G., Berger, J.O., 2010. Bayes and empirical-Bayes multiplicity adjustment in the variable-selection problem. Ann. Stat. 38 (5), 2587–2619.

Shahzad, S.J.H., Shahbaz, M., Ferrer, R., Kumar, R.R., 2017. Tourism-led growth hypothesis in the top ten tourist destinations: new evidence using the quantile-on-quantile approach. Tour. Manage 60 (June), 223–232.

Sims, C.A., 1972. Money, income, and causality. Am. Econ. Rev. 62, 540–552.

Stock, J.H., Watson, M.W., 2002. Forecasting using principal components from a large number of predictors. J. Am. Stat. Assoc. 97 (460), 1167–1179.

Tiwari, A.K., 2014. The frequency domain causality analysis between energy consumption and income in the United States. Economia Aplicada 18 (1), 51–67.

Tiwari, A.K., Albulescu, C., 2016. Renewable-to-total electricity consumption ratio: estimating the permanent or transitory fluctuations based on flexible Fourier stationarity and unit root tests. Renew. Sustain. Energy Rev. 57 (May), 1409–1427.

Trafimow, D., Marks, M., 2015. Editorial. Basic Appl. Soc. Psychol. 37 (1), 1–2.

Tukey, J.W., 1949. Comparing individual means in the analysis of variance. Biometrics 5 (2), 99–114.

Wasserman, L., 2000. Bayesian model selection and model averaging. J. Math. Psychol. 44 (1), 103.

Wasserstein, R.L., Lazar, N.A., 2016. The ASA's statement on p-values: context, process, and purpose. Am. Stat. 70 (2), 129–133.

Wolde-Rufael, Y., 2006. Electricity consumption and economic growth: a time series experience for 17 African countries. Energy Policy 34 (10), 1106–1114.

Wooldridge, 2002. Econometric Analysis of Cross Section and Panel Data. The MIT Press, Cambridge, Massachusetts. London, England.

Zellner, A., 1986. On assessing prior distributions and Bayesian regression analysis with g-prior distributions. In: Zellner, A. (Ed.), Bayesian Inference and Decision Techniques: Essays in Honor of Bruno de Finetti. Edward Elgar Publishing Limited, Amsterdam, pp. 389–399.

Zellner, A., Siow, A., 1980. Posterior odds ratio for selected regression hypotheses. Bernardo, J.M., DeGroot, M.H., Lindley, D.V., Smith, A.F.M. (Eds.), Bayesian Statistics, 1, Valencia University Press, Valencia, pp. 585–603.

Zellner, A., Siow, A., 1984. Basic Issues in Econometrics. University of Chicago Press, Chicago.

Further Readings

Lee, J., Strazicich, M.C., 2003. Minimum Lagrange multiplier unit root test with two structural breaks. Rev. Econ. Stat. 85, 1082–1089.

Lee, J., Strazicich, M.C., 2004. Minimum LM unit root test with one structural break. Working Paper no. 04–17, Department of Economics, Appalachian State University.

Marques, A., Fuinhas, J., Menegaki, A.N., 2014. Interactions between electricity generation sources and economic activity in Greece: a VECM approach. Appl. Energy 132, 24–46.

Narayan, P.K., Liu, R., 2011. Are shocks to commodity prices persistent? Appl. Energy 88 (1), 409–416.

Siedler, T., Schupp, J., Spiess, K., Wagner, G.G., 2009. The German socio-economic panel (SOEP) as reference data set. Schmollers Jahrbuch, 129, Duncker & Humblot, Berlin, pp. 367–374.

Tiwari, A.K., Albulescu, C.T., 2016. Oil price and exchange rate in India: fresh evidence from continuous wavelet approach and asymmetric, multi-horizon Granger-causality tests. Appl. Energy 179, 272–283.

Yıldırıma, S., Özdemira, B.K., Doğana, B., 2013. Is there a persistent inflation in OECD energy prices? Evidence from panel unit root tests. Procedia Econ. Fin. 5, 809–818.

Current Issues in Time-Series Analysis for the Energy-Growth Nexus (EGN); Asymmetries and Nonlinearities Case Study: Pakistan

Muhammad Shahbaz

Montpellier Business School, Montpellier, France

Chapter Outline

1 Introduction

The EGN has come under intense scrutiny in the last four decades (Apergis and Tang, 2013; Belke et al., 2011; Chontanawat et al., 2008; Jumbe, 2004; Khan et al., 2014; Menegaki, 2014; Ozturk, 2010; Payne, 2009; Shahbaz et al., 2017a). This issue is important because energy consumption drives the wheels of economic growth. A significant and sharp increase in the demand for energy can be attributed to the following factors: (1) the promotion of economic growth (this applies particularly for the emerging economies) and (2) the maintenance of living standards (this applies particularly for the developed nations). A considerable number of empirical studies investigating the EGN shed light on four different hypotheses (Growth, Conservation, Feedback, and Neutrality[1]). This ambiguity in empirical results may be due to

---

[1] To save space, a detailed presentation of the four hypotheses is provided in Chapter 1.

the asymmetries existent in the EGN. This reveals the complexity of the relationship between energy consumption and economic growth. These asymmetries are the outcome of a complex economic system, which generates the macroeconomic variables under consideration in the EGN. It is very important to take them into account, if we want the EGN analysis to generate reliable empirical results, helpful in designing comprehensive economic policies for the maintenance of the long-run economic growth.

1.1 The EGN in Pakistan

Pakistan has implemented numerous economic reforms in fiscal, external affairs, and energy sectors over time with the aim to maintain its macroeconomic performance. These reforms have not only affected its macroeconomic performance, but also created the ground for possible asymmetries in the variables' trends. In turn, these may affect the association between energy consumption and economic growth. Under such circumstances, the linear empirical investigation provides inconclusive empirical results (Shahbaz et al., 2017a). This implies the importance of recognizing asymmetries, while investigating the relationship between energy consumption and economic growth.

This chapter contributes to the existing time-series EGN literature in the following three ways:

1. The augmented production function is employed to investigate the association between energy consumption and economic growth by considering oil prices, capital, and labor using a nonlinear framework.
2. The unit root properties of the energy consumption, economic growth, oil prices, capital, and labor are investigated by applying linear and nonlinear unit root tests.
3. Finally, the NARDL approach developed by Shin et al. (2014) is applied for examining the nonlinear effect of energy consumption, oil prices, capital, and labor on economic growth.

The current empirical analysis deals with the asymmetric cointegration between economic growth and its determinants. Furthermore, economic growth is positively and negatively affected by the positive and negative shocks in energy consumption. Oil prices affect economic growth negatively and positively through their positive and negative shocks. Positive shocks in capital add to economic growth, but negative shocks in capital reduce it. Also, the contribution of positive and negative shocks of labor onto economic growth can have either an increasing or a decreasing (negative) effect.

The rest of this chapter is organized as follows: Section 2 reviews relevant up-to-date literature. Section 3 hosts data collection, the methodology, and modeling. Section 4 deals with nonlinear autoregressive distributive lag (NARDL) bounds testing approach for symmetric cointegration. Section 5 deals with results and their interpretation. Finally, Section 6 concludes the chapter with policy implications.

2 Literature Review

2.1 The Up-to-Date Evolution of Bivariate Analysis in the EGN

In 1978, Kraft and Kraft (1978) applied a bivariate model to investigate the EGN over the period between 1947 and 1974 for the United States. They found that energy consumption causes real GNP in the Granger sense. Later on, Abosedra and Baghestani (1991) reinvestigated the association between both variables by applying Granger (1969) causality approach. Their findings confirmed the results reported by Kraft and Kraft (1978). Conversely, Zarnikau (1997) applied the Granger (1969) causality test to reexamine the relationship between the energy demand (proxied by the Divisia energy index) and the real GDP growth for the US economy. The results provided support for the feedback effect, that is, the bidirectional causality between both variables. Payne (2008) considered the EGN relationship through the Toda and Yamamoto (1995) causality approach and reported the neutral effect between both variables. The Markov-switching causality approach was applied by Fallahi (2011) to observe the EGN in the case of United States. This model reports the causality results regime-wise, such as in 1971–75, 1977–82, 1989–95, and 2001–05. The empirical exercise revealed support for the feedback effect, that is, the energy use causes growth and vice versa in 1971–75, but no causality is found between both variables for the rest of regimes.

Using Pakistani data, Riaz and Stern (1984) studied the EGN by utilizing energy supply and demand functions. They reported that real GDP growth increases energy use and thus, energy use increases real GDP growth. Aqeel and Butt (2001) used the bivariate model to examine causality between energy sources (coal, petroleum, electricity consumption, and the consumption of natural gas and real GDP growth). They applied Granger (1969) cointegration and Hsiao Granger causality tests. Their results confirmed the existence of cointegration. The empirical exercise of causality unveiled that total energy consumption Granger causes economic growth. Petroleum consumption is the cause of real GDP growth. Neutrality exists between natural gas consumption and economic growth (coal consumption and real GDP growth), while in the short run, causality between electricity consumption and economic growth is bidirectional. Yang (2000) used data on growth in energy use and real GDP growth to investigate the causality relationship between the variables by applying Engle–Granger (1987) cointegration approach. He reported that variables are found to be cointegrated, as well as that a feedback effect exists between energy consumption and economic growth in Taiwan's economy. For the Turkish economy, Altinay and Karagol (2004) examined the causal association between growth in energy consumption and real GDP growth by utilizing Hsiao's version of the Granger causality approach and the Perron (1997) test with structural break points. They indicated the presence of structural breaks, showing the impact of macroeconomic policies on growth in the energy use and real GDP. Their analysis validated the existence of a neutral effect between the variables.

Lee and Chang (2005) used the bivariate model to analyze the relationship between energy use and real GDP growth. They applied tests developed by Perron (1997) and Gregory and Hansen (1996) respectively. Their empirical outcome revealed that cointegration is present. Real GDP growth causes energy use and vice versa, energy use causes real GDP growth in the Granger sense, for the Taiwan's economy. Furthermore, real GDP growth has a positive and significant effect on energy demand. Lee and Chang (2007) applied both linear and nonlinear models to observe the outcome of energy use on domestic output growth in the case of Taiwan. Their results indicated the relationship between energy use and domestic output growth that is of an inverted U-shape. This means that at initial levels of economic development, energy demand is increased with a boost to economic growth and starts declining after a threshold level in real GDP per capita is reached. Furthermore, the linear model confirms the existence of the feedback effect between both variables. Dhungel (2008) examined the cointegration and causality relationships between both variables by using the Johansen and Juselius (1990) cointegration method and the vector error correction model (VECM) Granger causality approaches. The empirical evidence unveiled that both series are linked in the long run and causality is found running from energy to economic growth in Nepal. For Tanzania, Odhiambo (2009) used the ARDL bounds testing to observe the long-run relationship between growth in energy use and income per capita growth. The cointegration between the variables is evidenced once energy and electricity consumption were used as dependent variables. The causality analysis by Granger (1969) approach indicated that total energy demand is the cause of income per capita growth in the Granger sense. Paul and Uddin (2011) looked into the EGN for Bangladesh by applying an innovative accounting approach. Their results showed that growth in real GDP is not led by growth in energy use, but shocks in real GDP growth adversely affect energy use growth. In the case of Indonesia, Arifin and Syahruddin (2011) inspected the causality between energy consumption measured by energy sources (renewable and nonrenewable), and income per capita growth by applying Toda and Yamamoto (1995) Granger causality test. They documented that growth in income per capita is Granger caused by renewable energy.

Using the Pakistan economic data, Liew et al. (2012) examined whether sectoral economic growth leads energy consumption by applying the Johansen and Juselius (1990) and pair-wise Granger causality approaches. They found that the cointegration relationship is valid for the long run. Moreover, their empirical exercise exposed the bidirectional causality between the agricultural growth and energy consumption growth. The neutral effect exists between the industrial growth and energy consumption and a similar finding exists for the services growth and energy consumption. Zaman et al. (2012) explored the impact of total energy consumption from agriculture, industrial, and service sectors using the bivariate models. They applied Johansen and Juselius (1990), error-correction model, and an innovative accounting approach for causality analysis. Their results suggest that total energy consumption has a negative impact on industrial growth, while population and agricultural growth negatively affect total energy use. They further found that the feedback effect is confirmed. Using data of the Croatian's economy, Borozan (2013) explored the relationship between both variables. The results of Vector Auto-Regressive (VAR) Granger causality test revealed that real GDP is Granger caused by total energy use, also confirmed by the impulse response function test.

2.2 The Evolution of the Multivariate Framework in the EGN

We find that the aforementioned studies focused on applying the bivariate model to test the EGN, but ignored the role of capitalization, labor, and other potential variables. This implies that findings of these studies may be biased due to the exclusion of pertinent variables. The aforementioned variables such as capital and labor play an important role in the production function and both are determinants of economic growth and energy consumption. Other variables such as employment, government consumption expenditures, development spending, consumer prices, energy prices, oil consumption, and exchange rate may affect domestic production and energy consumption. For example, Yu et al. (1988) applied Granger (1969) and Sims (1972) causality techniques to investigate the linkages among energy use, total employment, and nonfarm employment. Their empirical evidence showed that energy consumption Granger causes the nonfarm employment (a neutral effect is present between both variables by Sims causality test).

Mahmud (2000) applied the partial equilibrium model to investigate the effect of energy and nonenergy inputs on the manufacturing sector. The findings revealed that shocks in energy prices reduce capital investment and increase the cost of production in the manufacturing sector, while energy and nonenergy inputs are not possible substitutes. In the case of Greece, Hondroyiannis et al. (2002) explored the causality relationship among energy use, energy price, and growth by employing Johansen and Juselius (1990) cointegration for the long run. The VECM Granger causality test is employed for the causal relationship among the series. They found cointegration and that growth causes energy and vice versa, namely that energy causes growth.

Narayan and Smyth (2005) applied the trivariate model to test the causal relationship among energy use, income, and employment by using the Australian time-series data. They found cointegration and noted that employment and income cause energy use in the Granger sense. For Pakistan, Mushtaq et al. (2007) studied the impact of agricultural growth and agricultural energy prices on the agricultural energy consumption (oil, electricity, and natural gas consumption). Their empirical exercise indicated the occurrence of the long-run association and that agricultural growth Granger causes oil and electricity consumption in the agriculture sector. Salim et al. (2008) employed the energy demand function by incorporating energy prices and real GDP growth for Bangladesh, China, India, Malaysia, Thailand, and Pakistan. Their results showed that cointegration is present and energy use

Granger causes real GDP growth and energy prices, in the short run, in Pakistan. Yu et al. (2008) investigated energy demand by incorporating energy prices and per capita income growth in the trivariate framework for the Chinese economy by applying the innovative accounting approach. They found the negative (positive) outcome of energy prices (per capita income growth) on energy consumption. Wesseh and Zoumara (2012) employed energy consumption, employment, and growth to examine their relationship by applying nonparametric bootstrapped causality test in the case of Liberia. They confirmed the presence of cointegration by the ADRL bounds testing analysis. They found out that energy use and employment boost the economic growth process and enhance the domestic production. The bootstrapped causality findings indicated that growth is the cause of employment and that a bidirectional relationship is valid between energy and economic growth in the Granger sense. Adom (2013) used the energy demand function to test the EGN by applying a time-varying approach in the case of Ghana. The results suggested that economic growth adds to energy demand and energy prices reduce it, but the impact varies with regime shifts.

The empirical findings reported by Granger (1969) and Sims (1972) may be biased due to low explanatory power. These standard approaches failed to detect the causality from other channels and provided contradictory findings (Asafu-Adjaye, 2000). Given the limitation and discrepancies in the traditional cointegration and causality tests, Iqbal (1986) applied the Zellner's iterative method to examine the impact of energy consumption (all types of energy), capital, and labor on a manufacturing sector in Pakistan. He noted that energy consumption, capital, and labor are substitutes, while the relationship between natural gas consumption and electricity consumption reveals complementarity. Chishti and Mahmud (1990) reinvestigated the relationship among energy, nonenergy inputs and a large manufacturing sector using an aggregate Divisia index. They reported that energy consumption, capital, and labor are major determinants of growth in the large manufacturing sector. Their empirical evidence unveiled that energy and capital have a complementary relationship, while labor and energy are substitutes. Stern (1993) used the Divisia energy index as a measure of energy consumption to explore the EGN. He used the multivariate framework by including capitalization and employment in the EGN. The empirical evidence indicated that the unidirectional causal relationship runs from energy consumption to economic growth. Stern (2000) incorporated capital and labor as contributing factors to energy consumption and economic growth. The production function is utilized to investigate the EGN in the US economy. He applied Johansen (1991) for the long run and the VECM Granger causality approaches for the causality linkages. The empirical evidence showed the positive effect of the energy consumption, capital, and labor on the output growth, that is, shocks in energy consumption, capital, and labor reduce the output growth and thus, economic growth. The empirical results revealed the neutral effect between energy use and output growth.

Using Pakistani data, Alam and Butt (2002) reinvestigated the direction of causality between both variables by incorporating capital and labor as supplementary determinants of energy use and output growth. They employed Johansen and Juselius (1990) test for cointegration and the VECM for the causality relationship. They noted the existence of cointegration between the series. Their empirical exercise confirmed the feedback effect between both variables. Furthermore, capital causes energy consumption and economic growth and labor causes economic growth in the Granger sense for the short run. Oh and Lee (2004) applied energy demand (income and energy prices) and production (energy consumption, capital, and labor) function in the multivariate framework to verify the EGN in the case of Korea. Their findings indicated that energy use is led by growth in income per capita. Soytas et al. (2007) applied the Toda and Yamamoto (1995) and the variance decomposition approaches to reassess the causal relationship between energy consumption and economic growth by including capital, labor, and carbon emissions in the multivariate regression model for the United States. They found that the neutral effect is validated for energy consumption and economic growth. Similarly, Payne (2009) reinvestigated the causality between energy sources and income per capita growth by applying the Toda and Yamamoto (1995) causality approach. He found that energy sources do not contribute to income per capita growth. Kaplan et al. (2011) reinvestigated the causality between energy and growth using the multivariate versions of energy demand and the neoclassical production functions by adding energy prices, capital, and labor. They applied Johansen and Juselius (1990) for the long run and the VECM Granger test for the causality research. Their findings confirmed cointegration among the variables for both models. Moreover, they found bidirectional causation between the series.

Similarly, Shahiduzzaman and Alam (2012) applied the supply-side production function to look into the EGN by adding capital and labor for Australia. The Johansen and Juselius (1990) test for cointegration as well as the Toda and Yamamoto (1995) Granger causality tests was applied. Their results supported the occurrence of cointegration. They noted that energy use, capital, and labor add to real GDP growth. The findings of Toda and Yamamoto (1995) Granger causality showed that real GDP growth is the cause of energy use and the energy use is the cause of real GDP growth in the Granger sense. Using the US data, Gross (2012) reinvestigated the correlation between energy consumption and economic growth by including energy prices, trade, and capitalization in the multivariate framework. The empirical evidence revealed that economic growth has a positive impact on energy consumption. The feedback effect exists in the short run, but a neutral effect is valid for the long run between both variables and is validated by the VECM Granger causality test. Using Swedish data, Stern and Enflo (2013) assessed the causality between energy (Divisia energy index) and output, using the bivariate and multivariate production functions. They found that the output Granger causes energy consumption, but the reverse is not true by using the bivariate model. In the multivariate model, incorporating capital and labor, their empirical evidence indicated that energy use is the cause of economic growth in Granger sense. They have also used the energy demand function and noted that energy prices and economic growth cause energy use in the Granger sense. Shahbaz et al. (2012) investigated the impact of energy use measures by renewable and nonrenewable energy sources on real GDP growth. They confirmed the long-run association and that the feedback relationship exists between consumption of energy sources and economic growth. Ahmed et al. (2013) used the trivariate framework to examine the relationship between energy consumption and economic growth. Their empirical evidence provides that energy consumption plays an important role by stimulating economic growth. Yildirim et al. (2014) investigated the association between energy consumption and economic growth by using the bivariate framework for N-11 countries. They found that the neutral effect exists between energy consumption and economic growth in Pakistan. Ahmed et al. (2015) revisited the EGN in Pakistan and found that economic growth leads energy consumption. In the case of United States, Arora and Shi (2016) applied the augmented production function for the investigation of linkages between energy consumption and economic growth by adding capital and labor as additional determinants of economic growth. Their empirical results indicate that energy consumption plays a vital role in boosting the economic growth like capital and labor. Shahbaz et al. (2016) augmented the production function by adding financial development as an additional determinant of economic growth and energy consumption. Their empirical evidence reported that financial development strengthens the EGN. For the Turkish economy, Pata and Terzi (2017) noted that energy consumption is the main stimulator of economic growth.

Considering the important role played by asymmetries in the EGN, Arac and Hasanov (2014) have applied linear and nonlinear empirical approaches to examine the asymmetric relationship between energy consumption and economic growth. Their empirical evidence indicates that positive and negative shocks in energy consumption, positively and negatively affect economic growth, but the impact of negative shocks that stems in the energy consumption has a dominant effect over that of a positive shock. Shahbaz et al. (2017a) employed the classical production function to examine the asymmetric association between energy consumption and economic growth by applying the NARDL developed by Shin et al. (2014) for the Indian economy. Their empirical results confirmed the presence of asymmetries and cointegration as well. Economic growth is positively and negatively affected by the negative and positive shocks that occur in energy consumption. Capital (with positive and negative shocks) and labor (through a positive shock) also have a positive effect on economic growth.

3 The Modeling, Data, and Methodology

The inconclusive nature of the production function always provides a motivation to researchers for investigating the relationship between energy consumption and economic growth. In doing so, researchers applied different empirical approaches on the production function by incorporating additional determinants of economic growth, but empirical results on energy-growth are still ambiguous (Shahbaz et al., 2017a). Policy makers are not facilitated in the designing of comprehensive economic and energy policies for sustainable long-run economic growth. Pakistan has been facing an energy crisis for the last two decades and is satisfying domestic energy demand by importing oil.[2] This shows that oil price shocks in international market not only affect the EGN, but also the whole macroeconomic performance in Pakistan. In doing so, we have added oil prices as an additional factor of domestic production affecting economic activity. Oil prices affect economic growth via the supply-side and demand-side channels. The supply-side hypothesis entails that oil prices play a vital role in domestic production and a rise in oil prices increases the cost of production. This rise in the cost of production leads firms to lower output and increase production prices, which also increase inflation (Shahbaz et al., 2017b; Tang et al., 2010). According to the demand-side hypothesis, shocks in the oil prices affect consumption and investment activities. Oil prices slow down economic activity by lowering demand for labor and as a result of that affect real wages. Oil price shocks affect inflation via the cost of production and through that, the economic activity and economic growth are affected too (Ftiti et al., 2016; Shahbaz et al., 2017b). The general form of the augmented production function is modeled as follows:

$$Y_t = f(E_t, OP_t, K_t, L_t) \qquad (7.1)$$

The log-linear specification is used for the empirical purpose, following Shahbaz and Lean (2012). They argued that a log-linear specification is appropriate for attaining efficient and reliable empirical results compared to the simple linear specification. Variables were also transformed into per capita units except for oil prices. Following Shahbaz and Lean, and later on Shahbaz et al. (2017a), all variables were transformed into natural-log. The empirical equation of the augmented production function is modeled as follows:

$$\ln Y_t = \beta_0 + \beta_1 \ln E_t + \beta_2 \ln OP_t + \beta_3 \ln K_t + \beta_4 \ln L_t + \varepsilon_t \qquad (7.2)$$

where ln indicates natural-log, Y_t stands for economic growth measured by real GDP per capita, E_t stands for energy consumption, OP_t stands for oil prices, capital measured by the gross fixed capital formation and labor are shown by K_t and L_t respectively, while ε_t is an error term with a normal distribution.

The study covers the period of 1985Q$_I$–2016Q$_{IV}$. The World Development Indicators (CD-ROM, 2017) is used to collect data on real GDP (in local currency, constant 2010); energy consumption (kg of oil equivalent); gross fixed capital formation (in local currency, constant 2010); and labor force. The data on oil prices are obtained from Pakistan Energy Year Book (2017).[3] Total population is used to convert all the variables into per capita unit except for oil prices.

---

[2] Pakistan is basically an oil-dependent country.
[3] Oil prices transform into real terms by deflating inflation.

4 The NARDL Bounds Testing Approach for Asymmetric Cointegration

The presence of asymmetries in energy consumption, oil prices, capital, labor, and economic growth leads us to apply nonlinear cointegration approach to examine the asymmetric cointegration (long-run relationship) between the variables. In doing so, we choose a multivariate NARDL cointegration test originated by Shin et al. (2014) to examine the long-run relationship between the variables. This approach captures the asymmetries and nonlinearities that stem in time-series data. The NARDL approach differentiates the long-run and short-run asymmetric impact of energy consumption, oil prices, capital, and labor on the economic growth. The VECM or smooth transition model suffers from the convergence problem, if proliferation of estimates exists. The NARDL provides efficient empirical analysis by solving the issue of estimate proliferation. This test does not require that all the variables should be integrated at the same order of integration. The NARDL test is applicable, if all the variables are integrated at I(1), or the variables have a flexible order of integration. In the presence of asymmetries and nonlinearities, the flexibility of the integrating order is important (for more details, see Hoang et al., 2016). This approach also solves the problem of multicollinearity with the help of the appropriate lag-length selection of the variables (Shin et al., 2014). The empirical equation of the production function is modeled in Eq. (7.3) following the NARDL framework introduced by Shin et al. (2014):

$$\Delta Y_t = \alpha_0 + \rho Y_{t-1} + \theta_1^+ E_{t-1}^+ + \theta_2^- E_{t-1}^- + \theta_3^+ OP_{t-1}^+ + \theta_4^- OP_{t-1}^- + \theta_5^+ K_{t-1}^+ + \theta_6^- K_{t-1}^- + \theta_7^+ L_{t-1}^+$$

$$+ \theta_8^- L_{t-1}^- + \sum_{i=1}^{p} \alpha_1 \Delta Y_{t-i} + \sum_{i=0}^{q} \alpha_2 \Delta E_{t-i}^+ + \sum_{i=0}^{q} \alpha_3 \Delta E_{t-i}^- + \sum_{i=0}^{q} \alpha_4 \Delta OP_{t-i}^+ + \sum_{i=0}^{q} \alpha_5 \Delta OP_{t-i}^- \quad (7.3)$$

$$+ \sum_{i=0}^{q} \alpha_6 \Delta K_{t-i}^+ + \sum_{i=0}^{q} \alpha_7 \Delta K_{t-i}^- + \sum_{i=0}^{q} \alpha_7 \Delta L_{t-i}^+ + \sum_{i=0}^{q} \alpha_8 \Delta L_{t-i}^- + D_t + \mu_t$$

where α_i indicates the short-run estimates in Eq. (7.3) and long-run estimates are shown by θ_i with $i = 1, \dots, 8$. This shows that the short-run analysis intends to examine the immediate impacts of exogenous variable changes (i.e., energy consumption, oil prices, capital, and labor) on the dependent variable (i.e., economic growth). On the contrary, time reaction and speed of adjustment toward long-run equilibrium level are measured by the long-run analysis. The Wald test is applied to test for the presence of asymmetries in the long run $\left(\theta = \theta^+ = \theta^-\right)$ and the short run $\left(\alpha = \alpha^+ = \alpha^-\right)$, as well as for all the variables (i.e., energy consumption, oil prices, capital, and labor). Y_t is the economic growth, E_t indicates energy consumption, OP_t is the oil prices, K_t shows capital, and L_t is the labor. We also incorporate D_t as a dummy variable, which captures the effect of the structural break and is determined by Kim and Perron (2009) unit root test. Coefficients p and q are used to show the optimal lag length, not only for the dependent variable (Y_t), but also for the independent variables $\left(E_t, OP_t, K_t, L_t\right)$, employing the Akaike information criterion (AIC), due to its superior explanatory properties. The explanatory variables are decomposed into positive and negative partial sums as shown in Eq. (7.4):

$$x_t^+ = \sum_{j=1}^{t} \Delta x_j^+ = \sum_{j=1}^{t} \max\left(\Delta x_j, 0\right) x_t^- = \sum_{j=1}^{t} \Delta x_j^- = \sum_{j=1}^{t} \min(\Delta x_j, 0), \qquad (7.4)$$

with x_t indicating E_t, OP_t, K_t, and L_t.

We follow the bounds test, that is, the joint test of all the lagged levels of the regressors proposed by Shin et al. (2014), to examine the presence of the long-run cointegration, while accommodating asymmetries. We use two tests, the t-statistic developed by Banerjee et al. (1998) and the F-statistic originated by Pesaran et al. (2001). Using the t-statistic, we follow the null hypothesis: $\theta = 0$ against the alternative hypothesis: $\theta < 0$. The F-statistic follows the null hypothesis: $\theta^+ = \theta^- = \theta = 0$. The rejection of the null hypothesis of no cointegration indicates the presence of the long-run relationship between economic growth and its determinants, and vice versa. The asymmetric estimates for the long run are estimated following the relationships $L_{mi^+} = \theta^+/\rho$ and $L_{mi^-} = \theta^-/\rho$. These estimates for the long run, with respect to the positive and negative shocks in the energy consumption, oil prices, capital, and labor, quantify the association between the variables for the long-run equilibrium. The asymmetric dynamic multiplier effects are measured as shown in Eq. (7.5):

$$m_h^+ = \sum_{j=0}^{h} \frac{\partial Y_{t+j}}{\partial E_t^+}, \quad m_h^- = \sum_{j=0}^{h} \frac{\partial Y_{t+j}}{\partial E_t^-}, \quad m_h^- = \sum_{j=0}^{h} \frac{\partial Y_{t+j}}{\partial OP_t^+}, \quad m_h^- = \sum_{j=0}^{h} \frac{\partial Y_{t+j}}{\partial OP_t^-}, \quad m_h^- = \sum_{j=0}^{h} \frac{\partial Y_{t+j}}{\partial K_t^+},$$

$$m_h^- = \sum_{j=0}^{h} \frac{\partial Y_{t+j}}{\partial K_t^-}, \quad m_h^- = \sum_{j=0}^{h} \frac{\partial Y_{t+j}}{\partial L_t^+}, \quad m_h^- = \sum_{j=0}^{h} \frac{\partial Y_{t+j}}{\partial L_t^-} \quad \text{for } h = 0, 1, 2..... \qquad (7.5)$$

where if $h \to \infty$, then $m_h^+ \to L_{mi^+}$ and $m_h^- \to L_{mi^-}$.

The asymmetric response of economic growth to positive and negative shocks in energy consumption, oil prices, capital, and labor is shown by the dynamic multipliers. These multiplier estimates show the dynamic adjustments from the initial toward the new equilibrium between the variables in the system, following a variation affecting the system.

5 Results and Discussion

Table 7.1 reports the descriptive statistics and pair-wise correlations. The empirical evidence indicates that the volatility in oil prices is larger than the economic growth volatility. Energy consumption is less volatile than the volatility in capitalization and labor has less volatility compared to oil prices, economic growth, energy consumption, and capitalization. The Jarque–Bera test is also applied to test whether the variables follow a normal distribution or not. The results are reported in Table 7.1 and we find that the null hypothesis of a normal distribution is rejected. This implies that the distribution of all the variables is not independent and identical. We consider the distribution to be symmetric, if the data

Table 7.1: Descriptive statistics and pair-wise correlation.

Variables	ln Y_t	ln E_t	ln OP_t	ln K_t	ln L_t
Mean	9.3133	4.6898	2.4793	7.4505	2.6414
Median	9.2623	4.7083	2.2656	7.4203	2.6186
Maximum	9.6063	4.8496	3.3973	7.7528	2.7657
Minimum	9.0066	4.4389	1.4755	7.1917	2.5751
Standard deviation	0.1734	0.1098	0.5619	0.1378	0.0642
Skewness	0.1042	−0.6773	0.2888	0.5815	0.5065
Kurtosis	1.8072	2.4611	1.6564	2.7340	1.7390
Jarque–Bera	7.8186	11.3371	11.4077	7.5935	13.9529
Probability	0.0200	0.0034	0.0033	0.0224	0.0009
ln Y_t	1.0000				
ln E_t	0.4492	1.0000			
ln OP_t	−0.1244	0.2588	1.0000		
ln K_t	0.4863	0.2106	0.2729	1.0000	
ln L_t	0.0069	0.2307	0.2609	−0.1085	1.0000

distribution provides a bell-shaped curve. The results provided for skewness and kurtosis show the presence of a potential asymmetry in the distribution of time-series data. This leads us to apply the asymmetric ARDL-modeling (NARDL) for empirical analysis rather than symmetric ARDL-modeling. The NARDL approach to cointegration is helpful in solving the issue of nonnormality by capturing the presence of asymmetries stemming in time-series data (Shin et al., 2014). The pair-wise correlation analysis reveals the positive correlation between energy consumption and economic growth. Oil prices are inversely correlated with economic growth. A positive correlation exists between capital (or labor) and economic growth. Oil prices, capital, and labor are positively correlated with energy consumption. A positive correlation of capital and labor occurs with oil prices, but labor is negatively correlated with capital.

Traditional unit root tests such as the DF (Dickey and Fuller, 1979), the ADF (Dickey and Fuller, 1981), the PP (Phillips and Perron, 1988), the KPSS (Kwiatkowski et al., 1992), the ADF–GLS (Elliot et al., 1996) and the N–P (Ng and Perron, 2001) may provide ambiguous empirical results. These unit root tests may tend toward acceptance of the null hypothesis when it is false and vice versa, due to their weak explanatory properties. Furthermore, these unit root tests ignore the importance of structural breaks occurring in time-series data. The presence of structural breaks in the series may cause traditional unit root tests to provide vague empirical results. This issue is solved by applying the Z–A unit root test (Zivot and Andrews, 1992). This test provides superior empirical results containing information about the unknown single structural break occurring in the series. The results are reported in Table 7.2. We find that the economic growth, energy consumption, oil prices, capital, and labor have the unit root problem in the presence of structural breaks. These breaks are

Table 7.2: Unit root analysis.

Variables	Z–A Unit Root Test		K–P Unit Root Test		
	t-Statistic	Time Break	t-Statistic	Probability	Time Break
$\ln Y_t$	−3.247 (1)	1991Q$_\text{IV}$	−3.4779 (1)	0.3993	2003Q$_\text{I}$
$\ln E_t$	−4.025 (2)	2007Q$_\text{II}$	−3.0811 (3)	0.6391	1991Q$_\text{I}$
$\ln O_t$	−4.283 (1)	1998Q$_\text{I}$	−3.5585 (1)	0.3546	2003Q$_\text{I}$
$\ln K_t$	−3.368 (2)	2008Q$_\text{III}$	−2.6083 (2)	0.8660	1991Q$_\text{I}$
$\ln L_t$	−2.373 (1)	2006Q$_\text{II}$	−2.1683 (2)	0.8686	2013Q$_\text{I}$
$\Delta \ln Y_t$	−5.430 (2)**	1993Q$_\text{III}$	−5.7470 (2)*	0.0001	1992Q$_\text{II}$
$\Delta \ln E_t$	−7.041 (2)*	2007Q$_\text{IV}$	−6.9013 (3)*	0.0000	2008Q$_\text{I}$
$\Delta \ln O_t$	−7.963 (1)*	2005Q$_\text{III}$	−6.8085 (1)*	0.0000	1998Q$_\text{I}$
$\Delta \ln K_t$	−6.093 (2)*	2006Q$_\text{I}$	−6.4395 (2)*	0.0000	2005Q$_\text{I}$
$\Delta \ln L_t$	−4.568 (2)**	2001Q$_\text{II}$	−6.9575 (2)*	0.0000	2005Q$_\text{II}$

* and ** show significance at 1% and 5% levels of significance respectively.

related to the economic policies implemented into the energy market to sustain long-run economic growth. For example, in 1991, the government of Pakistan initiated economic reforms by introducing a policy package for the expansion of economic growth by using the free-market mechanism. The central point of these reforms was the disinvestment in public enterprises, deregulation, as well as denationalization. These reforms encouraged the private sector, which affected total factor productivity and hence economic growth (Looney, 1992). After first differencing, all the variables have been found as stationary. This leads us to conclude that economic growth, electricity consumption, oil prices, capital, and labor are integrated at I(1).

For testing whether nonlinearity is present in the variables or not, we apply the BDS test developed by Brock et al. (1988). Table 7.3 shows the results of the BDS test and we find that the null hypothesis of i.i.d. (independently and identically distributed) has been rejected. It implies the nonnormal distribution of data which shows the presence of nonlinearity. We may note that all the variables have a nonlinear behavior. The presence of nonlinearity in the variables renders the unit root analysis ambiguous. This issue is solved by applying nonlinear unit root tests developed by Bierens (1997) and Breitung (2000). The results are reported in Table 7.4 and we find that all the variables are found to be nonstationary in the presence of nonlinearity confirmed by Bierens (1997). Similarly, the results of the Breitung (2000) unit root test corroborated that all the variables have a unit root problem at levels, in the presence of nonlinearity. This confirms that all the variables are integrated at I(1) in the absence and the presence of nonlinearity.

The presence of nonlinearity leads us to the application of the nonlinear (asymmetric) ARDL approach developed by Shin et al. (2014) as all the variables have a nonlinear behavior. In doing so, the general to specific approach is applied following Greenwood-Nimmo et al. (2013). The lag length for the appropriate model is selected based on AIC. The appropriate

Table 7.3: BDS test for nonlinearity.

Dimension	BDS Statistic	Standard Error	z-Statistic	Probability
		$\ln Y_t$		
2	0.2003	0.0039	51.2741	0.0000
3	0.3379	0.0062	54.4014	0.0000
4	0.4332	0.0073	58.5949	0.0000
5	0.5002	0.0077	64.9521	0.0000
6	0.5477	0.0074	73.812	0.0000
		$\ln E_t$		
2	0.2048	0.0055	37.1813	0.0000
3	0.3482	0.0087	39.6414	0.0000
4	0.4482	0.0104	42.7088	0.0000
5	0.5173	0.0109	47.1505	0.0000
6	0.5647	0.0106	53.2128	0.0000
		$\ln O_t$		
2	0.1829	0.0042	42.6033	0.0000
3	0.3025	0.0067	44.6288	0.0000
4	0.3800	0.0080	47.4339	0.0000
5	0.4295	0.0082	51.8140	0.0000
6	0.4655	0.0079	58.6840	0.0000
		$\ln K_t$		
2	0.1817	0.0067	26.8869	0.0000
3	0.3026	0.0107	28.0414	0.0000
4	0.3812	0.0129	29.5359	0.0000
5	0.4302	0.0135	31.8459	0.0000
6	0.4574	0.0130	34.9531	0.0000
		$\ln L_t$		
2	0.2020	0.0047	42.8200	0.0000
3	0.3400	0.0074	45.4885	0.0000
4	0.4353	0.0088	49.0898	0.0000
5	0.5018	0.0092	54.5111	0.0000
6	0.5487	0.0088	62.0642	0.0000

Table 7.4: Nonlinear unit root analysis.

Variables	Bierens Unit Root Test		Breitung Unit Root Test	
	t-Statistic	Probability Values	t-Statistic	Probability Values
$\ln Y_t$	-2.2508	0.4940	0.0081	0.9684
$\ln E_t$	-1.4527	0.9080	0.0175	0.9970
$\ln OP_t$	-3.6630	0.1290	0.0165	0.9429
$\ln K_t$	-2.0808	0.6120	0.0073	0.1920
$\ln L_t$	-3.4330	0.2260	0.0076	0.1911

lag length contributes to the more accurate estimation and dynamic multipliers (Katrakilidis and Trachanas, 2012). The results of NARDL approach are reported in Table 7.5. We find that energy consumption, oil prices, capital, and labor explain economic growth by 76.58% ($R^2 = 0.7658$). This shows that the contribution of energy consumption, oil prices, capital, and labor are responsible for 76.58% of the variation and the rest (namely, 23.42%) is explained by the error term in the production function. The absence of autocorrelation in the empirical model is confirmed by the Durbin Watson (DW) test statistic which is 2.2. This implies that the considered variables, that is, energy consumption, oil prices, capital, and labor in the production function explain economic growth without autocorrelation. Additionally, we find that the empirical model follows the normal distribution. There is no problem of serial correlation and white heteroscedasticity. There is absence of the autoregressive conditional and the functional form of the empirical formulation is well constructed. This shows the reliability and consistency of the empirical model under the NARDL framework.

Under the ARDL framework, the empirical results confirm that the calculated FPSS statistic is statistically significant at 1% level of significance, which indicates that the upper critical bounds are calculated to be less than the FPSS statistic. This confirms the existence of cointegration among energy consumption, oil prices, capital, labor, and economic growth for the period of 1985Q$_1$–2015Q$_4$. The presence of asymmetry in the long run and the short run is investigated by applying the Wald test. Accounting for nonlinearity and asymmetry is very important in the production function estimation, while studying the association among energy consumption, oil prices, capital, labor, and economic growth. The *t*-statistic of BDM originated by Banerjee et al. (1998) also corroborates the presence of cointegration between the variables at 1% level of significance. We apply the NARDL *F*-statistic (FPSS) developed by Shin et al. (2014) and find that the energy consumption, oil prices, capital, labor, and economic growth have long-run asymmetric cointegration in the case of Pakistan. It implies that the asymmetries and nonlinearities are important, while investigating the production function by considering energy consumption and oil prices as additional determinants.

The long-run and short-run asymmetric impact of energy consumption, oil prices, capital, and labor on economic growth are reported in Table 7.5. We find that the positive shock existing in the energy consumption has both positive and significant impact on economic growth. A negative shock in energy consumption reduces economic growth insignificantly though. It implies that an increase in the energy consumption plays its role in stimulating economic growth. We may conclude that energy consumption has a positive effect on economic growth. This empirical evidence is consistent with the existing studies such as Zaman et al. (2011), Ahmed et al. (2013, 2015), and Shahbaz et al. (2016) who also report that energy consumption enhances domestic production and hence stimulates economic growth. A positive shock in oil prices reduces economic growth, but a negative shock in oil prices increases economic growth. This shows that a rise in oil prices decreases economic growth and the impact of a positive and a negative shock on growth is according to our

Table 7.5: Nonlinear autoregressive distributive lag empirical analysis.

Variables	Coefficient	Standard Error	t-Statistic
Constant	2.3482*	(0.4112)	[5.7104]
$\ln Y_{t-1}$	−0.2603*	(0.0456)	[−5.7119]
$\ln E_{t-1}^{+}$	0.1675*	(0.0354)	[4.7314]
$\ln E_{t-1}^{-}$	−0.0643	(0.0552)	[−1.1654]
$\ln OP_{t-1}^{+}$	−0.0167*	(0.0041)	[−4.1095]
$\ln OP_{t-1}^{-}$	0.0097**	(0.0038)	[2.5275]
$\ln K_{t-1}^{+}$	0.0559*	(0.0150)	[3.7362]
$\ln K_{t-1}^{-}$	−0.0301*	(0.0086)	[−3.5250]
$\ln L_{t-1}^{+}$	0.3128*	(0.0834)	[3.7519]
$\ln L_{t-1}^{-}$	−1.8712*	(0.4717)	[−3.9672]
$\Delta \ln Y_{t-1}$	0.5481*	(0.0770)	[7.1185]
$\Delta \ln E_{t-1}^{+}$	0.4925*	(0.0866)	[5.6859]
$\Delta \ln K_{t}^{+}$	0.1445*	(0.0230)	[6.2888]
$\Delta \ln OP_{t}^{+}$	−0.0404*	(0.0079)	[−5.1069]
$\Delta \ln L_{t-1}^{+}$	−18.4892*	(3.6589)	[−5.0532]
$\Delta \ln Y_{t-1}$	0.2120*	(0.0753)	[2.8141]
$\Delta \ln L_{t-1}^{+}$	12.5604*	(3.9859)	[3.1512]
$\Delta \ln K_{t-1}^{+}$	−0.0916*	(0.0262)	[−3.4999]
$\Delta \ln E_{t-1}^{+}$	−0.2554*	(0.0875)	[−2.9190]
$\Delta \ln OP_{t-1}^{+}$	0.0240*	(0.0085)	[2.8241]
Long-run Results and Diagnostic Analysis			
$L_{\ln E}^{+}$	0.6436*	(0.0753)	
$L_{\ln E}^{-}$	−0.2470	(0.2104)	
$W_{\ln E}$	15.8205*	(0.2239)	
$L_{\ln OP}^{+}$	−0.0642*	(0.0100)	
$L_{\ln OP}^{-}$	0.0373*	(0.0124)	
$W_{\ln OP}$	39.7908*	(0.0161)	
$L_{\ln K}^{+}$	0.2147*	(0.0385)	
$L_{\ln K}^{-}$	−0.1158*	(0.0251)	
$W_{\ln K}$	45.0498*	(0.0493)	
$L_{\ln L}^{+}$	1.2015*	(0.2321)	
$L_{\ln L}^{-}$	−7.1881*	(1.2515)	
$W_{\ln L}$	36.3336*	(1.3919)	
R^2	0.7658		
χ_{Norm}^2	2.5179		
χ_{SC}^2	1.2914		
χ_{HET}^2	0.3758		
χ_{ARCH}^2	0.3760		
χ_{FF}^2	1.9139		

Table 7.5: Nonlinear autoregressive distributive lag empirical analysis. (*cont.*)

Variables	Coefficient	Standard Error	*t*-Statistic
T_{BDM}	−5.7119*		
$F_{PSS-Nonlinear}$	4.2956*		
AIC	−8.7310		
SIC	−8.2785		
Hannan–Quinn	−8.5472		

99% upper (lower) bound with $k = 4$ is 5.06 (3.74). The 95% upper (lower) bound with $k = 6$ is 4.43 (3.15). The superscript "+" and "−" denote positive and negative cumulative sums, respectively. L^+ and L^- are the estimated long-run coefficients associated with positive and negative changes, respectively, defined by $\hat{\beta} = -\hat{\theta}/\hat{\rho}$. W_{LR} represents the Wald test for the null of the long-run symmetry for respective variable χ^2_{SC}, χ^2_{FF}, χ^2_{HET}, and χ^2_{Norm} denote Lagrange Multipliers (LM) tests for serial correlation, normality, functional form, and heteroscedasticity, respectively. *SE* stands for standard errors. * and ** indicate significance at 1% and 5% levels, respectively.

expectations. A rise in oil prices affects economic growth directly and indirectly. Directly, oil prices rise transmits into the cost of production which decreases the investment activities of firms and hence the domestic production is reduced. Indirectly, oil prices increase affects the exchange rate and inflation and thus economic activity and growth (Shahbaz et al., 2017a). This empirical evidence is consistent with existing studies in energy economics literature such as Jimnez-Rodrguez and Snchez (2005) and Ali (2016) for OECD countries, Farzanegan and Markwardt (2009) and Behmiri and Manso (2013) for sub-Saharan countries, Ftiti et al. (2016) who conclude that oil prices increases affect economic growth adversely.

Economic growth is positively affected by a positive shock that stems in capital and a negative shock in capital has a negative effect on economic growth. This shows that a positive shock that stems in capital stimulates fiscal investment in infrastructure development for the long run. This increases domestic production and hence, long-run sustainable economic growth. On the contrary, a negative shock occurring in capital reduces economic growth by lowering domestic production. Overall, capital is positively linked to economic growth. Similarly, existing studies in literature such as Mehta (2011) and Sahoo and Dash (2009) for India, Sahoo et al. (2010) for China, and Shahbaz et al. (2017b) for India also report that capital plays its significant role in stimulating economic activity in that it speeds up economic growth. The asymmetric relationship between labor and economic growth is interesting and statistically significant. Economic growth is positively and negatively affected by the positive and negative shocks that appear in labor. Labor positive and negative shocks on economic growth are estimated to be 0.3128 and −1.8712, respectively. This implies that a rise in labor force contributes to economic growth significantly by increasing consumption and investment activities. Pakistan's young population is almost 60% of total, which shows the economic dependence of Pakistan's economy on young population. In such circumstances, any adverse shock in labor force will not only decrease domestic production, but will also dismantle economic growth. These empirical findings are consistent with Shahbaz et al. (2017a,b), but they are opposite from Ismail et al. (2015) for India and Malaysia respectively.

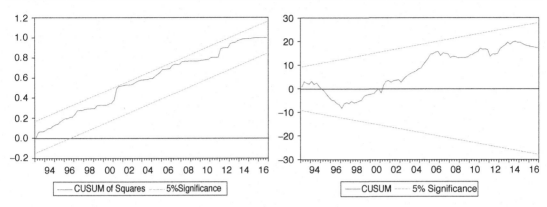

Figure 7.1: Stability.

For short-run analysis (Table 7.5), we find that a positive shock that stems in energy consumption has a positive and a significant impact on economic growth (coefficient is 0.4925). A lagged differenced positive shock that stems in energy consumption reduces economic growth by 0.2554 and it is statistically significant at 1% level. A differenced positive shock in oil prices has a negative impact on economic growth, but it is statistically significant at 1% level. Economic growth is positively affected by a lagged differenced positive shock in oil prices at 1% level of significance. Economic growth is positively stimulated by a differenced positive shock in capital. A positive shock in capital contributes to economic growth by 0.1455%, but a lagged differenced positive shock in capital reduces economic growth by 0.0916. These results are statistically significant at 1% level respectively. Differenced and lagged differenced positive shocks that appear in labor, have negative and statistically significant effects on economic growth. This shows that in the short run, a rise in labor force will not contribute to economic growth due to a mismatch between the supply and demand for labor, but it adjusts in the long run. The reliability of empirical findings is confirmed by applying CUSUM and CUSUMsq. The CUSUM and CUSUMsq are lying between the critical bounds at 5% level, which shows that empirical results are reliable and consistent (see Fig. 7.1).

We apply multiple dynamic adjustments. Results are shown in the plots with the cumulative dynamic multipliers (see Fig. 7.2). We find that these results indicate the pattern of adjustment in the production function in its new long-run equilibrium, due to positive and negative shocks that stem in energy consumption, oil prices, capital, and labor force respectively. The basis of dynamic multipliers for the empirical estimation is the best fitted NARDL fulfilling the AIC. The continuous black line and dashed black line show positive and negative capturing of the adjustments of the production function to positive and negative shocks in independent variables at given forecast horizons. The continuous red line (asymmetric curve) indicates the difference between the dynamic multipliers linked to positive and negative shocks, that is, $m_h^+ - m_h^-$. At a 95% confidence interval, the dotted red

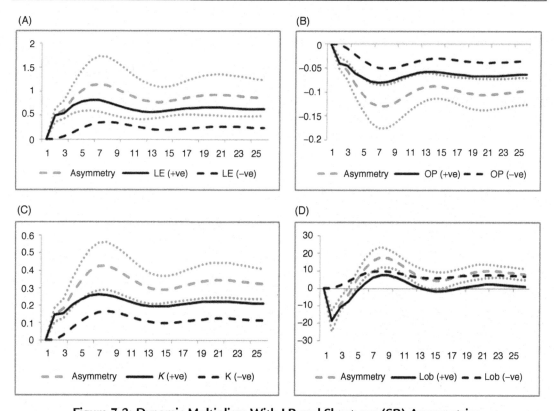

Figure 7.2: Dynamic Multipliers With LR and Short-run (SR) Asymmetries.
(A) Energy consumption; (B) Oil prices; (C) Capital; and (D) Labor. *Black (dotted) line* shows positive (negative) impact, while *dark gray lines (red lines* in web version) show asymmetry and confidence (upper and lower) bands.

lines (lower and upper bands) indicate the statistical significance at 95% confidence interval of asymmetry at any horizon *h*. The results are reported in Fig. 7.2. Overall, we find that energy consumption contributes to economic growth via stimulating economic activity and increasing the domestic production. From an initial point, the impact of a positive shock that stems in energy consumption has a dominant effect on economic growth compared to a negative shock in energy consumption. On a similar vein, the asymmetric response to shocks in energy consumption is statistically significant. The linkage between oil prices and economic growth is negative. This indicates that the positive shock in oil prices dominantly affects economic growth by lowering economic activity although negative shocks in oil prices increase economic growth, but at a lower magnitude. Oil prices rise retards economic growth, as the role of a positive shock that stems in oil prices dominates that of negative shocks in oil prices on economic growth for the long run (−0.0167 vs. 0.0097 in Table 7.5). Capital contributes to economic growth significantly. A positive shock in capital has a dominating positive effect on economic growth compared to a decline in economic growth, due to a negative shock in capital (0.0559 vs. −0.0301 in Table 7.5). Lastly, economic growth is positively and negatively affected by positive and negative shocks in labor, but a negative shock in labor has a negative and dominant effect on economic growth overwhelming that of a positive shock in labor. This shows that, due to a decline in economic activity, firms are ready to fire people, a fact that adversely affects domestic production and hence economic growth. In a recovery period, firms are reluctant to hire people due to an uncertain future. That's why a positive shock in labor even contributes to economic growth, but in a lower magnitude.

6 Conclusion and Policy Implications

The presence of the four competing hypotheses on the EGN is always the focus of interest and research for academicians, practitioners, and policy makers. These empirical findings are diverse due to the use of different data for different countries, data samples, and econometric approaches. The issue is still under consideration for implementing efficient empirical analysis that is able to provide comprehensive policy implications that foster sustainable economic development. For that purpose, we employed a production function to examine the relationship between energy consumption and economic growth by adding oil prices as additional factors affecting economic growth and energy consumption as well. The nonlinear unit root test is applied to confirm whether variables are integrated at I(0) or I(1). Considering the importance of asymmetries in time-series data, we apply the NARDL testing approach to test the asymmetric effect of energy consumption along with oil prices, capital, and labor on economic growth in Pakistan. The empirical evidence confirms the presence of symmetric and asymmetric cointegration among energy consumption, oil prices, capital, labor, and economic growth over the period of $1985Q_I$–$2016Q_{IV}$. Furthermore, a rise in energy consumption (positive shock) adds to economic growth via stimulating economic activity and energy consumption negative shock retards economic growth insignificantly. A rise (positive shock) and fall (negative shock) in oil prices decline and stimulate economic growth respectively.

The positive shocks in energy consumption have a substantial effect on domestic production and hence, on economic growth. This entails the importance of the efficient use of the existing energy sources for sustainable long-run economic growth. In such a situation, exploring new energy sources is also an appropriate solution to stimulate economic activity. Any reduction in energy supply will decline domestic output. This is confirmed by the evidence of the result of a negative shock in energy consumption. This suggests that the government should maintain a stable energy consumption rather than reducing energy supply. Thus the government should encourage the usage of energy efficient and savings technologies, not only in production activities, but also in the consumption activities by using proper electronic and print media campaigns. A negative relationship between oil prices and economic growth reveals the importance of using alternative sources of energy. Pakistan oil market is directly linked to the international market, which affects domestic oil prices if any shock in oil prices at international level. Pakistan is an oil-dependent country and a huge amount of its foreign reserves is consumed to importing oil for meeting domestic energy demand. This directly affects the exchange rate and weakens local currency, which also increases local inflation. This entails the clear need to explore new energy sources such as oil, which not only will save foreign reserves, but also will reduce dependence on the imported oil. This amount of foreign reserves can be used to import energy-efficient and environment-friendly technologies to stimulate domestic production.

The positive and negative effects of positive and negative shocks of capital on economic growth reveals the importance of capital for domestic production. This entails that capital plays a vital role in stimulating economic activity. This shows that government should focus more on establishing the infrastructure development for achieving long-run sustainable economic growth. There is a clear need for policy makers to understand the capital-growth nexus. A consistent improvement in capital will speedup economy by raising economic growth while a reduction in capital improvement will retard economic growth which is confirmed by the negative shock in capital. In such circumstances, the government should focus on increasing R&D expenditures, improving the quality of capital via conducting research on introducing energy efficient capital, which not only enhances domestic production, but also saves energy for future generations—a prerequisite for sustainable economic growth. Finally, a positive shock stems in labor that leads to economic growth. This shows that without labor, sustainable economic growth in the long run is impossible, because a negative shock in labor reduces economic growth. Therefore the government should invest in labor force to attain long-run economic growth by improving their technical efficiency via technical education. The agriculture sector absorbs a major portion of labor force and adopting technology in agriculture enhances its production much easier. In such a situation, government should disseminate the technical level at town level to educate farmers, that is, the related labor force for using energy-saving and growth-stimulating technology for agriculture production. This model can also be implemented in the industrial sector after a careful and comprehensive policy design. Improvements in technical education not only help labor force to increase their productivity, but also save energy wastage and environment from degradation.

References

Abosedra, S., Baghestani, H., 1991. New evidence on the causal relationship between United States energy consumption and gross national product. J. Energy Dev. 14, 285–292.

Adom, P.K., 2013. Time-varying analysis of aggregate electricity demand in Ghana: a rolling analysis. OPEC Energy Rev. 37, 63–80.

Ahmed, M., Riaz, K., Khan, A.M., Bibi, S., 2015. Energy consumption–economic growth nexus for Pakistan: taming the untamed. Renew. Sustainable Energy Rev. 52, 890–896.

Ahmed, W., Zaman, K., Taj, S., Rustam, R., Waseem, M., Shabir, M., 2013. Economic growth and energy consumption nexus in Pakistan. South Asian J. Global Bus. Res. 2, 251–275.

Alam, S., Butt, M.S., 2002. Causality between energy and economic growth in Pakistan: an application of cointegration and error-correction modeling techniques. Pac. Asian J. Energy 12, 151–165.

Ali, S., 2016. The impact of oil price on economic growth: test of Granger causality, the case of OECD countries. Int. J. Social Sci. Econ. Res. 1, 1333–1349.

Altinay, G., Karagol, E., 2004. Structural break, unit root, and the causality between energy consumption and GDP in Turkey. Energy Consumption 26, 985–994.

Apergis, N., Tang, C.F., 2013. Is the energy-led growth hypothesis valid? New evidence from sample of 85 countries. Energy Econ. 38, 24–31.

Aqeel, A., Butt, M.S., 2001. The relationship between energy consumption and economic growth in Pakistan. Asia-Pac. Dev. J. 8 (2), 101–110.

Arac, A., Hasanov, M., 2014. Asymmetries in the dynamic interrelationship between energy consumption and economic growth: evidence from Turkey. Energy Econ. 44, 259–269.

Arifin, J., Syahruddin, N., 2011. Causality relationship between renewable and nonrenewable energy consumption and GDP in Indonesia. Econ. Finance Indones. 59, 1–18.

Arora, V., Shi, S., 2016. Energy consumption and economic growth in the United States. Appl. Econ. 39, 3763–3773.

Asafu-Adjaye, J., 2000. The relationship between energy consumption, energy prices and economic growth: time series evidence from Asian developing countries. Energy Econ. 22, 615–625.

Banerjee, A., Dolado, J., Mestre, R., 1998. Error-correction mechanism tests for cointegration in a single-equation framework. J. Time Ser. Anal. 19, 267–283.

Behmiri, N.B., Manso, J.P., 2013. How crude oil consumption impacts on economic growth of Sub-Saharan Africa? Energy 54, 74–83.

Belke, A., Dobnik, F., Dreger, C., 2011. Energy consumption and economic growth: new insights into the cointegration relationship. Energy Econ. 33, 782–789.

Bierens, H.J., 1997. Testing the unit root with drift hypothesis against nonlinear trend stationarity, with an application to the US price level and interest rate. J. Econometr. 81, 29–64.

Borozan, D., 2013. Exploring the relationship between energy consumption and GDP: evidence from Croatia. Energy Policy 59, 373–381.

Breitung, J., 2000. The local power of some unit root tests for panel data. Baltagi, B. (Ed.), Nonstationary Panels, Panel Cointegration, and Dynamic Panels Advances in Econometrics, vol. 15, JAI, Amsterdam, pp. 161–178.

Brock, W.A., Dechert, W.D., Scheinkman, J., LeBaron, B., 1988. A test for independence based on the correlation dimension. Department of Economics, University of Chicago, USA.

Chishti, S., Mahmud, F., 1990. The demand for energy in the large-scale manufacturing sector of Pakistan. Energy Econ. 12, 251–254.

Chontanawat, J., Hunt, L.-C., Pierse, R., 2008. Does energy consumption cause economic growth? Evidence from a systematic study of over 100 countries. J. Policy Model. 30, 209–220.

Dhungel, K.R., 2008. A causal relationship between energy consumption and economic growth in Nepal. Asia-Pac. Dev. J. 15, 137–150.

Dickey, D., Fuller, W., 1979. Distribution of the estimators for autoregressive time series with a unit root. J. Am. Stat. Assoc. 74, 427–431.

Dickey, D., Fuller, W., 1981. Likelihood ratio statistics for autoregressive time series with a unit root. Econometrica 49, 1057–1072.

Elliot, G., Rothenberg, T.J., Stock, J.H., 1996. Efficient tests for an autoregressive unit root. Econometrica 64, 813–836.

Fallahi, F., 2011. Causal relationship between energy consumption (EC) and GDP: a Markov-Switching (MS) causality. Energy 36, 4165–4170.

Farzanegan, M.R., Markwardt, G., 2009. The effects of oil price shocks on the Iranian economy. Energy Econ. 31, 134–151.

Ftiti, Z., Guesmi, K., Teulon, F., Chouachi, S., 2016. Relationship between crude oil prices and economic growth in selected OPEC countries. J. Appl. Bus. Res. 32, 11–22.

Granger, C.W.J., 1969. Investigating causal relations by econometric models and cross-spectral methods. Econometrica 37, 424–438.

Greenwood-Nimmo, M., Shin, Y., van, Treeck T., 2013. The Decoupling of Monetary Policy From Long-Term Interest Rates in the U.S. and Germany. , Available from: http://ssrn.com/abstract=1894621.

Gregory, A.W., Hansen, B.E., 1996. Residual-based tests for cointegration in models with regime shifts. J. Econom. 70, 99–126.

Gross, C., 2012. Explaining the (non-) causality between energy and economic growth in the U.S.—a multivariate sectoral analysis. Energy Econ. 34, 489–499.

Hoang, T., Lahiani, A., Heller, D., 2016. Is gold a hedge against inflation? A nonlinear ARDL analysis. Econ. Model. 54, 54–66.

Hondroyiannis, G., Lolos, G., Papapetrous, E., 2002. Energy consumption and economic growth: assessing the evidence from Greece. Energy Econ. 24, 319–336.

Iqbal, M., 1986. Substitution of labor, capital and energy in the manufacturing sector of Pakistan. Empirical Econ. 11, 81–95.

Ismail, N.W., Rahman, H.S.W.A., Hamid, T.A.T.A., 2015. Does population aging affect economic growth in Malaysia? Prosiding Perkem 10, 205–210.

Jimnez-Rodrguez, R., Snchez, M., 2005. Oil price shocks and real GDP growth: empirical evidence for some OECD countries. Appl. Econ. 37, 201–228.

Johansen, S., 1991. Estimation and hypothesis testing of cointegration vectors in Gaussian vector autoregressive models. Econometrica 59 (6), 1551–1580.

Johansen, S., Juselius, K., 1990. Maximum likelihood estimation and inference on cointegration with applications to the demand for money. Oxford Bull. Econ. Stat. 52, 169–210.

Jumbe, C., 2004. Cointegration and Causality between electricity consumption and GDP: empirical evidence from Malawi. Energy Econ. 26, 61–68.

Kaplan, M., Ozturk, I., Kalyoncu, H., 2011. Energy consumption and economic growth in Turkey: cointegration and causality analysis. Rom. J. Econ. Forecast. 2, 30–41.

Katrakilidis, C., Trachanas, E., 2012. What drives housing price dynamics in Greece: new evidence from asymmetric ARDL cointegration. Econ. Model. 29, 1064–2069.

Khan, A., Khan, M.Z., Zaman, K., Arif, M., 2014. Global estimates of energy-growth nexus: application of seemingly unrelated regressions. Renew. Sustainable Energy Rev. 29, 63–71.

Kim, D., Perron, P., 2009. Unit root tests allowing for a break in the trend function at an unknown time under both the null and alternative hypotheses. J. Econom. 148, 1–13.

Kraft, J., Kraft, A., 1978. On the relationship between energy and GNP. J. Energy Dev. 3, 401–403.

Kwiatkowski, D., Phillips, P.C.B., Schmidt, P., Shin, Y., 1992. Testing the null hypothesis of stationarity against the alternative of a unit root. J. Econometr. 54, 159–178.

Lee, C.-C., Chang, C.-P., 2005. Structural breaks, energy consumption, and economic growth revisited: evidence from Taiwan. Energy Econ. 27, 857–872.

Lee, C.-C., Chang, C.-P., 2007. The impact of energy consumption on economic growth: evidence from linear and nonlinear models in Taiwan. Energy 32, 2282–2294.

Liew, V., Nathan, T., Wong, W., 2012. Are sectoral outputs in Pakistan led by energy consumption? Econ. Bull. 32, 2326–2331.

Looney, R.E., 1992. An assessment of Pakistan attempt at economic reforms. J. South Asian Middle East. Stud. 15, 1–28.

Mahmud, U.S.F., 2000. The energy demand in the manufacturing sector of Pakistan: some further results. Energy Econ. 22, 541–648.

Mehta, R., 2011. Short-run and long-run relationship between capital formation and economic growth in India. Indian J. Manage. Technol. 19, 170–180.

Menegaki, A., 2014. On energy consumption and GDP studies: a meta-analysis of the last two decades. Renew. Sustainable Energy Rev. 29, 31–36.

Mushtaq, K., Abbas, F.A., Ghafoor, A., 2007. Energy use for economic growth: cointegration and causality analysis from the agriculture sector of Pakistan. Pak. Dev. Rev. 46, 1065–1073.

Narayan, P.K., Smyth, R., 2005. Electricity consumption, employment and real income in Australia: evidence from multivariate Granger causality tests. Energy Policy 33, 1109–1116.

Ng, S., Perron, P., 2001. Lag length selection and the construction of unit root tests with good size and power. Econometrica 69, 1519–1554.

Odhiambo, N.M., 2009. Energy consumption and economic growth nexus in Tanzania: an ARDL bounds testing approach. Energy Policy 37, 617–622.

Oh, W., Lee, K., 2004. Energy consumption and economic growth in Korea: testing the causality relation. Energy Econ. 26, 973–981.

Ozturk, I., 2010. A literature survey on energy-growth nexus. Energy Policy 38, 340–349.

Pata, U.K., Terzi, H., 2017. A multivariate causality analysis between energy consumption and growth: the case of Turkey. Energy Sources B Econ. Plann. Policydoi: 10.1080/15567249.2016.1278484.

Paul, P.B., Uddin, G.S., 2011. Energy and output dynamics in Bangladesh. Energy Econ. 33, 480–487.

Payne, J.E., 2008. On the dynamics of energy consumption and output in the US. Appl. Energy 86, 575–577.

Payne, J.E., 2009. On the dynamics of energy consumption and output in the US. Appl. Energy 86, 575–577.

Perron, P., 1997. Further evidence on breaking trend functions in macroeconomic variables. J. Econometr. 80, 355–385.

Pesaran, M.H., Shin, Y., Smith, R., 2001. Bounds testing approaches to the analysis of level relationships. J. Appl. Econom. 16, 289–326.

Phillips, P.C.B., Perron, P., 1988. Testing for unit roots in time series regression. Biometrika 75, 335–346.

Riaz, T., Stern, N.H., 1984. Pakistan: energy consumption and economic growth. Pak. Dev. Rev. 23, 431–453.

Sahoo, P., Dash, R.K., 2009. Infrastructure development and economic growth in India. J. Asian Pac. Econ. 14, 351–365.

Sahoo, P., Dash, R.K., Nataraj, G., 2010. Infrastructure Development and Economic Growth in China. Institute of Developing Economies (IDE), Discussion Paper No. 261.

Salim, R.A., Rafiq, S., Hassan, A.F.M.K., 2008. Causality and dynamics of energy consumption and output: evidence from non-OECD Asian countries. J. Econ. Dev. 33, 1–18.

Shahbaz, M., Lean, H.H., 2012. Does financial development increase energy consumption? The role of industrialization and urbanization in Tunisia. Energy Policy 40, 473–479.

Shahbaz, M., Hoang, T.H.V., Mahalik, M.K., Roubaud, D., 2017a. Energy consumption, financial development and economic growth in India: new evidence from a nonlinear and asymmetric analysis. Energy Econ. 63, 199–212.

Shahbaz, M., Islam, F., Butt, M.S., 2016. Finance–growth–energy nexus and the role of agriculture and modern sectors: evidence from ARDL bounds test approach to cointegration in Pakistan. Global Bus. Rev., 1037–1059.

Shahbaz, M., Sarwar, S., Chen, W., Malik, M.N., 2017b. Dynamics of electricity consumption, oil price and economic growth: global perspective. Energy Policy 108, 256–270.

Shahbaz, M., Zeshan, M., Afza, T., 2012. Is energy consumption effective to spur economic growth in Pakistan? New evidence from bounds test to level relationships and Granger causality tests. Econ. Model. 29, 2310–2319.

Shahiduzzaman, M., Alam, K., 2012. Cointegration and causal relationship between energy consumption and output: assessing the evidence from Australia. Energy Econ. 34, 2182–2188.

Shin, Y., Yu, B., Greenwood-Nimmo, M., 2014. Modelling asymmetric cointegration and dynamic multipliers in a nonlinear ARDL framework. Festschrift in Honor of Peter Schmidt. Springer, New York, pp. 281–314.

Sims, A.C., 1972. Money, income, and causality. Am. Econ. Rev. 62, 540–552.

Soytas, U., Sari, R., Ewing, B.T., 2007. Energy consumption, income, and carbon emissions in the United States. Ecol. Econ. 62, 482–499.

Stern, D.I., 1993. Energy and economic growth in the USA: a multivariate approach. Energy Econ. 15, 137–150.

Stern, D.I., 2000. A multivariate cointegration analysis of the role of energy in the US macroeconomy. Energy Econ. 22, 267–283.

Stern, D.I., Enflo, K., 2013. Causality between energy and output in the long-run. Energy Econ. 39, 135–146.

Tang, W., Wu, L., Zhnag, Z.X., 2010. Oil price shocks and their short- and long-term effects on the Chinese economy. Energy Econ. 32, 3–14.

Toda, H.Y., Yamamoto, T., 1995. Statistical inference in vector autoregression with possible integrated process. J. Econometr. 66, 225–250.

Wesseh, P.K., Zoumara, B., 2012. Causal independence between energy consumption and economic growth in Liberia: evidence from a non-parametric bootstrapped causality test. Energy Policy 50, 518–527.

Yang, H.Y., 2000. A note on the causal relationship between energy and GDP in Taiwan. Energy Econ. 22, 309–317.

Yildirim, E., Sukruoglu, D., Aslan, A., 2014. Energy consumption and economic growth in the next 11 countries: the bootstrapped autoregressive metric causality approach. Energy Econ. 44, 14–21.

Yu, E.S.H., Chow, P.C.Y., Choi, J.Y., 1988. The relationship between energy and employment: a reexamination. Energy Syst. Policy 11, 287–295.

Yu, W., Ju'e, G., Youmin, X., 2008. Study on the dynamic relationship between economic growth and China energy based on cointegration analysis and impulse response function. China Popul. Resour. Environ. 18, 56–61.

Zaman, K., Khan, M.M., Saleem, Z., 2011. Bivariate cointegration between energy consumption and development factors: a case study of Pakistan. Int. J. Green Energy 8, 820–833.

Zaman, K., Khan, M.M., Saleem, Z., 2012. Bivariate cointegration between energy consumption and development factors: a case study of Pakistan. Int. J. Green Energy 8, 820–833.

Zarnikau, J., 1997. A reexamination of the causal relationship between energy consumption and gross national product. J. Energy Dev. 21, 229–239.

Zivot, E., Andrews, D., 1992. Further evidence of great crash, the oil price shock and unit root hypothesis. J. Bus. Econ. Stat. 10, 251–270.

Further Readings

Brock, W.A., Sayers, C., 1988. Is the business cycle characterized by deterministic chaos? J. Monetary Econ. 22, 71–90.

Brown, R.L., Durbin, J., Evans, J.M., 1975. Techniques for testing the constancy of regression relations over time. J. R. Stat. Soc. 37, 149–163.

Jayaraman, T.K., Choong, C.-K., 2009. Growth and oil price: a study of causal relationship in small Pacific Island countries. Energy Policy 37, 2182–2189.

CHAPTER 8

Panel Data Analysis in the Energy-Growth Nexus (EGN)

Can T. Tugcu

Akdeniz University, Antalya, Turkey

Chapter Outline

1 Introduction: The Benefits and Limitations of Using Panel Data in the EGN

Given that the data mining procedure is fairly formal, the most restrictive issue for conducting an econometric analysis in the EGN field is the lack of a sufficient number of observations. That is why panel data methodologies have been on the increase in the empirical side of the economics. For instance, one of the most important data sources of researchers in the EGN is the World Bank database. Thus, if one goes to the World Development Indicators database for obtaining nonrenewable energy consumption data, it can be realized that there exists continuous annual data for only 22 out of 217 countries for the period 1960–2014.

On the other hand, if you are looking for renewable energy consumption data for the same period, one discovers that the only available time span for this data set in the World Development Indicators is the 1990–2012 period. Under these circumstances, it is not possible to perform an inclusive econometric analysis that investigates the impact of nonrenewable energy consumption on economic growth; and it is not appropriate to accommodate a time-series analysis that deals with the consequences of renewable energy consumption on growth. However, the solution is not very far from that point. Through the pooling of all observations on a cross-section of economic units, the panel data approach has achieved a milestone in overcoming this problem. With its two-dimensional form, panel data studies allow energy researchers to make inclusive analyses using small sample-sized data sets. In addition to that, as Baltagi (2005) states, employing a panel data approach in energy economics has some additional benefits in terms of less collinearity, more degrees of freedom, the speed of adjustment to economic policy changes, controlling heterogeneity, and efficiency in the identification and measurement of economic issues.

As far as the limitations from using panel data in energy economics are concerned, these are basically twofold. The first one is the cross-sectional dependence phenomenon. High degrees of globalization or cross-unit relationships may give rise to the existence of this problem. In the case of panel data, which is cross-sectionally dependent, the estimation generally experiences efficiency loss and yields invalid test statistics. Second, the fact that panel data have a two-dimensional form causes the error-term of the analysis to contain individual-specific cross-section and time effects. If these effects are correlated to explanatory variables, the endogeneity problem occurs, and this constitutes the second limitation of panel data utilization in the EGN. Under the existence of the endogeneity problem, the estimated coefficients become inconsistent and upward-biased.

The rest of this chapter is organized as follows: besides Section 1, which explains why panel data analysis is useful; Section 2 describes the steps in panel data analysis; Section 3 illustrates the tools (steps) of the previous section, through an empirical example; and Section 4 offers conclusions.

2 The Steps of Panel Data Analysis in the EGN

Basically, focusing on the connection between the various types of panel data methodologies, this section draws a path that a researcher should follow in order to conduct a complete empirical analysis in the field of the EGN.

2.1 Testing for Cross-Sectional Dependence

As aforementioned, cross-sectional dependence is one of the most important diagnostics that a researcher should investigate before estimating any panel data models. The existence

H0: No cross-sectional problem exists in the data

Figure 8.1: The Null Hypothesis for the Cross-Sectional Dependence Test.

or absence of this problem determines the path that is to be followed next. If cross-sectional dependence is found in the data, other steps of the analysis have to be taken that will allow for cross-sectional dependence.

There are limited numbers of cross-sectional dependence tests that can be used for detecting the problem. These are the Breusch and Pagan (1980) LM test, Pesaran (2004) scaled LM test, Pesaran (2004) CD test, and Baltagi et al. (2012) bias-corrected scaled LM test. The general null hypothesis for these tests is that "no cross-sectional dependence exists in the data" (Fig. 8.1).

If the data set is composed of panel observations from a small number of cross-section units, then the Breusch and Pagan (1980) LM test can be acknowledged as the best choice. On the other hand, as the Breusch and Pagan (1980) LM test is not appropriate for panel data sets with a large number of cross-section units, Pesaran (2004) proposed a standardized version of the LM test. The scaled LM test is applicable for the panels under large time and cross-sectional settings. However, if the number of cross-section units is large, while the time dimension is not, the size distortion originating from the expected value of the correlation coefficients obtained from the unobserved individual-specific effects gets things worse.

To overcome this shortcoming of the scaled LM test, Pesaran (2004) formulated the CD test. The CD test has good properties for the panels with both small cross-sections and time dimensions. In addition to that, Baltagi et al. (2012) offered a bias correction for the scaled LM test of Pesaran (2004). Under the setting of fixed effects homogenous panels, the bias-corrected scaled LM test can be utilized in the panels with large cross-section and small time dimensions.

2.2 Testing for Stationarity

Within the panel unit root-testing framework, there are two generations of tests. The first generation of tests assumes that cross-section units are cross-sectionally independent; whereas the second generation of panel unit root tests relaxes this assumption and allows for cross-sectional dependence. In this context, it is possible to summarize the first and second generation of panel unit root tests that are often used in the EGN studies, as illustrated in Table 8.1.

Table 8.1: Panel unit root tests.

First Generation		Second Generation	
Nonstationarity tests Im et al. (2003) Levin et al. (2002) Choi (2001) Breitung (2000) Maddala et al. (1999)	Stationarity tests Hadri (2000)	Nonstationarity tests Pesaran (2007) Moon and Perron (2004) Bai and Ng (2004) Chang (2002)	Stationarity tests Bai and Ng (2005) Harris et al. (2005)

Among the first generation panel unit root tests, all the tests except for Hadri (2000), test the null hypothesis of a unit root. Furthermore, while the Levin et al. (2002), Breitung (2000), and Hadri (2000) tests assume that there is a common unit root process (e.g., cross-sections are homogeneous) across cross-section units, the Im et al. (2003), Choi (2001), and the Maddala et al. (1999) tests are based on heterogeneous cross-section formation. Finally, it is possible to utilize all the first-generation tests except for the Hadri (2000) test for unbalanced panels.

The second generation of panel unit root tests aims to overcome the shortcoming of cross-sectional dependence in the first-generation tests. With regards to this, all the tests except for the Bai and Ng (2005) and Harris et al. (2005) assume that there is a unit root in the data. The second-generation tests are based on the heterogeneity assumption. Accordingly, there is no common autoregressive (AR) structure in the series and the panels are heterogeneous.

2.3 Testing for Panel Cointegration in the EGN

The sufficient condition for analyzing the cointegration relationship in the panel analysis of the EGN is to prove that the variables in question are not level stationary (Erdem and Tugcu, 2012a). If one looks into the literature, one can find three groups of panel cointegration approaches that have been adopted for proving the long-run relationship between energy consumption and economic growth (the interested reader should also see Ozturk, 2010). These tests can be classified as follows (Table 8.2).

Baltagi (2005) states that, especially for small data sets, panel cointegration tests are more powerful than time series cointegration tests. In this sense, irrespective of their starting base (residual, likelihood, or error correction), the tests illustrated in Table 8.2 are applicable for

Table 8.2: Panel cointegration tests.

Residual-Based Tests	Likelihood-Based Tests	Error Correction-Based Tests
McCoskey and Kao (1998) Kao (1999) Pedroni (1999, 2004)	Larsson et al. (2001) Groen and Kleibergen (2003)	Westerlund (2007)

the investigation of the long-run relationship in the EGN. However, Groen and Kleibergen (2003)[1] and Westerlund (2007) cointegration methodologies have an advantage over the others, in terms of allowing for cross-sectional dependence. The main difference between these two tests is that Groen and Kleibergen (2003) methodology allows for the possibility of multiple cointegration relationship in the panels; whereas Westerlund (2007) does not.

Given that the cross-section units are independent, Gutierrez (2003) proved that residual-based tests have a better performance than the Larsson et al. (2001) in explaining the cointegration relationship. On the other hand, the Pedroni (1999, 2004) tests have more explanatory power than other residual-based panel cointegration tests, when the time dimension of the data set gets larger. That is, the McCoskey and Kao (1998) and the Kao (1999) tests perform better in panels, which exhibit small sample properties. Nevertheless, under the assumption that the cross-section units are heterogeneous, the Pedroni (1999, 2004) tests seem to be the best choice for the studies that deal with the investigation of the EGN.

2.4 Testing for Panel Causality in the EGN

One of the most popular tools utilized in the panel data analysis of the EGN, is the causality approach. If there is no theoretical restriction, the causality between energy consumption and economic growth can be investigated without requiring a precondition. In addition to that, although it is not an obligation to conduct a causality analysis following a cointegration methodology, evidence for a possible cointegration relationship indicates a causal link in at least one direction. There are two basic causality approaches that can be adopted for the EGN. The first one is the classical Granger causality test accompanied with a large stacked panel data set. Under the null of non-Granger causality, utilizing the classical Granger causality test is the simplest way for the detection of a possible causal link between energy consumption and economic growth. This method assumes that the panels are homogeneous, and does not allow any interconnections of the data between cross-section units.

The second panel causality approach that is applicable in this field was recently developed by Dumitrescu and Hurlin (2012). By assuming that cross-sections in the panel are heterogeneous, the test has good small sample properties, even in the presence of cross-sectional dependence. Although the cross-section units in the panel are found to be independent, utilizing the Dumitrescu and Hurlin (2012) test is convenient, as it is also capable of reporting the individual-specific causal linkages.

2.5 Panel Estimation in the EGN

According to Phillips and Moon (1999), there are four different cases that a researcher may be faced with, when performing a panel data analysis and are also depicted in Fig. 8.2: (1)

---

[1] This test is an augmented version of the Larsson et al. (2001), as it allows for cross-sectional dependence.

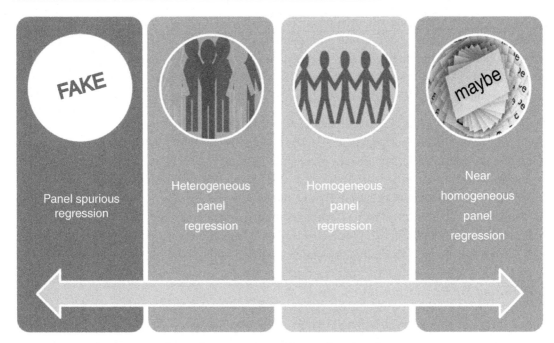

Figure 8.2: Four Cases in Panel Estimation.

panel spurious regression (noncointegrated panels), (2) heterogeneous panel cointegration regression, (3) homogeneous panel cointegration regression, and (4) near-homogeneous panel cointegration regression.

These cases encompass the selection procedure for the appropriate methodology toward a crucial matter: that of obtaining the most efficient, consistent, and unbiased estimation results. In a very recent study, Tiba and Omri (2017) have summarized the existing empirical panel data studies on the EGN in light of the cases stated earlier. The interested readers had better review the aforementioned study for the details.

2.5.1 Estimating noncointegrated panels

Starting from the first case shown in Fig. 8.2; if the variables in a panel do not have any integration order, coefficient estimates can be secured by using the pooled OLS that is assumed as one of the most efficient estimators. However, this may result in inconsistent findings, as the OLS ignores the unobservable cross-section (group) and time effects. To avoid this inconsistency, the fixed effects (FE) or the random effects (RE) models can be applied.

Theoretically, if the purpose of the panel data analysis is to reach a general result by using the data of randomly selected cross-section units from the whole sample, the RE model is an appropriate choice. However, if it is proposed to obtain a general result by employing the data

Table 8.3: Tests for the error-component formation.

RE Models	FE Models
Breusch and Pagan (1980) LM tests Honda (1985) LM tests	Moulton and Randolph (1989) F test

FE, Fixed effects; RE, random effects.

of a fixed group, the most suitable model is FE. However, in the case of endogeneity, the RE model results in biased and inconsistent estimates, whereas it performs better than the FE under the assumption of exogeneity. In this sense, the test statistic (hereafter, Hausman test) that was developed by Hausman (1978) and Hausman and Taylor (1981) can be utilized to provide evidence for the existence or absence of endogeneity. According to Baltagi (2005), under the null of no endogeneity (i.e., no correlation between unobservable effects and the independent variables), the Hausman test shows that the feasible generalized least squares estimator of the RE model is unbiased and consistent. On the other hand, the rejection of the null hypothesis forces to employ the fixed effect estimator (also known as the within estimator), that is, a component of the FE model.

After selecting the appropriate estimator, one should define the error-component formation of the analysis. In this context, the panel data approach accommodates two unobservable error-components, namely the group effects and time effects. There are several methods for deciding the error-component formation of the regression, depending on whether it is estimated by the FE or RE models. These methods can be illustrated as follows (Table 8.3).

Breusch and Pagan (1980) developed the LM test statistics to test the existence of group and time effects in the RE models. The LM1 statistic tests the significance of the group effects under the null of no group effects and the LM2 statistic tests the significance of time effects under the null of no time effects. If one of the null of the LM1 or LM2 is rejected, the model is called as one-way RE model; on the other hand, if both the nulls of LM1 and LM2 are rejected, then the model takes the form of two-way RE.

The Honda (1985) tests differ from the tests of Breusch and Pagan (1980) by being applicable in unbalanced panels and allowing the alternative hypotheses to be one-sided. On the other hand, under the nulls of no group and no time effects, the Moulton and Randolph (1989) F-test investigates the presence of the group and time effects in the FE models. If one of the nulls is rejected, the model becomes a one-way FE model. On the contrary, if both the nulls are rejected, then the model takes the form of a two-way FE.

As the FE, the feasible generalized least squares and the pooled OLS estimators are not capable in dealing with strong serial-correlation, heteroscedasticity, and multicollinearity problems, researchers are responsible for detecting and correcting those failures, if they exist (the interested reader might also see Erdem and Tugcu, 2011).

2.5.2 Estimating cointegrated panels

Due to the endogeneity and serial-correlation problems, the OLS estimators exhibit a nonnegligible bias in finite samples for estimating the cointegrated panels. Thus, it is vital to find an appropriate tool for achieving efficient outcomes. There are basically two approaches for the estimation of cointegrated panels in the field of the EGN. The first one is the fully modified OLS (FMOLS), which is an extended version of the Phillips and Hansen (1990) methodology. The second approach is the dynamic OLS (DOLS) that was originally developed by Saikkonen (1992) and Stock and Watson (1993). The FMOLS and the DOLS have been improved many times in light of the needs of panel data analysis and have assumed the form they have today. In this respect, it is possible to classify the related methodologies as shown in Fig. 8.3.

In contrast to the OLS estimators, each of the FMOLS and DOLS estimators shown in Fig. 8.3, has a power against endogeneity and serial-correlation, both in homogeneous and heterogeneous panels. Nevertheless, Phillips and Moon (1999) FMOLS estimators, as well

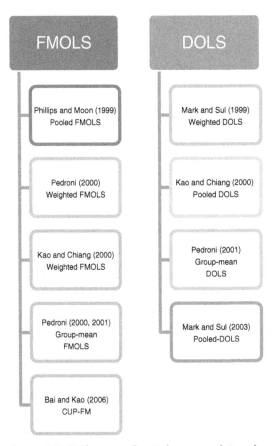

Figure 8.3: Estimators for Cointegrated Panels.

as Kao and Chiang (2000) and Mark and Sul (2003) DOLS estimators were proposed for homogeneous and/or near-homogeneous cointegrated panels.

Among the fully modified methodologies, the continuously updated fully modified (CUP-FM) estimator of Bai and Kao (2006) has an advantage over the others, in terms of considering the possible cross-sectional dependence. However, by Kao and Chiang (2000) and Mark and Sul (2003), the DOLS estimators are proposed to be superior to the FMOLS estimators, as the DOLS estimators are fully parametric and do not require any preestimation and nonparametric correction. In this context, assuming the cross-section units in the panel are independent, utilizing the DOLS methodology for estimating the cointegrated panels in the EGN field, may provide statistically more significant parameters compared to the FMOLS.

2.5.3 Estimating heterogeneous dynamic cointegrated panels: a panel autoregressive distributed lag approach

The Pesaran et al. (2001) bounds testing methodology in time series econometrics can be regarded as one of the most important milestones in the field of cointegration analysis. This methodology has allowed researchers to adopt a cointegration analysis with time series that are purely I(0), purely I(1) or mutually cointegrated (Erdem and Tugcu, 2012b). It is clear that the origins of this methodology can be traced in the seminal papers of Pesaran and Smith (1995) and Pesaran et al. (1999).

In the EGN studies that fulfill the autoregressive distributed lag conditions,[2] it is appropriate to utilize the mean-group (MG) and pooled mean-group (PMG) estimators, proposed by Pesaran and Smith (1995) and Pesaran et al. (1999) in performing a dynamic heterogeneous panel cointegration analysis. With the panels having large cross-section and time dimensions, the MG and PMG are able to produce consistent estimates. The main difference between these estimators is that, while both estimators allow for their intercepts, short-run coefficients, and error variances to differ freely across cross-section units, only the PMG constraints the long-run coefficients to be the same. That is, under the assumption of long-run slope homogeneity, although the PMG and the MG estimators are supposed to be consistent, the PMG estimator is also assumed to be efficient. In this respect, Pesaran et al. (1999) suggested employing a joint Hausman test statistic, which tests the null of long-run homogeneity. Accepting the null hypothesis supports the consistency and efficiency of the coefficients obtained from the PMG estimator rather than the MG estimator (Menegaki and Tugcu, 2017).

As the MG and PMG estimators do not consider the cross-sectional dependence, Pesaran (2006) proposed the common correlated effects (CCE) estimators, namely, common correlated effects mean-group (CCEMG) and common correlated effects pooled (CCEP)

---

[2] In the case of a panel data set that exhibits I(2) integration order, it is not possible to proceed with the ARDL analysis. Foremost, this methodology requires a statistically significant and negative error-correction parameter to be estimated as evidence for the cointegration relationship.

estimators, which can make allowance for the existence of cross-sectional dependence, as well as endogeneity and serial correlation, and yield consistent and efficient parameter estimates. Just as in the PMG estimator case, the CCEP estimator either constraints long-run coefficients to be the same, or allows the intercepts, short-run coefficients, and error variances to differ freely across cross-section units. Keeping the null of the long-run slope homogeneity, the CCEP estimator is assumed to produce consistent and efficient parameters. Besides, it is also proved by Pesaran (2006) that the CCEP estimator performs slightly better than the CCEMG estimator in panels with small cross-section and time dimensions.

3 An Empirical Example for the G7 Countries

3.1 The Model and the Data

The link between energy consumption and economic growth is generally investigated by following the classical and/or the augmented Cobb–Douglas production function. In this section, and for simplicity reasons, the whole analysis will be performed on a univariate model:

$$Y = F(E)$$

where Y and E represent nominal GDP and total final energy consumption, respectively.

The data set includes GDP in current US dollars and total final energy consumption measured as TJ (terra joules = 1012 joules) of the G7 countries for the period 1990–2012. While the GDP data were sourced from the World Development Indicators database, total final energy consumption data were obtained from the "Sustainable Energy for All" database of the World Bank. To achieve a smooth analysis, namely, for avoiding outliers and supporting stationarity of the variables, as well as obtaining moderate and acceptable coefficients, and to yield elasticity estimates rather than slope parameters, the data was transformed in natural logarithms.

3.2 Investigating Cross-Sectional Dependence

As aforementioned, cross-sectional dependence is one of the most important diagnostics that a researcher should investigate before performing a panel data analysis. In this context, the Breusch and Pagan (1980) LM test, Pesaran (2004) scaled LM test, Pesaran (2004) CD test, and Baltagi et al. (2012) bias-corrected scaled LM test were utilized [note: the interested reader might find useful to read (Tugcu and Tiwari, 2016) for a detailed explanation of the tests] and the findings were reported in Table 8.4.

Findings in Table 8.4 illustrate that the null of "no cross-sectional dependence" is rejected even at 1% level of significance. Thus, we need to proceed with tests and estimation techniques that can take account of cross-sectional dependence.

Table 8.4: Cross-sectional dependence test results.

Tests	Y	E
Breusch–Pagan LM	353.481 (0.00)	195.328 (0.00)
Pesaran-scaled LM	50.222 (0.00)	25.819 (0.00)
Pesaran CD	18.566 (0.00)	8.837 (0.00)
Bias-corrected scaled LM	50.063 (0.00)	25.660 (0.00)

Numbers in parentheses are *P*-values.

3.3 Investigating Stationarity

The existence of cross-sectional dependence requires a unit root test, which allows for cross-sectional dependence, while investigating stationarity in panel data. This prerequisite is materialized in the Pesaran (2007) methodology. Pesaran (2007) proposed a unit root test defined as cross-sectionally augmented IPS test (hereafter, CIPS) for investigating the existence of a unit root in a cross-sectionally dependent panel data set. The proposed test statistic can be formulated as follows:

$$\text{CIPS}(N,T) = N^{-1} \sum_{i=1}^{N} t_i(N,T)$$

where ti(*N,T*) is the cross-sectionally augmented Dickey–Fuller statistic for the *i*th cross-section unit.

As seen from Table 8.5, all variables are first-difference stationary. As it is clearly explained in Section 2.5, this result supports evidence for a possible cointegration relationship between the nominal GDP and total final energy consumption; and thus requires checking whether there exists a cointegration relationship.

3.4 Investigating Panel Cointegration

Westerlund (2007) proposed four panel cointegration test statistics with a bootstrapping option that test the existence of long-run relationships among integrated variables, even with

Table 8.5: Panel unit root test results.

Variables	CIPS
Y	−1.505
E	−2.086
ΔY	−3.406
ΔE	−5.648

Δ is the first difference operator.
The critical values for Y, E and ΔY are −2.12, −2.25, and −2.51 at 10, 5, and 1% levels of significance, respectively.
The critical values for ΔE are −2.10, −2.22, and −2.44 at 10, 5, and 1% levels of significance, respectively.

the existence of cross-sectional dependence. The underlying idea is to test for the absence of cointegration by determining whether the individual panel members are error-correcting. The considered error-correction model can be expressed as follows:

$$\Delta y_{i,t} = c_i + \alpha_{i,1} \Delta y_{i,t-1} + \alpha_{i,2} \Delta y_{i,t-2} + \ldots + \alpha_{i,p} \Delta y_{i,t-p} + \beta_{i,0} \Delta x_{i,t} + \beta_{i,1} \Delta x_{i,t-1} + \ldots + \beta_{i,p} \Delta x_{i,t-p}$$
$$+ \alpha_i \left(y_{i,t-1} - \beta_i x_{i,t-1} \right) + \mu_{i,t}$$

where α_i provides an estimate of the speed of error-correction toward the long-run equilibrium; $y_{i,t} = -\left(\beta_i/\alpha_i\right) \times x_{i,t}$ for that series i.

The first two test statistics (i.e., panel statistics) of Westerlund (2007) test H_0: $\alpha_i = 0$ for all i versus the alternative H_1: $\alpha_i < 0$ for all i, which indicates that rejection of H_0 should be taken as evidence of cointegration for the panel as a whole. The second two statistics (i.e., group mean statistics) test H_0: $\alpha_i = 0$ for all i against H_1: $\alpha_i < 0$ for at least one i, suggesting that rejection of H_0 should be taken as evidence of cointegration for at least one of the cross-sectional units. The two panel statistics are as follows:

$$P_r = \frac{\hat{\alpha}_i}{SE\left(\hat{\alpha}_i\right)}$$

$$P_\alpha = T\hat{\alpha}$$

where $\hat{\alpha}_i$ is the estimated value of the error correction parameter and $SE\left(\hat{\alpha}_i\right)$ is the conventional standard error of $\hat{\alpha}_i$. The two group mean statistics are as follows:

$$G_r = \frac{1}{N} \sum_{i=1}^{N} \frac{\hat{\alpha}_i}{SE\left(\hat{\alpha}_i\right)}$$

$$G_\alpha = \frac{1}{N} \sum_{i=1}^{N} \frac{T\hat{\alpha}_i}{\hat{\alpha}_i(1)}$$

where $\hat{\alpha}_i(1) = 1 - \sum_{j=1}^{P_i} \hat{\alpha}_{ij}$ and $SE\left(\hat{\alpha}_i\right)$ is the conventional standard error of $\hat{\alpha}_i$.

According to Westerlund (2007), in addition to be allowing for cross-sectional dependence, each test is able to accommodate individual-specific short-run dynamics, including serially correlated error terms, nonstrictly exogenous regressors, individual-specific intercept and trend terms, and individual-specific slope parameters. These tests also have good small-sample properties with small size distortions and high power relative to other popular residual-based panel cointegration tests (Erdem and Tugcu, 2012a) (Table 8.6).

Findings indicate that the null hypotheses are rejected in both cases, that is, final energy consumption is cointegrated to nominal GDP in G7 countries. With respect to this, it is

Table 8.6: Panel cointegration test results.

Tests	Statistics
P_r	$-9.544\ (0.00)$
P_α	$-14.385\ (0.00)$
G_r	$-3.743\ (0.00)$
G_α	$-21.282\ (0.00)$

Numbers in parentheses are *P*-values.
The average lag length selected by the AIC information criterion is 3.
The width of the Bartlett kernel window is set to 2.
Panel cointegration relationship contains a constant and deterministic trend.

necessary to follow a methodology that is capable of estimating a cointegrated panel that also suffers from cross-sectional dependence.

3.5 Coefficient Estimation From the Cointegrated Panel

Bai and Kao (2006) analyzed the limiting distribution of some panel estimators and formulated a CUP-FM estimator, which accounts for cross-sectional dependence in panel data. It is mentioned that the CUP-FM estimator has better small-sample properties than other estimators (e.g., OLS, two staged-FM, etc.). The CUP-FM estimator can be formulated in the following manner:

$$
\hat{\beta}_{\text{CUP-FM}} = \left[\sum_{i=1}^{n} \left(\sum_{t=1}^{T} y_{i,t}^{+}\left(\hat{\beta}_{\text{CUP-FM}}\right)\left(x_{i,t} - \tilde{x}_i\right)' - T\left(\lambda_i'\left(\hat{\beta}_{\text{CUP-FM}}\right)\Delta_{Fei}^{+}\left(\hat{\beta}_{\text{CUP-FM}}\right) + \Delta_{\mu ei}^{+}\left(\hat{\beta}_{\text{CUP-FM}}\right)\right)\right) \right]
$$
$$
\left[\sum_{i=1}^{n} \sum_{t=1}^{T} \left(x_{i,t} - \bar{x}_i\right)\left(x_{i,t} - \bar{x}_i\right)' \right]^{-1}
$$

where $y_{i,t}^{+} = y_{i,t} - (\lambda_i'\Omega_{Fei} + \Omega_{\mu ei})\Omega_{ei}^{-1}\Delta x_{i,t}$ is the transformed dependent variable of the fixed effect panel regression, which was modified for correcting endogeneity, and λ_i' stands for the estimated factor loadings. The CUP-FM is constructed by estimating parameters and the long-run covariance matrix and loadings recursively. Thus, the $\hat{\beta}_{\text{FM}}$, $\hat{\Omega}$, and λ_i' are estimated repeatedly, until convergence is reached (Bai and Kao, 2006).

According to the estimation result in Table 8.7, 1% increase in final energy consumption raises nominal GDP by 1.033%.

Table 8.7: The CUP-FM estimation results.

Dependent Variable: ln *Y*	Statistics
ln*E*	1.033 (8.579)

Number in parenthesis is *t*-statistic.

Table 8.8: Panel causality test results.

Cross-Section Units	Null Hypotheses	
	E Does Not Homogeneously Cause Y	Y Does Not Homogeneously Cause E
The whole panel sample	2.441 (0.04)	2.608 (0.02)
Canada	1.833 (0.17)	0.066 (0.79)
France	2.746 (0.09)	0.310 (0.57)
Germany	0.655 (0.41)	8.373 (0.00)
Italy	4.484 (0.03)	6.050 (0.01)
Japan	6.359 (0.01)	0.416 (0.51)
United Kingdom	1.087 (0.29)	3.812 (0.05)
United States	0.822 (0.36)	0.192 (0.66)

Numbers in parentheses are *P*-values.
Lag length was set to 1.

3.6 Investigating Panel Causality

Dumitrescu and Hurlin (2012) proposed a panel causality test based on the individual Wald statistic of Granger noncausality, averaged across the cross-section units (Table 8.8). The testing procedure allows considering the heterogeneity of causal relationships and the heterogeneity of the regression model used for testing Granger causality. The linear panel regression model followed by Dumitrescu and Hurlin (2012) is as follows:

$$Y_{i,t} = \alpha_i + \sum_{j=1}^{J} \lambda_i^j Y_{i,t-j} + \sum_{j=1}^{J} \beta_i^j E_{i,t-j} + \varepsilon_{i,t}$$

where Y is the nominal GDP and E is total final energy consumption.

Dumitrescu and Hurlin (2012) state that "a homogeneous specification of the relationship between the variables x and y does not allow to interpret causality relationships, if any individual from the sample has an economic behavior different from that of the others". Thus, they propose an average Wald statistic, which tests the null that "no homogeneous causal relationship exists for any of the cross-section units," namely, $[H_0 : \beta_i = 0, (i = 1,..,N)]$, against the alternative hypothesis that the causal relationships occur for at least one subgroup of the panel, namely, $[H_1 : \beta_i = 0, (i = 1,..N_1); \beta_i \neq 0, (i = N_1 + 1, N_1 + 2,.., N)]$. Rejection of the null hypothesis with $N_1 = 0$ indicates that x Granger causes y for all i, whereas rejection of the null hypothesis with $N_1 > 0$ provides evidence that the regression model and the causal relationships vary from one individual or the sample to another. Under these circumstances, the average of individual Wald statistic generated by Dumitrescu and Hurlin (2012) can be formulated in the following manner:

$$W_{N,T}^{\text{Hnc}} = \frac{1}{N} \sum_{i=1}^{N} W_{i,T}$$

where $W_{i,T}$ is the individual Wald statistic for the ith cross-section unit.

Findings illustrate that the feedback hypothesis holds, as evidence of a bidirectional causality is present between total final energy consumption and nominal GDP in the G7 countries for the period 1990–2012. Moreover, country-specific estimates reveal that there exists a unidirectional causal link running from total final energy consumption to nominal GDP in France; whereas causality is also running from nominal GDP to total final energy consumption in Germany and the United Kingdom. That is, the growth hypothesis is supported for France, whereas the conservation hypothesis holds for Germany and the United Kingdom. Finally, rejection of the null hypotheses for Italy results in a bidirectional causality between total final energy consumption and nominal GDP, which provides evidence for the validity of the feedback hypothesis.

4 Conclusions

This chapter has been dedicated to shed light on the ordinary challenges that a researcher may be faced with, while performing a panel data analysis in the field of the EGN. By avoiding complicated econometric material, this chapter has attempted to draw a simple path to be followed, without getting stuck in complex statistics and formulae. However, performing an econometric analysis and transferring it into a scientific paper is quite different than learning how to conduct that analysis. A scientific study is established on a theoretical basis. Empirical analysis, on the other hand, is just a tool that can make that theory more apparent. Thus, potential researchers should not neglect the fundamental link between the theoretical and empirical sides of the energy economics field. In this sense, a detailed and extensive review of literature may help to provide the necessary knowledge in terms of the theoretical and the statistical components and the interpretations of panel data analyses in the field of EGN.

References

Bai, J., Kao, C., 2006. On the estimation and inference of a panel cointegration model with cross-sectional dependence. In: Baltagi, B.H. (Ed.), Panel Data Econometrics: Theoretical Contributions and Empirical Applications. Elsevier, Amsterdam.

Bai, J., Ng, S., 2004. A PANIC attack on unit roots and cointegration. Econometrica 72 (4), 1127–1177.

Bai, J., Ng, S., 2005. A new look at panel testing of stationarity and the PPP hypothesis. In: Rothenberg, T.J., Andrews, D., Stock, J. (Eds.), Identification and Inference in Econometric Models: Essay. Cambridge University Press, New York.

Baltagi, B., 2005. Econometric Analysis of Panel Data, third ed. John Wiley & Sons Ltd, United Kingdom.

Baltagi, B.H., Feng, Q., Kao, C., 2012. A Lagrange multiplier test for cross-sectional dependence in a fixed effects panel data model. J. Econom. 170 (1), 164–177.

Breitung, J., 2000. The local power of some unit root tests for panel data. Baltagi, B. (Ed.), Nonstationary Panels, Panel Cointegration, and Dynamic Panels, Advances in Econometrics, 15, JAI, Amsterdam, pp. 161–178.

Breusch, T.S., Pagan, A.R., 1980. The Lagrange multiplier test and its applications to model specification in econometrics. Rev. Econ. Stud. 47 (1), 239–253.

Chang, Y., 2002. Nonlinear IV unit root tests in panels with cross-sectional dependency. J. Econom. 110 (2), 261–292.

Choi, I., 2001. Unit root tests for panel data. J. Int. Money Finance 20 (2), 249–272.

Dumitrescu, E.I., Hurlin, C., 2012. Testing for Granger non-causality in heterogeneous panels. Econ. Model. 29 (4), 1450–1460.

Erdem, E., Tugcu, C.T., 2011. Investigating the macroeconomic and qualitative dynamics of urban economic growth: evidence from the most productive Turkish cities. Int. J. Bus. Soc. Sci. 2, 136–145.

Erdem, E., Tugcu, C.T., 2012a. New evidence on the relationship between economic freedom and growth: a panel cointegration analysis for the case of OECD. Glob. Econ. J. 12 (3), 1–18.

Erdem, E., Tugcu, C.T., 2012b. Higher education and unemployment: a cointegration and causality analysis of the case of Turkey. Eur. J. Educ. 47 (2), 299–309.

Groen, J.J.J., Kleibergen, F., 2003. Likelihood-based cointegration analysis in panels of vector error-correction models. J. Bus. Econ. Stat. 21 (2), 295–318.

Gutierrez, L., 2003. On the power of panel cointegration tests: a Monte Carlo comparison. Econ. Lett. 80 (1), 105–111.

Hadri, K., 2000. Testing for stationarity in heterogeneous panel data. Econom. J. 3 (2), 148–161.

Harris, D., Leybourne, S., McCabe, B., 2005. Panel stationarity tests for purchasing power parity with cross-sectional dependence. J. Bus. Econ. Stat. 23 (4), 395–409.

Hausman, J.A., 1978. Specification tests in econometrics. Econometrica 46 (6), 1251–1271.

Hausman, J.A., Taylor, W.E., 1981. Panel data and unobservable individual effects. Econometrica 49 (6), 1377–1398.

Honda, Y., 1985. Testing the error components model with non-normal disturbances. Rev. Econ. Stud. 52 (4), 681–690.

Im, K.S., Pesaran, M.H., Shin, Y., 2003. Testing for unit roots in heterogeneous panels. J. Econom. 115 (1), 53–74.

Kao, C., 1999. Spurious regression and residual-based tests for cointegration in panel data. J. Econom. 90 (1), 1–44.

Kao, C., Chiang, M.H., 2000. On the estimation and inference of a cointegrated regression in panel data. Baltagi, B.H. (Ed.), Advances in Econometrics: Nonstationary Panels, 15, Panel Cointegration and Dynamic Panels, pp. 179–222.

Larsson, R., Lyhagen, J., Löthgren, M., 2001. Likelihood-based cointegration tests in heterogeneous panels. Econom. J. 4 (1), 109–142.

Levin, A., Lin, C.F., Chu, C.S.J., 2002. Unit root tests in panel data: asymptotic and finite-sample properties. J. Econom. 108 (1), 1–24.

Maddala, G.S., Wu, S., Liu, P., 1999. Do panel data rescue purchasing power parity (PPP) theory?". In: Krishnakumar, J., Ronchetti, E. (Eds.), Panel Data Econometrics: Future Directions. Elsevier.

Mark, N.C., Sul, D., 2003. Cointegration vector estimation by panel DOLS and long-run money demand. Oxf. Bull. Econ. Stat. 65 (5), 655–680.

McCoskey, S., Kao, C., 1998. A residual-based test of the null of cointegration in panel data. Econom. Rev. 17 (1), 57–84.

Menegaki, A.N., Tugcu, C.T., 2017. Energy consumption and Sustainable Economic Welfare in G7 countries; a comparison with the conventional nexus. Renew. Sustain. Energy Rev. 69, 892–901.

Moon, H.R., Perron, B., 2004. Testing for a unit root in panels with dynamic factors. J. Econom. 122 (1), 81–126.

Moulton, B.R., Randolph, W.C., 1989. Alternative tests of the error components model. Econometrica 57, 685–693.

Ozturk, I., 2010. A literature survey on energy–growth nexus. Energy Policy 38 (1), 340–349.

Pedroni, P., 1999. Critical values for cointegration tests in heterogeneous panels with multiple regressors. Oxf. Bull. Econ. Stat. 61 (S1), 653–670.

Pedroni, P., 2004. Panel cointegration: asymptotic and finite sample properties of pooled time series tests with an application to the PPP hypothesis. Econom. Theory 20 (03), 597–625.

Pesaran, M.H., 2004. General diagnostic tests for cross section dependence in panels. Cambridge Working Papers in Economics No: 0435. Faculty of Economics, University of Cambridge.

Pesaran, M.H., 2006. Estimation and inference in large heterogeneous panels with a multifactor error structure. Econometrica 74 (4), 967–1012.

Pesaran, M.H., 2007. A simple panel unit root test in the presence of cross-section dependence. J. Appl. Econom. 22 (2), 265–312.

Pesaran, M.H., Shin, Y., Smith, R.P., 1999. Pooled mean group estimation of dynamic heterogeneous panels. J. Am. Stat. Assoc. 94 (446), 621–634.

Pesaran, M.H., Shin, Y., Smith, R.J., 2001. Bounds testing approaches to the analysis of level relationships. J. Appl. Econom. 16 (3), 289–326.

Pesaran, M.H., Smith, R., 1995. Estimating long-run relationships from dynamic heterogeneous panels. J. Econom. 68 (1), 79–113.

Phillips, P.C., Hansen, B.E., 1990. Statistical inference in instrumental variables regression with I (1) processes. Rev. Econ. Stud. 57 (1), 99–125.

Phillips, P.C., Moon, H.R., 1999. Linear regression limit theory for nonstationary panel data. Econometrica 67 (5), 1057–1111.

Saikkonen, P., 1992. Estimation and testing of cointegrated systems by an autoregressive approximation. Econom. Theory 8 (01), 1–27.

Stock, J.H., Watson, M.W., 1993. A simple estimator of cointegrating vectors in higher order integrated systems. Econometrica 61, 783–820.

Tiba, S., Omri, A., 2017. Literature survey on the relationships between energy, environment and economic growth. Renew. Sustain. Energy Rev. 69, 1129–1146.

Tugcu, C.T., Tiwari, A.K., 2016. Does renewable and/or non-renewable energy consumption matter for total factor productivity (TFP) growth? Evidence from the BRICS. Renew. Sustain. Energy Rev. 65, 610–616.

Westerlund, J., 2007. Testing for error correction in panel data. Oxf. Bull. Econ. Stat. 69 (6), 709–748.

Further Reading

Mark, N., Sul, D., 1999. A computationally simple cointegration vector estimator for panel data. Ohio State University Manuscript.

Pedroni, P., 2000. Fully modified OLS for heterogeneous cointegrated panels. Adv. Econom. 15, 93–130.

Pedroni, P., 2001. Purchasing power parity tests in cointegrated panels. Rev. Econ. Stat. 83 (4), 727–731.

CHAPTER 9

Testing for Causality: A Survey of the Current Literature

Nicholas Apergis

University of Piraeus, Piraeus, Greece

Chapter Outline

1 Introduction

Many tests of the Granger-type causality have been derived and implemented to test the direction of causality for time series (Granger, 1969; Geweke et al., 1983; Sims, 1972). These tests are based on investigating the validity of null hypotheses and are formulated as zero restrictions on the coefficients of the lags of a subset of the variables. The goal of this chapter is to provide a review survey on the most significant contributions to the literature in relevance to causality testing.

2 Bivariate Causality Tests in Time Series

2.1 Causality Tests Without Cointegration

With the aim of testing for the presence of bivariate causality, first, we need to introduce a Vector Autoregressive (VAR) model. In particular, we introduce a bivariate VAR model with two-period lag (actually, the number of appropriate lags must be determined through an informational optimization criterion, such as the Akaike criterion). We assume the presence of two endogenous variables y_t and x_t:

$$y_t = a_{10} + b_{11}y_{t-1} + b_{12}y_{t-2} + b_{13}x_{t-1} + b_{14}x_{t-2} + \varepsilon_{y_t}$$
$$x_t = a_{20} + b_{21}y_{t-1} + b_{22}y_{t-2} + b_{23}x_{t-1} + b_{24}x_{t-2} + \varepsilon_{x_t}$$

where ε_s are the error terms in the two equations. In matrix form, the above model yields:

$$\begin{pmatrix} y_t \\ x_t \end{pmatrix} = \begin{pmatrix} a_{10} \\ a_{20} \end{pmatrix} + \begin{pmatrix} b_{11} & b_{12} \\ b_{21} & b_{22} \end{pmatrix} \begin{pmatrix} y_{t-1} \\ x_{t-1} \end{pmatrix} + \begin{pmatrix} b_{13} & b_{14} \\ b_{23} & b_{24} \end{pmatrix} \begin{pmatrix} y_{t-2} \\ x_{t-2} \end{pmatrix} + \begin{pmatrix} \varepsilon_{y_t} \\ \varepsilon_{x_t} \end{pmatrix}$$

Needless to say, before any testing for causality, unit root testing should have identified whether the variables x and y should be in levels or in first differences.

The variable x does not cause the variable y, if the null hypothesis H_0: $b_{13} = b_{14} = 0$ cannot be rejected. Similarly, the variable y does not cause the variable x, if the null hypothesis H_0: $b_{21} = b_{22} = 0$ cannot be rejected. The causality tests are performed in both directions through a system of equations. To test the above null hypotheses, we employ Wald tests, in the same fashion as a restriction of the null hypothesis would be tested with the assistance of the F-test. If we reject the null hypothesis, then causality is established. The F-statistic is calculated as follows:

$$F = \left[\text{SSE}_0 - \text{SSE}\right] / \text{SEE} \times \left[N - 2k - 1\right] / k$$

where SSE_0 denotes the sum of squared residuals for a model with restrictions ($b_{13} = b_{14} = b_{21} = b_{22} = 0$), SSE denotes the sum of squared residuals for the model without any restrictions, k is the number of lags, and N denotes the number of observations. Under the null hypothesis, the F-statistic follows an F-distribution with k degrees of freedom in the numerator and $N - 2k - 1$ degrees of freedom in the denominator (Box 9.1).

2.2 Causality Tests With Cointegration

In the case where the variables y and x are cointegrated, an Error Correction VAR (ECVAR) model should be employed. The ECVAR model can distinguish the long- and short-run relationship between the variables and can identify sources of causality that cannot be detected by the usual Granger causality test. In particular:

$$y_t = a_{10} + b_{11}y_{t-1} + b_{12}y_{t-2} + b_{13}x_{t-1} + b_{14}x_{t-2} + c_1 z_{t-1} + \varepsilon_{y_t}$$
$$x_t = a_{20} + b_{21}y_{t-1} + b_{22}y_{t-2} + b_{23}x_{t-1} + b_{24}x_{t-2} + c_2 z_{t-1} + \varepsilon_{x_t}$$

where zs represent the residuals from the cointegrating vector (i.e., the error-correction term). These terms also represent the speed of adjustment, namely the speed of returning

BOX 9.1 A Published Case Study

The STUDY: Sbaouelgi Jihène and Boulila Ghazi, 2013: The causality between income inequality and economic growth: empirical evidence from the Middle East and North Africa region. Asian Economic and Financial Review 3, 668–682.

Data

They use four indicators of income inequality: (1) the Gini index, (2) the openness rate, (3) the secondary school enrolment rate, and (4) the gross fixed capital formation as a percentage of GDP. Concerning economic growth, the growth rate of GDP per capita is employed, with the data sources being the Word Development Indicators database from the World Bank (2011). Nine MENA countries are used, spanning the period 1960–2011 on an annual basis.

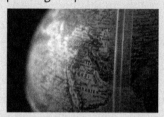

Causality results

Table 9.1: Short-run Granger causality tests.

Countries and Variables	Null Hypothesis	
	Income Inequality → Growth	Growth → Income Inequality
Algeria		
GDP, Gini	0.435	0.210
GDP, GCF	1.442	0.854
GDP, Openness	1.001	0.664
GDP, School	0.331	0.008
Egypt		
GDP, Gini	0.526	0.052
GDP, GCF	14.565**	0.025
GDP, Openness	4.625	1.063
GDP, School	0.630	1.536
Jordan		
GDP, Gini	0.185	0.032
GDP, GCF	0.740	0.027
GDP, Openness	1.705	0.236
GDP, School	2.506	1.131
Mauritania		
GDP, Gini	0.277	0.407
GDP, GCF	0.119	0.017
GDP, Openness	1.204	1.454
GDP, School	2.172	8.596**

** $P \leq 0.05$. All variables are expressed in first differences in logs.

Causality conclusions

According to the findings in Table 9.1, the evidence illustrates that causation is present in the case of Egypt where it turns out to be unidirectional, running from income inequality to growth, while in the case of Mauritania it runs from growth to income inequality. These results hold at the 5% significance level.

to the equilibrium path when a shock hits the two variables. In this new case, causality can be tested both on short-run and long-run bases. In terms of the short-run analysis, the procedure is similar to the case shown in Section 1.1. In terms of the long-run analysis, it is the statistical significance of coefficients c_1 and c_2 that determines the presence of long-run causality. More specifically, if the coefficient c_1 turns out to be statistically significant, then the variable x causes the variable y in the long run. Similarly, if the coefficient c_2 turns out to be statistically significant, then that variable y causes the variable x in the long run as well.

Overall, the term "causality" does not mean that one variable causes the other variable, but it simply means that there is a correlation between the informational content of one variable and the previous values of another variable. Causality does not provide information, however, on the sign of the impact or how long it will last. However, Dufour and Renault (1998) are the first to present a theoretical framework, referred as long (or short) horizon noncausality, which allows researchers to disentangle potentially different causality relations over different forecast horizons. They provide definitions and a set of conditions which ensure the equivalence between standard Wiener–Granger type one-step-ahead noncausality and noncausality at any forecast period. Their methodology on the conditions on noncausality between two variables at a forecast horizon greater than one involves examining the statistical significance of multilinear zero restrictions on the coefficients of the VAR parameters. However, more details on this methodology are provided later in the text (Box 9.2).

BOX 9.2 A Published Case Study

The STUDY: Wadjamsse B. Djezou, 2014. The democracy and economic growth nexus: empirical evidence from Côte d'Ivoire. The European Journal of Comparative Economics 11, 251–266.

Data

Data are on an annual basis, spanning the period from 1960 to 2012. The economic performance is proxied by real GDP per capita, while data are obtained from the World Bank's World Development Indicators database. Data on democracy and regime durability were obtained from the Polity IV database. Regime durability indicates the number of years as the last substantive change in authority characteristics, that is, a measure of the durability of the regime's authority pattern. It is defined as a 3-point move in a country's democracy score. Democracy scores are based on the premise that a mature and internally coherent democracy is an institutional framework in which a political participation is unrestricted, that is, open and fully competitive, an executive recruitment is elective, and the constraints on the chief executive are substantial. Higher values represent stronger democracy with a range of -10 to 10.

Causality results

Table 9.2: Granger causality results.

Dependent Variable	Source of Causation		Long-run Causality
	Short-run Causality		
	Democracy Score	GDP	
Democracy score	–	0.426[0.65]	−3.52[0.00]***
GDP	0.068[0.93]	–	

The null hypothesis is of no causality. All variables are expressed in first differences. Figures in brackets denote *P*-values.
*** $P \leq 0.01$.

Causality conclusions

The results of causality tests are reported in Table 9.2. They document that there is a long-run causality running from GDP to democracy as the estimated coefficient of the lagged error correction term is not only negative, but also statistically significant in the democracy equation. This is quite a fundamental result as it highlights the economic performance of the country that impacts its democratic status and not the reverse. In other words, past information on a country's economic performance does permit a better prediction of the level of democratization in that country. This suggests that poor countries should first of all eliminate poverty before discussing about political freedom (or election). Thus economic growth, through strong institutions, is a precondition for democratization.

3 Multivariate Causality Tests in Time Series

3.1 Causality Without Cointegration

Next, we extend the bivariate causality tests into the multiple variable setting in the VAR scheme. For $t = 1,\ldots, T$, the n-variable VAR model can be represented as follows:

$$\begin{pmatrix} x_{1t} \\ x_{2t} \\ \vdots \\ x_{nt} \end{pmatrix} = \begin{pmatrix} a_{10} \\ a_{20} \\ \vdots \\ a_{n0} \end{pmatrix} + \begin{pmatrix} b_{11}(L) & \cdots & b_{1n}(L) \\ b_{21}(L) & \cdots & b_{2n}(L) \\ & \vdots & \\ b_{n1}(L) & \cdots & b_{nn}(L) \end{pmatrix} \begin{pmatrix} x_{1(t-1)} \\ x_{2(t-1)} \\ \vdots \\ x_{n(t-1)} \end{pmatrix} + \begin{pmatrix} \varepsilon_{1t} \\ \varepsilon_{2t} \\ \vdots \\ \varepsilon_{nt} \end{pmatrix}$$

where $(x_{1t},\ldots, x_{nt})$ is the n-variable vector stationary time series at time t, L is the backward operation, where $L(x_t) = x_{t-1}$, a_{i0} are intercept parameters, and $b_{ij}(L)$ are polynomials in the lag operator L:

$b_{ij}(L) = b_{ij}(1)L + b_{ij}(2)L^2 + \ldots + b_{ij}(p)L^p$ and $\varepsilon_t = (\varepsilon_{1t},\ldots, \varepsilon_{nt})'$ is the disturbance vector obeying the assumptions of the standard linear regression model.

Each equation in the VAR model is expected to have the same lag length for each variable and the regressors are identical in all equations. In other words, a uniform order p is chosen across all the lag polynomials $b_{ij}(L)$ in the VAR model according to a certain criterion, such as the Akaike's Information Criterion (AIC). To test causality, that is, with causality running from x_{2t} to x_{1t}, with the null hypothesis pertaining that the terms $b_{ij}(L)$ are significantly different from zero. Similarly, this applies for the remaining causality cases. First, we run regressions for each equation without any restrictions on the parameters, and we obtain the residual covariance matrix represented with the symbol Σ. Next, we run the regressions with the restriction imposed by the null hypothesis and obtain the restricted residual covariance matrix represented as Σ_0. Then, we get the likelihood ratio statistics suggested by Sims (1980):

$$(T - c)(\log |\Sigma_0| - \log |\Sigma|)$$

where T is the number of observations, c is the number of parameters estimated in each equation of the unrestricted system, and $\log|\Sigma_0|$ and $\log|\Sigma|$ are the natural logarithms of the determinant of restricted and unrestricted residual covariance matrix, respectively. This test statistic follows an asymptotic χ^2 distribution with the degrees of freedom being equal to the number of restrictions on the coefficients in the system.

3.2 Causality With Cointegration

In case of the presence of cointegration across the multiple variables, the causality test involves specifying a multivariate pth order vector error correction model (VECM) as follows:

$$\begin{pmatrix} x_{1t} \\ x_{2t} \\ \vdots \\ x_{nt} \end{pmatrix} = \begin{pmatrix} a_{10} \\ a_{20} \\ \vdots \\ a_{n0} \end{pmatrix} + \begin{pmatrix} b_{11}(L) & \cdots & b_{1n}(L) \\ b_{21}(L) & \cdots & b_{2n}(L) \\ & \vdots & \\ b_{n1}(L) & \cdots & b_{nn}(L) \end{pmatrix} \begin{pmatrix} x_{1(t-1)} \\ x_{2(t-1)} \\ \vdots \\ x_{n(t-1)} \end{pmatrix} + \begin{pmatrix} c_1 z_{t-1} \\ c_2 z_{t-1} \\ \vdots \\ c_n z_{t-1} \end{pmatrix} + \begin{pmatrix} \varepsilon_{1t} \\ \varepsilon_{2t} \\ \vdots \\ \varepsilon_{nt} \end{pmatrix}$$

where again z_{t-1} is the lagged error-correction term derived from the long-run cointegrating relationship. The presence of short-run causality is investigated again through the employment of Wald testing in which if the null hypothesis is rejected, then causality is established. However, it is not easy to test for long-run causality just by looking at the significance of the error-correction term.

It is also worth mentioning here (this also applies in the bivariate case) that Pesaran and Pesaran (1997) recommended the cumulative sum of recursive residuals (CUSUM) and the CUSUM of square (CUSUMSQ) tests proposed by Brown et al. (1975) to assess the parameter constancy. Moreover, Hansen (1992) recommended three tests for parameter stability: the supremum F test (SupF), the Mean F test (MeanF), and L_C. These tests have the same null hypothesis that the parameters are stable. When the calculated probability values are greater than 0.05, then the null hypothesis is accepted (Box 9.3).

BOX 9.3 A Published Case Study

The STUDY: Dritsakis, N., Varelas, E., Adamopoulos, A., 2006. The main determinants of economic growth: an empirical investigation with Granger causality analysis for Greece. European Research Studies IX, 47–58.

Data

Annual data on real GDP (adjusted by the GDP deflator, real gross fixed capital formation, adjusted by the GDP deflator, and exports, measured by the real revenues of exports are obtained from the International Financial Statistics database. In addition, FDI data, measured by foreign direct investments adjusted by the GDP deflator, are also obtain from the World Bank database, spanning the period 1960–2002, while all variables measured in logarithms.

Granger causality results

Given the presence of cointegration, the analysis tests for Granger causality through the error-correction mechanism (Table 9.3).

Granger causality conclusions

There is a bidirectional causal relationship between foreign direct investments and real GDP, a unidirectional causal relationship between exports and real GDP, running from exports to real GDP, and finally, a unidirectional causal relationship between investments and real GDP.

Table 9.3: Granger causality results.

Dependent Variable	Testing Hypothesis	*F*-Test
	GDP	
	Exports do not cause GDP	40.323[0.00]
	Investments do not cause GDP	32.894[0.00]
	FDI do not cause GDP	36.171[0.00]
	Exports	
	GDP does not cause Exports	1.970[0.42]
	Investments	
	GDP does not cause Investments	33.652[0.00]
	FDI	
	GDP does not cause FDI	31.895[0.00]

Figures in brackets denote *P*-values. All variables are expressed in differences in logs. *** $P \leq 0.01$.

4 Alternative Causality Test Approaches in Time Series

Toda and Yamamoto (1995) and Dolado and Lutkepohl (1996) proposed a methodological approach that is applicable irrespective of the integration and cointegration properties of the system of the involved variables. This methodology involves using a modified Wald statistic for testing the significance of the parameters of the VAR model. The estimation of a VAR($s + d_{max}$) guarantees the asymptotic χ^2 distribution of the Wald statistic, where s is the lag length in the system and d_{max} is the maximal order of integration in the model. As lagged dependent variables appear in each equation of the aforementioned causal models, their presence is expected to purge serial correlation among the error terms. The traditional F tests and its Wald test counterpart to determine whether some parameters of a stable VAR model are jointly zero, are not valid for nonstationary processes, as the test statistics do not have a standard distribution (Toda and Phillips, 1994).

Hill (2007) introduced a sequential multihorizon noncausality test. His testing approach is based on Wald type test statistics under the null hypothesis of joint zero parameter linear restrictions. He introduces a VAR framework of order p at horizon h. This framework yields:

$$Z_{t+h} = a + \sum_{k=1}^{p} 1_k^{(h)} Z_{t+1-k} + v_{t+h}$$

where Z is an m-vector stationary process with $m \geq 2$, $1_k^{(h)}$ are matrix-valued coefficients, v is a zero mean $m \times 1$ vector of a white noise process, and a is the intercept. Causality occurs at any horizon if it occurs at horizon 1. In the case that some or all variables are nonstationary, the above equation is extended in the Toda and Yamamoto (1995) model by adding d extra lags to the VAR process. Then, a Wald testing procedure of linear zero restrictions is employed to test for 1-step ahead noncausality. Hill (2007) has also developed a parametric bootstrap methodology that simulates small sample P-values.

Finally, we examine causality through the generalization proposed by Dufour et al. (2006). Given that certain studies have displayed that, in multivariate models, where a vector of auxiliary variables Z is used in addition to the variables of interest x_1 and x_2, it is possible that x_1 does not cause x_2 in the Granger sense (one period ahead), but can still help to predict x_2 several periods ahead. Such a generalization allows for the possibility of distinguishing between short-run and long-run causalities. The statistical procedure in Dufour et al. (2006) tested noncausality at various horizons in the context of finite-order VAR models. In such models, the noncausality restrictions at horizon 1 take the form of relatively simple zero restrictions on the coefficients of the VAR. However, at higher horizons, noncausality restrictions are generally nonlinear, taking the form of zero restrictions on multilinear forms in the coefficients of the VAR. When applying standard test statistics, such as Wald-type criteria, such forms can easily lead to asymptotically singular covariance matrices, so that the standard asymptotic theory would not apply to such statistics. Consequently, they recommend simple tests that can be implemented only through linear regression methodologies. These tests are based on considering multiple-horizon VAR, that is, $(p; h)$-autoregressions, where the parameters of interest can be estimated by linear approaches. Restrictions of non-causality at various horizons may then be tested through simple Wald-type criteria after taking into account the fact that such autoregressions involve autocorrelated errors that are orthogonal to the regressors. The correction for the presence of autocorrelation in the errors may then be performed by using a heteroscedastic autocorrelation consistent covariance matrix estimator. Given the presence of a large number of parameters that could alleviate the unreliability of asymptotic approximations, the use of finite-sample procedures turns out to be crucial.

The concept of autoregression at horizon h and the relevant notations yield a VAR(p) process of the form:

$$Z_t = a_t + \sum_{k=1}^{p} 1_k Z_{t-k} + v_t$$

where $Z_t = (z_{1t}, z_{2t}, \ldots, z_{mt})'$ is a random vector, a_t is a deterministic trend, and v_t is a white noise process of order 2 with a nonsingular variance–covariance matrix Ω. The most common specification for a_t assumes that a_t is a constant vector, although other deterministic trends, such as seasonal dummies, could also be considered. The VAR(p) is an autoregression at horizon 1. This autoregressive form can be generalized to allow for projection at any horizon h given the information available at time t. Hence, the observation at time $t + h$ can be computed recursively from the above equation and yields:

$$Z_{t+h} = a_t^{(h)} + \sum_{k=1}^{p} 1_k^{(h)} Z_{t+1-k} + \sum_{j=0}^{h-1} d_j b_{t+h-j}$$

where $d_0 = I_m$ and $h < T$. The appropriate formulae for the coefficients $1_k^{(h)}$ and $a_t^{(h)}$ are given in Dufour and Renault (1998), and the d_j matrices are the impulse-response coefficients of the process. The above equation is called an autoregression of order p at horizon h or a $(p; h)$-autoregression. Within this framework, the hypothesis is that a variable z_{jt} does not cause another one, say z_{it}, at horizon h. The restrictions related to that hypothesis take the form:

$$H_0^{(h)} : 1_{ijk}^{(h)} = 0, k = 1, \ldots, p$$

Thus the null hypothesis takes the form of a set of zero restrictions on the coefficients of the matrix $1_k^{(h)}$. Under the null hypothesis of noncausality at horizon h from z_{jt} to z_{it}, the asymptotic distribution of the Wald statistic is $x^2(p)$. To get an appropriate distribution, researchers need to take into account that the prediction error follows an MA($h - 1$) process. To that end, the approach uses the Newey–West procedure, which gives an automatically positive-semi-definite variance–covariance matrix. The provided Gaussian asymptotic distribution may not be very reliable in finite samples, especially if a VAR system is considered with a large number of variables and/or lags.

Due to autocorrelation, a larger horizon may also affect the size and the power of the test. An alternative to using the asymptotic chi-square distribution lies in using Monte Carlo test methodologies (Dufour, 2006) or bootstrap methodologies (Boxes 9.4 and 9.5).

BOX 9.4 Published Case Study on the Toda–Yamamoto Approach

The STUDY: Alimi, S.R., Ofonyelu, C.C., 2013. Toda–Yamamoto causality tests between money market interest rates and expected inflation: the Fisher hypothesis revisited. European Scientific Journal 9, 125–142.

Data

Annual time series data on nominal interest rates, inflation, and effective exchange rates for Nigeria are obtained from the Annual Report and Statements of Accounts published by the Central Bank of Nigeria, spanning the period 1970–2011. The analysis makes use of money market interest rates as nominal interests, as along with inflation can be used as proxies for expected inflation. In addition, the analysis employs the US six-month London Interbank Rate (USRATE), obtained from the World Economic Outlook Publication Report, as a proxy for the foreign interest rate. Finally, all variables are expressed in percentages.

Causality results

Given the uncertain results generated by the unit root tests, the Toda–Yamamoto results are reported in Table 9.4.

Causality conclusions

The findings illustrated in Table 9.4 show that there is only a unidirectional causality between inflation and the nominal interest rate, running from inflation to nominal interest rates, while there is no other causality evident, which provides empirical support to the Fisher hypothesis.

Table 9.4: Toda–Yamamoto causality results.

Null Hypothesis	χ^2 test	P-value	Results
Inflation does not cause interest rates	4.352	0.04	Reject the null
Interest rates do not cause inflation	0.162	0.69	Accept the null
Exchange rates do not cause interest rates	0.254	0.61	Accept the null
Interest rates do not cause exchange rates	1.583	0.21	Accept the null
US rate does not cause interest rates	0.884	0.35	Accept the null
Interest rates do not cause the US rate	0.086	0.77	Accept the null

BOX 9.5 Published Case Study on the Dufour, Pelletier, and Renault Causality Approach

The STUDY: Dufour, J.M., Bixi, J., 2016. Multiple Horizon Causality in Network Analysis: Measuring Volatility Interconnections in Financial Markets. Available at SSRN: https://ssrn.com/abstract=2745341.

Data

They study the crisis-sensitive volatility network in the US stock market. They are also interested in examining whether their volatility connectedness measures can reflect the underlying market systemic risk that plays an important role in the recent global financial crisis. Given that the volatility in stock markets is latent, they need a volatility proxy. The well-known VIX index, which has been widely accepted as a market volatility index by financial practitioners, is calculated from implied volatilities of the S&P 500 index options. It is sensitive to market turmoils. For each firm, they also exploit the information in their respective option contracts and thus, they use implied volatility, rather than using realized volatility estimated from stock intraday prices, for the quantities they are dealing which are more comparable to market indices. Volatility or implied volatility is sensitive to "terrifying news" in financial markets. The stock implied volatilities are inevitably contaminated by shocks in financial markets as risks are traded on markets. Nevertheless, implied volatility is still an excellent proxy to study the high-dimensional market volatility network.

Similar to the VIX index for the S&P 500 stock composite, in this paper the S&P 100 components implied volatilities are constructed with their respective at-the-money option contracts with 30-day maturity. This implied volatility measures the expected volatility of the underlying stock over the next 30 days. Therefore, they only consider the option contracts with 30-day maturity. The date range of the database is from 01/01/1996 to 08/31/2015. The companies whose IPO dates are after 01/01/2000 are dropped off, such that they can examine the two most important crises in the US stock market (i.e., the IT Bubble Burst and the Financial Crisis of 2007–09). The remaining full sample is from 20/08/1999 to 31/08/2015. There are missing values on some dates for some companies and, thus, they take linear interpolations to impute the missing values to get completed time series processes for estimations. They end up with 90 companies in the final sample, $N = 90$. The Industry Group classification for each node is from the North American Industry Groups database from MorningStar, LLC.

Causality results

This part reports only a fraction of the entire findings spectrum (mainly due to space availability issue). In particular, Table 9.5 reports the top 10 influential firms and their respective sector at different forecast horizons, $h = 1, 2, 3, 4, 5$. Given the forecast horizon h, they obtain the summary statistics of each row of the firm-wise causality table. The causality table is estimated by the full data sample (20/08/1999–31/08/2015). Nodes are the firms of selected S&P 100 components. For each firm i, they provide the median value of the entries, while for each given forecast horizon h, they sort the tickers by their median values and identify the top 10 influential firms. Moreover, Table 9.6 illustrates summary statistics of causality measures from each financial firm to other financial firms. The causality table is estimated by the full data sample (20/08/1999–31/08/2015). Nodes are the firms of selected S&P 100 components. For each

Table 9.5: Causality at different horizons.

Rank	h = 1 Sector	h = 1 Ticker	h = 2 Sector	h = 2 Ticker	h = 3 Sector	h = 3 Ticker	h = 4 Sector	h = 4 Ticker	h = 5 Sector	h = 5 Ticker
1	F	BAC	T	CSCO	T	CSCO	T	CSCO	T	CSCO
2	C AAPL		C	AAPL	C	AAPL	C	AAPL	C	AAPL
3	T	CSCO	F	C	F	AIG	F	AIG	F	AIG
4	F	C	F	AIG	F	C	F	C	F	C
5	F	BK	F	GS	F	GS	F	GS	F	GS
6	F	AIG	I	GE	I	GE	I	GE	I	GE
7	F	MET	F	MS	F	JPM	F	JPM	F	JPM
8	C	F	F	JPM	C	F	C	F	C	F
9	F	JPM	F	MET	T	IBM	T	IBM	T	IBM
10	F	MS	C	F	F	MET	T	EMC	T	EMC

B, Basic materials; C, consumer goods; F, financial; H, healthcare; I, industrial goods; S, services; T, technology; U, utilities.

Table 9.6: Summary statistics of causality measures from each financial firm to other financial firms.

Ticker	Median	Mean	Minimum	25%	75%	Maximum
BAC	0.42	1.21	0.00	0.04	1.66	5.33
MS	0.30	1.19	0.00	0.04	2.21	4.83
GS	0.25	0.43	0.00	0.00	0.82	1.70
BK	0.06	0.91	0.00	0.00	0.98	5.75
WFC	0.03	0.49	0.00	0.00	0.38	3.96
ALL	0.01	0.30	0.00	0.00	0.37	1.46
SPG	0.01	0.54	0.00	0.00	0.08	5.63
AXP	0.00	0.43	0.00	0.00	0.24	2.78
C	0.00	0.42	0.00	0.00	0.53	2.77
AIG	0.00	0.14	0.00	0.00	0.12	0.74
COF	0.00	0.39	0.00	0.00	0.12	4.63
JPM	0.00	0.10	0.00	0.00	0.15	0.51
MET	0.00	0.31	0.00	0.00	0.58	1.32
USB	0.00	0.11	0.00	0.00	0.03	0.55

financial firm i, it reports the minimum value, the maximum value, the mean value and the quantiles [25%, 50% (median), and 75%] of the entries in its "OUT" vector truncated within the financial sector. The reported values are 100 times of the raw values, and are kept with two digits. It sorts the tickers by their median values and identifies the top 3 influential firms in the financial sector.

Causality conclusions

Table 9.5 reports the top 10 influential firms at different forecast horizons, h = 1, 2, 3, 4, 5, to take spillover effects into account. The firms and their orders in the list of top 10 influential firms are slightly different at different forecast horizons. For instance, in the case of only taking direct effects into account (*h* = 1), the most influential financial firm is BAC and 7 out of 10 most influential firms belong to the financial sector; in the case of taking direct and indirect effects into account (*h* = 5), the most influential financial firm becomes AIG and only 4 out of 10 most influential firms is from the financial sector. The technology firms are actually influential. In the case of *h* = 5, 4 out 10 most influential firms belong to the technology sector and the top two influential firms come from the technology sector, if the Apple Inc. is considered as a technology firm. In short, measuring a static network that only characterizes direct effects in an economic network is far from enough to fully understand all interconnections and indirect effects. In contrast, directly measuring direct and indirect effects with the causality tables at different forecast horizons can provide "dynamic" pictures of interconnections in the S&P 100 network with different effect-radius. In many cases, what is truly important is firm's total effect (direct effect and indirect effect) rather than just its direct effect.

5 Asymmetric Causality

Given that a number of research studies consider that the impact of a positive shock is similar to the impact of a negative shock, this part of the survey considers the literature of the presence of an asymmetric structure in terms of causality. To this end, the literature considers the asymmetric causality behavior through the consideration of cumulative sums of positive and negative shocks. Hatemi-J (2012) considered such an asymmetric causality approach, along with a bootstrap simulation approach with leverage adjustments that generates the appropriate critical values. This methodological approach has the advantage that it is not necessarily based on datasets coming from a normal distribution.

Asymmetry here implies that positive and negative shocks could generate different causal effects. Let us consider again the bivariate model introduced in Section 1.1:

$$y_t = a_{10} + b_{11} y_{t-1} + b_{12} y_{t-2} + b_{13} x_{t-1} + b_{14} x_{t-2} + \varepsilon_{y_t}$$
$$x_t = a_{20} + b_{21} y_{t-1} + b_{22} y_{t-2} + b_{23} x_{t-1} + b_{24} x_{t-2} + \varepsilon_{x_t}$$

After estimating the above two equations and getting the corresponding residuals, we define as $\varepsilon_{y_t}^+ = \max(\varepsilon_{y_t}, 0)$, $\varepsilon_{y_t}^- = \min(\varepsilon_{y_t}, 0)$, $\varepsilon_{x_t}^+ = \max(\varepsilon_{x_t}, 0)$, and $\varepsilon_{x_t}^+ = \min(\varepsilon_{x_t}, 0)$ their corresponding positive and negative shocks. Based on those definitions, the above system of equations yields:

$$y_t = a_{10} + \sum_{i=1}^{t} \varepsilon_{y_t}^+ + \sum_{i=1}^{t} \varepsilon_{y_t}^-$$

and

$$x_t = a_{20} + \sum_{i=1}^{t} \varepsilon_{x_t}^+ + \sum_{i=1}^{t} \varepsilon_{x_t}^-$$

Next, the positive and negative shocks of each variable can be defined in a cumulative form as:

$$y_t^+ = \sum_{i=1}^{t} \varepsilon_{y_i}^+$$

$$y_t^- = \sum_{i=1}^{t} \varepsilon_{y_i}^-$$

$$x_t^+ = \sum_{i=1}^{t} \varepsilon_{x_i}^+$$

and

$$x_t^- = \sum_{i=1}^{t} \varepsilon_{x_i}^-$$

In the next step, the analysis investigates the causal relationship between the above components. For example, in testing for causal relationship between positive cumulative shocks, the analysis assumes that $z_t^+ = (y_t^+, x_t^+)$ and that a VAR(p) model is used:

$$z_t^+ = b_0 + B_1 z_{t-1}^+ + \cdots + B_p z_{t-1}^+ + v_t^+$$

where z_t^+ is the 2×1 vector of the variables, b_0 is the 2×1 vector of constant terms, and v_t^+ is the 2×1 vector of error terms. The matrix B_r is a 2×2 matrix of parameters for lag order r, with $r = 1, \ldots, p$. The analysis will test the null hypothesis that the kth element of z_t^+ does not Granger cause the ωth element of z_t^+ or in other words:

H_0: the row ω, column k element in B_r equals zero.

To test for causality, the analysis makes use of a Wald test by assuming that

$$Z = \left(z_1^+, \ldots, z_T^+ \right) \text{ being a } (n \times T) \text{ matrix}$$
$$D = (b_0, B_1, \ldots, B_p) \text{ being a } [n \times (1 + np)] \text{ matrix}$$

$$\Lambda_t = \begin{pmatrix} 1 \\ z_t^+ \\ z_{t-1}^+ \\ \vdots \\ z_{t-p+1}^+ \end{pmatrix} \text{ being a } \left[(1 + np) \times 1 \right] \text{ matrix}$$

$$\Lambda = \left(\Lambda_0, \ldots, \Lambda_{T-1} \right) \text{ being a } \left[(1 + np) \times T \right] \text{ matrix}$$

$$\delta = \left(v_1^+, \ldots, v_T^+ \right) \text{ being a } (n \times T) \text{ matrix}$$

The analysis defines the VAR(p) model as: $Z = D\Lambda + \delta$ and the null hypothesis of noncausality is H_0: $C\beta = 0$ and is tested through the following Wald test:

$$(C\beta)'[C((\Lambda'\Lambda)^{-1} \times S_u)C']^{-1}(C\beta)$$

where $\beta = \text{vec}(D)$ and vec shows the column stacking operator, x represents the Kronecker product, and C is a $p \times n(1 + np)$ indicator matrix with all elements equal to one for restricted parameters and zeros for the remaining parameters. Finally, S_u is the variance–covariance matrix of the unrestricted VAR model (Box 9.6).

BOX 9.6 Published Case on Asymmetric Causality Case

The STUDY: Yildirim, S., Özdemir, B.K., Doğan, B., 2013. Financial development and economic growth nexus in emerging European economies: new evidence from asymmetric causality. International Journal of Economics and Financial Issues 3, 710–722.

Data

Given that both the stock and bond markets in the emerging European economies have not developed adequately, the analysis uses two different indicators, representing the activities and mediating dimension of the banking sector. These indicators are the ratio of M2 to GDP (denoted by FD1) and the liquid liabilities to GDP ratio (denoted by FD2). Economic growth is based on real GDP changes. Data are obtained from the International Financial Statistics database. Quarterly data are used for 10 countries: Bulgaria, Croatia, Hungary, Latvia, Lithuania, Poland, Romania, Russia, Turkey, and Ukraine. The time span is 1990–2012. Finally, all variables are expressed in natural logarithms.

Causality results

Due to space limits, a part of Table 9.7 is reported.

Table 9.7: Asymmetric causality results.

Bulgaria	Test Results
fdi$^+$ does not cause economic growth$^+$	1.072(6)
fdi$^-$ does not cause economic growth$^-$	2.023(5)
fdi$^-$ does not cause economic growth$^+$	1.368(5)
fdi$^+$ does not cause economic growth$^-$	35.132(5)
economic growth$^+$ does not cause fdi$^+$	4.187(6)
economic growth$^-$ does not cause fdi$^-$	22.716(5)
economic growth$^-$ does not cause fdi$^+$	6.420(5)
economic growth$^+$ does not cause fdi$^-$	8.704(5)

Critical values at 1% are: 38.042, 23.712, 21.700, 40.604, 50.993, 35.294, 31.789, 24.162, respectively.

Causality conclusions

The analysis uses the broad measure of the money stock to GDP ratio (M2 to GDP) as the financial development indicator in Table 9.7. This simple indicator helps measure the degree of monetization in an economy, and it is expected that the increases in M2 would be higher than GDP growth if financial deepening is occurring. At first glance, the impact of economic growth on financial development is more prominent. This pattern can be interpreted under the context of supply leading hypothesis. Further elaboration indicates that it is commonly observed that causality runs from negative growth shock to negative shocks in financial development. The feedback hypothesis or bidirectional causality does not exist.

6 Linear Panel Causality

We start with the Pooled Mean Group (PMG) estimator recommended by Pesaran et al. (1999). The PMG estimator is based on the Autoregressive Distributed Lag (ARDL) $(p, q, q,...., q)$ model which is described as:

$$y_{it} = a_i + \sum_{j=1}^{p} \lambda_{ij} y_{i,t-j} + \sum_{j=0}^{q} \delta_{ij} x_{i,t-j} + v_{it}$$

where x_{it} is a vector of explanatory variables (regressors) for group i, a_i represents fixed effects, λ_{ij} are the coefficients of the lagged dependent variables, and δ_{ij} are coefficient vectors. We reparameterize the above equation and get:

$$\Delta y_{it} = \varphi_i y_{i,t-1} + \beta_i' x_{it} + \sum_{j=1}^{p-1} \lambda_{ij}^* \Delta y_{i,t-j} + \sum_{j=0}^{q-1} \delta_{ij}^* \Delta x_{i,t-j} + a_i + v_{it}$$

$$i = 1,2,...,N, t = 1,2,...,T$$

where $\varphi_i = -\left[1 - \Sigma_{j=1}^{p} \lambda_{ij}\right], \beta_i = \Sigma_{j=0}^{q} \delta_{ij}, \lambda_{ij}^* = -\Sigma_{m=j+1}^{p} \lambda_{im}$, for $j = 1,..., p-1, \delta_{ij}^* = -\Sigma_{m=j+1}^{q} \delta_{im}$ for $j = 1,....,q-1$.

Pesaran et al. (1999) assumed that the ARDL $(p, q, q,...., q)$ model is stable, if the roots of the following equation:

$$1 - \sum_{j=1}^{p} \lambda_{ij} = 0$$

lies outside the unit circle. This assumption ensures that $\varphi_i < 0$ and, thus, there is a long-run relationship between y_{it} and x_{it} defined by:

$$y_{it} = -[\beta_i / \varphi_i x_{it} + \eta_{it}]$$

where η_{it} is a stationary process and the long-run coefficients $\theta_i = -\beta_i/\varphi_i = \theta$ are the same across the group. Therefore when a cointegrating relationship between x_{it} and y_{it} is established, namely as:

$$y_{it} = a_{0i} + a_{1i} x_{it} + a_{2i} z_{it} + v_{it}$$

with z representing another control variable, the ARDL (1,1,1) equation yields:

$$y_{it} = a_{0i} + a_{1i} x_{it} + a_{2i} x_{i,t-1} + a_{3i} z_{it} + a_{4i} z_{i,t-1} + \lambda_i y_{i,t-1} + v_{it}$$

and

$$x_{it} = b_{0i} + b_{1i}y_{it} + b_{2i}y_{i,t-1} + b_{3i}z_{it} + b_{4i}z_{i,t-1} + \mu_i y_{i,t-1} u_{it}$$

while the error-correction equations yield:

$$\Delta y_{it} = \varphi_1(y_{i,t-1} - \theta_1 x_{it} - \theta_2 z_{it}) - a_{2i}\Delta x_{it} - a_{3i}\Delta z_{it} + v_{it}$$

and

$$\Delta x_{it} = \varphi_2(x_{i,t-1} - \theta_3 y_{it} - \theta_4 z_{it}) - a_{4i}\Delta y_{it} - a_{5i}\Delta z_{it} + u_{it}$$

We can evaluate the null hypothesis of non-causality as $a_{2i} = 0$ and $b_{2i} = 0$.

We can also analyze causal empirical relationships by using the panel data causality testing methodology, developed by Holtz-Eakin et al. (1988). This methodology is closely related to an approach proposed by Anderson and Hsiao (1981). The test involves estimation of error-correction equations as below:

$$\Delta \ln y_{it} = b_1(\ln y_{i,t-1} - c_1 \Delta \ln x_{it}) + a_{1i} + v_{1it}$$
$$\Delta \ln x_{it} = b_2(\ln x_{i,t-1} - c_2 \Delta \ln y_{it}) + a_{2i} + v_{2it}$$

where a denote the time effects or alternatively the time trend. The parameters b_1 and b_2 denote the error-correction terms. The question of whether or not x causes y can be tested through the hypothesis:

$b_1 = c_1 = 0, H_0 : x$ does not Granger cause y in the long run

$b_2 = c_2 = 0, H_0 : y$ does not Granger cause x the long run.

The rejection of the first null hypothesis and the acceptance of the second null hypothesis is interpreted as causality from x to y, while the rejection of the second null hypothesis and the acceptance of the first null hypothesis are interpreted as causality in the reverse direction. If both hypotheses are rejected, then there is no feedback between the two variables. Assuming that the residuals of the level equation are serially uncorrelated, the values of y lagging two periods or more can be used as instruments in the first-differenced equation. The estimation equation and moment conditions can be estimated by first-differenced General Method of Moments (GMM), which was developed by Arellano and Bond (1991). However, conventional GMM estimations exhibit a major drawback if the explanatory variables display persistence over time. By persistence, we mean that the mean reversion process of the explanatory variable is very slow. In this case, their lagged levels may be rather poor instruments for their differences. Therefore researchers should use the system GMM estimator that was introduced by Blundell and Bond (1998), which combines the regression equation in first differences, instrumented with the lagged levels of the regressors, with the regression equation in the levels, instrumented with lagged the differences of the regressors. However, a drawback of the approach is that it ignores the heterogeneity problem in the cross-sections.

The above problem led to the development of the panel causality test by Hurlin and Venet (2001). They make use of a panel Granger model where for each individual *i* and for all *t* = 1,..., *T* the model yields:

$$y_{it} = \sum_{k=1}^{r} \gamma i^k y_{i,t-1} + \sum_{k=0}^{r} \beta_i^k x_{i,t-k} + v_{it}$$

where the autoregressive coefficients (*k*) and the regression coefficients slopes *i*(*k*) are assumed to be constant for all *k* = 1,..., *r*. The autoregressive coefficients are also assumed to be the same for all units, while the regression coefficients slopes can vary across individuals. Due to this, there are four types of causality relationships proposed by Hurlin and Venet (2001) to take account of the heterogeneity in the underlying processes. The testing procedure involves the combination of various sets of tests that make use of the *F*-test and when the null hypothesis of the first case is rejected, we could then proceed to the second case and even further, depending on the results. But when the null hypothesis for any case could not be rejected, the tests can then be ended. The four causality tests are therefore stated below. The Homogenous Noncausality (HNC) case is the first test, which implies testing whether a particular variable is not causing another one in all the cross-sections of the samples. If the computed *F*-statistic is significant, then the HNC hypothesis is rejected, that is, implying that there is causality in at least one member of the panel, and then we continue testing the HNC. In contrast, if the HNC hypothesis is accepted, then there is not any causality relationship across any member of the panel and the testing process will not proceed further. The second case is testing the Homogenous Causality (HC) hypothesis and we test whether one variable is causing another across all entities, that is, countries of the sample. If the *F*-statistic is not significant, then the HC hypothesis is accepted, implying that there is causality across all members of the panel and further testing will not be necessary. But when the HC hypothesis is rejected, then there is no causality relationship in at least one member of the panel. In the third case, we consider that when the HC hypothesis is rejected, there is no HC and then we move on to the heterogeneity tests to determine which of the members of the panel exhibits a causal relationship. Heterogeneous Causality (HEC) is the third case and the implication of this is that there are causal relationships that exist in at least one individual, and causality could rise to a maximum of *N* individuals. Finally, the fourth case, the Heterogeneous Noncausality (HENC) hypothesis implies that for not less than one individual, and for at most *N* − 1 individuals, there is causality across them. The rejection of the HENC hypothesis implies that the statistic is significant; therefore, there is causality for the individual under consideration. The second test is tested the joint hypothesis of no causality for a subgroup of individuals in the panel. In this case, the slope coefficients of all lags across the individuals of the subgroups are constrained to zero. If the *F*-statistic is significant, this implies the rejection of the HENC hypothesis for the subgroup under consideration and we can conclude that causality exists for this subgroup of panel members (Babajide, 2010).

Next, we consider the panel causality test introduced by Dumitrescu and Hurlin (2012). This test is a simple version of the Granger (1969) noncausality test for heterogeneous panel data models with fixed coefficients, while it takes into consideration two dimensions of heterogeneity: the heterogeneity of the regression model used to test the Granger causality and the heterogeneity of the causality relationships. We consider the following linear model:

$$y_{it} = a_i + \sum_{k=1}^{K} \gamma i^k y_{i,t-k} + \sum_{k=1}^{K} \beta_i^k x_{i,t-k} + v_{it}$$

where x and y are two stationary variables observed for N individuals in T periods. Also, $\beta_i = \left(\beta_i^1, \ldots, \beta_i^k \right)'$ and the individual effects a_i are assumed to be fixed in the time dimension. We assume that the lag orders of K are identical for all cross-section units of the panel. We also allow the autoregressive parameters γ_i^k and the regression coefficients β_i^k to vary across groups. Under the null hypothesis, it is assumed that there is no causality relationship for any of the units of the panel. This assumption is called the HNC hypothesis, which is defined as:

$$H_0 : \beta_i = 0, \quad \text{for} \quad i = 1, \ldots, N$$

The alternative is specified as the HENC hypothesis. Under this hypothesis, we allow for two subgroups of cross-section units. There is a causality relationship from x to y for the first one, but it is not necessarily based on the same regression model. For the second subgroup, there is no causality relationship from x to y. We consider a heterogeneous panel data model with fixed coefficients (in time) for this group. This alternative hypothesis is described by the following two parts:

$$H_1 : \beta_i = 0 \quad \text{for } i = 1, \ldots, N_1$$

and

$$\beta_i \neq 0 \quad \text{for } i = N_1 + 1, \ldots, N$$

It is also assumed that β_i may vary across groups and there are $N_1 < N$ individual processes with no causality from x to y. N_1 is unknown, but it provides the condition $0 \leq N_1/N < 1$. The following average statistic HNC $W_{N,T}^{\text{HNC}}$ is proposed, which is related with the null HNC hypothesis, as follows:

$$W_{N,T}^{\text{HNC}} = 1/N \sum_{i=1}^{N} W_{i,T}$$

where $W_{i,T}$ denotes the individual Wald statistics for the ith cross-section unit corresponding to the individual test: H_0: $\beta_i = 0$.

We define $Z_i = [e, Y_i, X_i]$ as a $(T \times 2K + 1)$ matrix, where e denotes a $(T, 1)$ vector and $Y_i = \left[y_i^1, \ldots, y_i^k \right]$, $X_{i=} \left[x_i^1, \ldots, x_i^k \right]$. $\theta_i = (\alpha_i \ \gamma_i' \ \beta_i')$ is the vector of parameters of the model. In addition, $R = [0, I_K]$ is a $(K, 2K + 1)$ matrix. For each $i = 1, \ldots, N$, the Wald statistic $W_{i,T}$ corresponding to the individual test H_0: $\beta_i = 0$ is defined as:

$$W_{it} = \theta_i' R' [\sigma_i^2 R(Z_i' Z_i)^{-1} R']^{-1} R \theta_i$$

Under the null hypothesis of noncausality, each individual Wald statistic converges to a chi-squared distribution with K degrees of freedom for $T \to \infty$.

Next, a third panel causality approach has been developed by Kònya (2006) which allows accounting for both cross-sectional dependence and heterogeneity issues. It is based on Seemingly Unrelated Regressions (SUR) systems and Wald tests with specific bootstrap critical values. This particular methodology enables testing for Granger causality on each individual panel member separately, by taking into account the potential contemporaneous correlation across entities (e.g., countries, regions, sectors, firms, banks) involved in the sample. This panel causality approach examines the relationship between y and x and can be studied using the following bivariate finite order vector autoregressive (VAR) model as before:

$$y_{it} = a_{1i} + \sum_{s=1}^{ly_1} b_{1is} y_{i,t-s} + \sum_{s=1}^{lx_1} c_{1is} x_{i,t-s} + v_{1it}$$

and

$$x_{it} = a_{2i} + \sum_{s=1}^{ly_2} b_{2is} y_{1,t-s} + \sum_{s=1}^{lx_2} c_{2is} x_{i,t-s} + v_{2it}$$

where the index i $(i = 1, \ldots, N)$ denotes one entity, that is, country, the index t $(t = 1, \ldots, T)$ the period, s is the lag, and ly_1, lx_1, ly_2, and lx_2 denote the lag lengths. The error terms, v_{1it} and v_{2it} are considered to be white noises (i.e., they have zero means, constant variances, and are individually serially uncorrelated) and may be correlated with each other for a given entity, that is, country.

In one entity, that is, country i, there is one-way Granger causality running from x to y, if in the first equation not all c_{1i} are zero, but in the second equation all b_{2i} are zero. There is one-way Granger causality from y to x, if in the first equation all c_{1i} are zero, but in the second equation not all b_{2i} are zero. There is two-way Granger causality between y and x, if neither all b_{2i} nor all γ_{1i} are zero. Finally, there is no Granger causality between y and x, if all b_{2i} and γ_{1i} are zero. The OLS estimators of the parameters are consistent and asymptotically efficient. This suggests that the $2N$ equations in the system can be estimated one-by-one, in any preferred order, while the entire system for all entities, that is, countries, can be estimated by the SUR procedure to take into account contemporaneous correlations within the systems (in the presence of contemporaneous correlation, the SUR estimator is more efficient than the OLS estimator). Following Kònya (2006), specific bootstrap Wald critical values are used to implement Granger causality. This methodological procedure has several advantages: (1) it does not assume that the panel is homogeneous, so it is possible to test for Granger-causality on each individual panel member separately, (2) this approach does not require pretesting for unit roots and cointegration, though it still requires the specification of the lag structure, and (3) this panel Granger causality approach allows researchers to detect for how many and for which members of the panel there exists one-way Granger causality, two-way Granger causality, or no Granger causality. However, Konya's testing procedure has a drawback for the panel datasets, if the number of cross-sections (N) is not reasonably smaller than time periods (T), because the SUR estimator is only feasible for panels with large T and small N.

The approaches described above can control for heterogeneity, but they are not capable of accounting for cross-sectional dependence. To overcome this problem, Emirmahmutoglu and Kose (2011) developed a panel causality methodology that accounts for cross-country heterogeneity irrespective of whether the variables of interest are nonstationary or cointegrated. In addition to this flexibility, because the critical values for panel statistics are derived from bootstrap distributions, this methodology also considers the cross-section dependency. This approach considers the following VAR model for each cross-section:

$$y_{it} = a_i + B_{1i}y_{i,t-1} + \ldots + B_{pi}y_{i,t-pi} + \ldots + A_{p+d,i}y_{i,t-pi-di} + v_{it}$$

where y_{it} is a vector of endogenous variables, a_i denotes the p dimensional vector of fixed effects, p_i is the optimal lag(s), and d_i is the maximum integration degree of the variables. The null hypothesis of no-Granger causality against the alternative hypothesis of Granger causality is tested by imposing zero restrictions on the first p parameters. The so-called modified Wald statistic has the asymptotic chi-square distribution with p degrees of freedom. To test the Granger noncausality hypothesis for the panel, the Fisher statistic is developed and defined as:

$$\lambda = -2\sum_{i=1}^{N} \ln(\pi_i)$$

where π_i is the probability corresponding to the individual modified Wald statistic. The Fisher statistic has an asymptotic chi-square distribution with $2N$ degrees of freedom. However, the limit distribution of the Fisher test statistic is no longer valid in the presence of cross-section dependency. To accommodate for cross-section dependency in the panel, Emirmahmutoglu and Kose (2011) suggest obtaining an empirical distribution of the panel statistic using the bootstrap methodology. For more details on the bootstrap methodology, the reader should refer to Emirmahmutoglu and Kose (2011) (Boxes 9.7 and 9.8).

BOX 9.7 Published Case in Panel Granger Causality by Pesaran et al. (1999)

The STUDY: Costantini, V., Martini, C., 2009. Causality Between Energy Consumption and Economic Growth: a Multi-sectoral Analysis Using Non-stationary Cointegrated Panel Data. Working Paper, No. 102/09, Department of Economics, University of Rome III.

Data

They obtained information on 71 countries, divided into two groups: OECD, with 26 countries, and NO-OECD, with 45 countries. The countries included in the OECD group are quite homogeneous, whereas those in the NO-OECD group are quite heterogeneous, both with respect to their development level and their policy settings. The dataset combines several sources. For the energy sectors, they obtained data from the IEA publications on OECD and NO-OECD energy balances, containing annual data on energy final consumption for the entire economy and for its main sectors, that is, industry, commerce and public services, transport, and residential sectors, all expressed in terms of kg of oil equivalent. All information on economic performance across the different sectors is taken from the World Bank dataset on World Development Indicators. More specifically, they consider gross domestic product, the value added of the industry and services, household's final consumption expenditures, all considered in terms of per capita constant 2000 US$. For the transport sector, they employed GDP as the economic dimension. Data on energy prices are provided by IEA statistics on energy prices and taxes (quarterly) for OECD countries, only for the period 1978–2005. They obtained data for the whole energy sector, along with the four specific end use sectors and considered four different energy prices: total energy price, total industry price, total household price, and total gasoline price (all expressed in terms of constant 2000 US$ per ton). They use the total industry price both for the industrial and the service sector even though many contributions affirm that the best price variable for services is the cost of electricity. In this dataset, electricity prices are often missing or are not complete throughout the time period, thus, consistently reducing the number of observations. Although they are aware that they could specify energy sectors with prices even for NO-OECD countries through the general Consumer Price Index as a proxy of energy prices, but they opted to adopt sector-specific energy prices to obtain more accurate estimates of price elasticities, as CPI does not account homogenously for energy services across all countries. For bivariate models, data availability allows considering the period 1970–2005 for the full sample and the NO-OECD sample, whereas for the case of OECD countries, the time series cover the period 1960–2005.

For multivariate models including energy prices, they have a reduced sample with only OECD countries, spanning the period 1978–2005. Considering the wide divergence among countries, both in the energy sectors and in economic performance, they considered per capita levels and thus they transformed all data into natural logarithms, given the high variance in levels between developed and developing countries.

Panel Granger causality results

Having estimated the VECM across all sectors and distinct subsamples, the analysis performs simple Wald F-tests on the significance of the coefficients, evaluating three different Granger causality relationships: a short-run causality, testing the significance of the coefficients related

Table 9.8: Panel causality tests.

Dependent Variable	Short-run	Long-run	Strong Causality		
	GDP	EC	ECT	GDP-ECT	EC-ECT
GDP	–	3.15*	−1.75	–	4.00
EC	4.06**	–	−9.67***	18.53***	–

EC, Energy consumption; *ECT*, error correction term. * Significant at 10% level. ** Significant at 5% level. *** Significant at 1% level. The heteroschedasticity of the error terms is corrected by using White robust standard errors both in periods (White period system robust covariances) and in cross-sections (coefficient covariance method: White cross-section system robust). The method for iteration control for GLS and GMM weighting specifications is to iterate weights and coefficients sequentially to convergence. To correct for possible autocorrelation we use the Newey–West estimator of the weighting matrix in the GMM criterion.

to the lagged economic and energy variables, a long-run causality related to the coefficient for the error-correction term, and a strong causality to test whether the sources of causation are jointly significant. The results of the VECM with two simultaneous equations for the analysis of the causal relationships between energy consumption and economic growth are reported in Table 9.8. This Table reports results in terms of the Wald *F*-test on the coefficients. Again, due to space limitations, only a part of this table is reported (the full sample results).

Granger causality conclusions

When the bivariate VECM model is performed on the whole economy, a bidirectional short-run causality and a unidirectional long-run relationship, where the economic growth is the driver of energy consumption and not vice versa is established. In addition, the negative sign of the estimated speed of adjustment coefficients are in accordance with the convergence toward long-run equilibrium. The larger the value of that coefficient, the stronger is the response of the variable to the previous period's deviation from long-run equilibrium, if any.

BOX 9.8 Panel Granger Causality on Published Case by Dumitrescu and Hurlin (2012)

The STUDY: Zeren, F., Ari, A., 2013. Trade openness and economic growth: a panel causality test. International Journal of Business and Social Science, 4 317–324.

Data

Their study investigates causality between trade openness and economic growth in the case of the G7 countries (i.e., Germany, France, Canada, Japan, Italy, US, and UK), spanning the period 1970–2011. Economic growth is measured using per capita GDP with constant 2000 US$, and trade openness is measured exports plus imports as a share of GDP. The data used in the paper are obtained from the World Development Indicators database provided by the World Bank. Both variables are expressed in their natural logarithms.

Panel Granger results

Table 9.9: Panel Granger causality results.

Tests Openness	Trade Openness → Economic Growth	Economic Growth → Trade
$W_{N,T}$	6.811*	5.937*
$Z_{N,T}$	12.351*	9.519*
$\hat{Z}_{N,T}$	3.411*	2.588*

* Significant at 5%.

Panel Granger causality conclusions

When the test statistics in Table 9.9 are compared to the bootstrap critical values in Table 9.9 by Dumitrescu and Hurlin (2012), it is inferred that these test statistics are statistically significant. Thus a bidirectional causality relationship exists between trade openness and economic growth. For the case of the G7 countries there exists a bidirectional causality relationship.

7 Nonlinear and Nonparametric Causality

This section describes the available nonlinear tests for exploring Granger causality. There are two main facts justifying the employment of such nonlinear testing procedures. First, standard linear Granger causality tests have extremely low power in detecting certain kinds of nonlinear relationships (Brock, 1991; Gurgul and Lach, 2009). Second, as the traditional linear approach is based on testing the statistical significance of suitable parameters only in terms of the mean equation, causality in higher-order structures (i.e., causality in variance) cannot be explored (Diks and Degoede, 2001). The application of nonlinear causality approaches may be a solution to this problem, because it allows exploring complex dynamic links between the variables of interest.

First, we refer to the approach proposed by Diks and Panchenko (2006). To this end, we define for $t = 1, 2, \ldots$, the $L_x + L_y + 1$-dimensional vector $W_t = [x_{t-L_x}^{L_x}, y_{t-L_y}^{L_y}, Y_t)$. The null hypothesis that x_t does not Granger cause y_t may be written in terms of density functions:

$$f_{x,y,z}(x, y, z) = f_{x,z}(x, z) f_{z|x,y}(z \mid x, y) = f_{x,z}(x, z) f_{z|y}(z \mid y)$$

where $f_x(z)$ stands for the probability density function of the random vector X at point z, $x = x_{t-B_x}^{L_x}, y = y_{t-L_y}^{L_y}, z = y_t$ for $t = 1, 2, \ldots$ Next, we can define the correlation integral $C_W(\varepsilon)$ for the multivariate random vector W by the following expression:

$$C_W(\varepsilon) = P[\| W_1 - W_2 \|] \le \varepsilon = \iint I(\| s_1 - s_2 \|) \le \varepsilon \, f_W(s_1) \, f_W(s_2) \, ds_2 \, ds_1$$

where W_1 and W_2 are independent with distributions in the equivalence class of the distribution of W, I denotes the indicator function (equal to one, if the condition in brackets holds true, otherwise is equal to zero), $\|x\| = \sup\{|x_i|: i = 1,..., d_W\}$ denotes the supremum norm (d_W is the dimension of the sample space W) and $\varepsilon > 0$. Hiemstra and Jones (1994) claimed that the null hypothesis in Granger's causality test implies that for every $\varepsilon > 0$:

$$C_{x,y,z}(\varepsilon)/C_{x,y}(\varepsilon) = C_{y,z}(\varepsilon)/C_y(\varepsilon)$$

They recommend calculating sample versions of correlation integrals and then testing whether the left-hand- and right-hand-side ratios differ significantly or not. They propose the use of the following formula as an estimator of the correlation integral:

$$C_{W,n}(\varepsilon) = 2/[n(n-1)] \times \sum_{i<j} \sum I_{ij}^W$$

where $I_{ij}^W = I\left(\left\|W_i - W_j\right\|\right) < \varepsilon$. In terms of the expected value and density functions, the authors managed to test their recommendation based on the asymptotic theory of the F-test statistic. Furthermore, they present some advice concerning the proper way of choosing the bandwidth according to the sample size. This adaptation is helpful in reducing the bias of the test, which is one of the serious problems which arise for long time series. The performance of the modified test is also based on the same lags for each pair of time series analyzed. They generate the following statistic for testing for causality:

$$T(\varepsilon) = (n-1)/[n(n-2)] \times \sum_i \dot{f}_{x,y,z}(x_i,y_i,z_i)\dot{f}_y(y_i) - \dot{f}_{x,y}(x_i,y_i)\dot{f}_{y,z}(y_i,z_i)$$

where n is the sample size, and $\dot{f}_w(w)$ is a local density estimator of a d_w. They prove that under strong mixing (strong mixing refers to the condition introduced by Rosenblatt (1971)) and has to do with the central limit theorem for "weakly dependent" random variables. It has received considerable importance in terms of the probability theory, due to its tractability in the derivation of the asymptotic properties for various functions in relevance to dependent random variables, for example, members of the important class of linear stochastic processes are strongly strong mixing, provided they are based on innovation random variables which have Lebesgue-integrable characteristic functions, while certain AR(1) processes do not represent strong mixing cases, the above test statistic satisfies:

$$\sqrt{n}[T(\varepsilon)-q]/S_n \to N(0,1)$$

where $\to$ denotes convergence in distribution, q denotes lags, and S_n is the estimator of the asymptotic variance of T. We can follow Diks and Panchenko's suggestion to implement a one-tailed version of the test, rejecting the null hypothesis, if the left-hand-side of the above expression is too large. As the statistic diverges to positive infinity under the alternative hypothesis, a calculated statistic greater than 1.28 implies the rejection of the null hypothesis at the 10% level of significance. In their test, the value of the bandwidth plays an important role in making a decision on nonlinear causality. Since a bandwidth value smaller (larger) than one generally results in larger (smaller) P-values (Bekiros and Diks, 2008), usually, the bandwidth value is equal to one.

This part examines nonlinear causality through the Smooth Transition Autoregressive (STAR) modeling approach. To this end, the following univariate STAR model is described based on the methodological approach proposed by Teräsvirta (1994):

$$y_t = a_{10} + b_1 w_t + (b_{20} + b_2 w_t) F(y_{t-d}) + u_t$$

where $w_t = (y_{t-1}, \ldots, y_{t-p})$, $u_t \to N(0, \sigma_u^2)$ and a $F(y_{t-d})$ denotes a transition function, while d is an unknown delay parameter. There are two choices for the transition function that are based on the following logistic function:

$$F(y_{t-d}) = \left[1 + \exp(-\gamma_L (y_{t-d} - c_L))\right]^{-1} \quad \text{with } \gamma_L > 0$$

and the following exponential function:

$$F(y_{t-d}) = \left[1 + \exp(-\gamma_E \left(y_{t-d} - c_E\right)^2)\right]^{-1} \quad \text{with } \gamma_E > 0$$

The above transition functions yield logistic STAR (LSTAR) and exponential STAR (ESTAR) models, respectively. In the case of LSTAR, two regimes can be considered, depending on the small and large values of the transition variable relative to the threshold parameter, c_L. This type of models can be appropriate to model business cycle asymmetries where expansion and contraction periods have different dynamics. In other words, the LSTR specification accounts for asymmetric realizations, in the sense that the two regimes are associated with small and large values of the transition variable relative to the threshold value. In contrast, the regimes in ESTAR models are subject to small and large absolute values of the transition function relative to c_E. The ESTAR transition function is symmetric around the threshold parameter, while the values close to c_E differ. The STAR-based test of Granger causality can be performed through the additive smooth transition regression model, presented with reference to Skalin and Teräsvirta (1996):

$$y_t = a_{10} + b_1 w_t + (b_{20} + b_2 w_t) F(y_{t-d}) + \delta_1 v_t + (\delta_{20} + \delta_2 v_t G(x_{t-e}) + u_t$$

where $v_t = (x_{t-1}, \ldots, x_{t-q})'$ and $G(.)$ shows the transition function, while e is an unknown delay parameter. The noncausality hypothesis is H_0: $G0$ and $\delta_{1i} = 0$, and $i = 1, \ldots, q$.

In case that nonlinear or threshold cointegration has been detected, we may test the transmissions using threshold error-correction modeling approaches (TECM). The TECM can be presented as follows (Enders and Siklos, 2001):

$$\Delta Y_t = a + \rho_1 Z_{t-1}^+ + \rho_2 Z_{t-1}^- + \sum_{i=1}^{n_1} \delta_i \Delta y_{t-i} + \sum_{i=1}^{n_2} \theta_i \Delta x_{t-i} + \varepsilon_t$$

where $Y_t = (y_t, x_t)'$, $Z_{t-1}^+ = h_t \hat{g}_{t-1}$, $Z_{t-1}^- = (1-h_t)\hat{g}_{t-1}$, such that $h_t = 1$, if $\hat{g}_{t-1} \geq \psi$, $h_t = 0$, if $\hat{g}_{t-1} < \psi$ and ε is a white noise disturbance. Through the system, the Granger causality tests are examined by testing whether all the coefficients of Δy_{t-i} and Δx_{t-i} are statistically different from zero, based on a standard F-test and if the ρ coefficients of the error-correction are also significant. Granger causality tests are very sensitive to the selection of the lag length, and thus the appropriate lag lengths can be determined through the Akaike criterion.

Next, we may test for Granger noncausality through the nonparametric test proposed by Nishiyama et al. (2011). The test statistic is constructed based on moment conditions for causality in the mean. To apply the test, the Nadaraya–Watson nonparametric estimator of moments is needed. Let $z(x_t, y_t)$ be a sample of T observations on dependent random variables in $\mathbb{R} \times \mathbb{R}$, with a joint distribution function F. Suppose now we need to test Granger noncausality in the mean from x_{t-1} to y_t. This corresponds to testing the null hypothesis:

$$H_0 : \Pr\{E[u_t \mid Z_{t-1}] = 0\} = 1$$

against the alternative hypothesis:

$$H_1 : \Pr\{E[u_t \mid Z_{t-1} = 0]\} < 1$$

where $u_t = y_t - E[y_t | y_{t-1}]$. If the null hypothesis is true, then past changes in x_t cannot affect the conditional mean of y_t. The authors have illustrated that the above null and alternative hypotheses can be rewritten in terms of unconditional moment restrictions:

$$H_0 : \Pr\{E[u_t f(y_{t-1})h(Z_{t-1})] = 0\} < 1$$

against the alternative hypothesis:

$$H_1 : \Pr\{E[u_t f(y_{t-1})h(Z_{t-1})] = 0\} < 1$$

where $h(z)$ is any function in the Hilbert space $s_r^\perp$ that is orthogonal to the Hilbert L_2 space: $s_r = \{s(.)|E[s(y_{t-1})^2] < \infty$. As $E[u_t f(y_{t-1}) h(Z_{t-1})]$ is unknown, we can use a nonparametric approach to estimate it. According to these authors, and through the Nadaraya–Watson methodology, we can estimate this conditional mean. To test the null hypothesis against the alternative hypothesis, the authors use the following test statistic:

$$\hat{S}_T = \sum_{i=1}^{k_T} w_i \alpha_i^2$$

where $\alpha_i = 1/\sqrt{T} \Sigma u_t \hat{f}(y_{t-1}) \hat{h}(Z_{t-1})$, while w_i is a nonnegative weighting function, such as $w_i = 0.9^i$. The above test statistic depends on the sample size. The authors have illustrated that, under the null hypothesis, the statistic converges in distribution to:

$$\sum_{i=1}^{\infty} w_i \varepsilon_i^2, \text{as } T \to \infty, \quad \text{where } \varepsilon_i \text{ are i.i.d. } N(0,1).$$

In other words, for a given summable positive sequence of weights $\{w_i\}$, the test statistic is pivotal and is asymptotically distributed as an infinite sum of weighted chi-squares. The main advantage of this test is that the simulation is very simple and the critical values are not dependent on the data.

Finally, we can also test whether past changes in x_t can affect the conditional distribution of y_t. The null hypothesis is defined when the distribution of y_t is conditional on its own past and past changes in x_t are equal to the distribution of y_t conditional on their own past only, almost everywhere. This is similar to test the conditional independence between y_t and past changes in x_t conditionally on the past y_t. According to Florens and Mouchart (1982) Florens and Mouchart (1982) and Florens and Fougère (1996), this is also a test of Granger noncausality in distribution, as opposed to the tests of Granger noncausality in mean mentioned above. This allows researchers to capture the dependence due to both low and high-order moments and quantiles. Furthermore, Granger causality tests provide useful information on whether knowledge of past changes in x_t can improve short-run forecast of movements in y_t. A new nonparametric test statistic is considered, proposed by Bouezmarni and Taamouti (2014). The test is based on a comparison of the conditional distribution functions using an L_2 metric. In case researchers are interested in testing Granger noncausality in the distribution from x_{t-1} to y_t, the following null hypothesis is tested:

$$H_0 : \Pr\{E[y_t \mid y_{t-1}, x_{t-1}] = F(y_t \mid y_{t-1})]\} = 1$$

against the alternative hypothesis:

$$H_1 : \Pr\{E[y_t \mid y_{t-1,}x_{t-1}] = F(y_t \mid y_{t-1})]\} < 1$$

where $F(y_t|y_{t-1}, x_{t-1})$ is the conditional distribution function of y_t given y_{t-1} and x_{t-1}, and $F(y_t|y_{t-1})$ is the conditional distribution function of y_t given only y_{t-1}. If the null hypothesis is true, then the past changes in x_t cannot affect the conditional distribution of y_t. As $F(y_t|y_{t-1}, x_{t-1})$ and $F(y_{t+1}|y_t)$ are unknown, a nonparametric approach is used to estimate them. If we denote $\bar{U}_{t-1} = (y_{t-1}, x_{t-1})'$ and $\bar{u} = (y, x)'$ then the Nadaraya–Watson estimator of the conditional distribution of y_t given y_{t-1} and x_{t-1} is defined as:

$$\hat{C}_{h_1}(y_t \mid \bar{u}) = \left[\sum_{t=2}^{T} K_{h_1}(\bar{u} - \bar{U}_{t-1})I_{Ay_t}(y_t)\right] \bigg/ \left[\sum_{t=2}^{T} K_{h_1}(\bar{u} - \bar{U}_{t-1})\right]$$

where $K_{h_1}(.) = h_1^{-2}K(./h_1)$, $K(.)$ stands for a kernel function, $h_1 = h_{1,T}$ is a bandwidth parameter, and $I_{Ay_t}(.)$ is an indicator function, which is defined on the set $A_{y_t} = [y_t, +\infty]$. Similarly, the Nadaraya–Watson estimator of the conditional distribution of y_t given y_{t-1} is defined as:

$$\hat{C}_{h_2}(y_t \mid y) = \left[\sum_{t=2}^{T} K_{h_2}^*(y - y_{t-1})I_{A_{y_t}}(y_t)\right] \bigg/ \left[\sum_{t=2}^{T} K_{h_2}^*(y - y_{t-1})\right]$$

where $K_{h_2}^*(.) = h_2^{-1}K^*(./h_2)$, for $K^*(.)$ a different kernel function, $h_2 = h_{2,T}$ is a different bandwidth parameter. To test the null hypothesis against the alternative hypothesis, the authors recommend the following test statistic:

$$\hat{A} = (1/T)\sum_{t=2}^{T}\left\{\hat{C}_{h_1}\left(y_t|\bar{u}\right) - \hat{C}_{h2}\left(y_t|y\right)\right\}^2 w(\bar{U}_{t-1})$$

where $w(.)$ is a nonnegative weighting function of the data $\bar{U}_{t-1}$ for $2 \leq t \leq T$. The test statistic $\hat{A}$ is close to zero, if conditionally on y_{t-1}, the variables y_t and x_{t-1} are independent, and it diverges in the opposite case. The authors have established the asymptotic distribution of the nonparametric test statistic. They show that the test is asymptotically pivotal under the null hypothesis and follows a normal distribution. As the distribution of the test statistic is valid only asymptotically, for finite samples they suggest using a local bootstrap version of the test statistic. The simple resampling from the empirical distribution will not conserve the conditional dependence structure in the data, thus, it is important to use the local smoothed bootstrap suggested by Paparoditis and Politis (2000). The latter improves quite a lot the finite sample properties (size and power) of the test. As optimal bandwidths are not available, they have considered the bandwidths $h_1 = c_1 T^{-1/4.75}$ and $h_2 = c_2 T^{-1/4.25}$ for various values of c_1 and c_2. The empirical power of the test also performs quite well (Boxes 9.9 and 9.10).

BOX 9.9 Published Paper on Nonparametric Causality

The STUDY: Muhtaseb, B.M.A., Daoud, H.E., 2017. Tourism and economic growth in Jordan: evidence from linear and nonlinear frameworks. International Journal of Economics and Financial Issues 7, 214–223.

Data

Data for Jordan are on a quarterly basis, spanning the period 1998–2015, and are on real GDP in US\$ and at constant 2005 prices and on real international tourism receipts expressed in constant US\$. GDP data were obtained from the Central Bank of Jordan, while those on the tourism variable from the World Travel and Tourism Council.

Granger causality results and conclusions

Table 9.10 presents the results from the nonlinear causality testing between tourism and economic growth. The DP test indicates that there is a bidirectional nonlinear causality.

Table 9.10: The DP test.

Null Hypothesis	Statistic	*P*-value
International tourism receipts does not cause economic growth	1.71*	0.0015
Economic growth does not cause international tourism receipt	1.43*	0.0023

* Statistically significant at 1%.

BOX 9.10 Published Paper on Noncausality Through the Nonparametric Test Proposed by Nishiyama et al. (2011)

The STUDY: Bekiros, S., Gupta, R., 2015. Predicting Stock Returns and Volatility Using Consumption Aggregate Wealth Ratios: A Nonlinear Approach. Working Paper, No. 2015-05, Department of Economics, University of Pretoria.

Data

The value adjusted CRSP index (CRSP-VW), obtained from the Center for Research in Security Prices, is deflated by the personal consumption expenditure chain type price deflator (2009 = 100) to provide real stock prices. Stock returns are computed as the real log returns (rcrspr), and its volatility (rcrspv) as the squared values of the returns. The data span ranges from 1952 to 2013, it is on a quarterly basis and is obtained from Sydney Ludvigson's website. Their goal is to determine whether the consumption aggregate wealth ratio can cause both real stock returns and their volatility.

Causality results and conclusions

The results are reported in Table 9.11. As it can be seen, the consumption-based wealth ratio is found to cause both real stock returns and their volatility.

Table 9.11: Nonlinear causality test.

Dependent Variable	Test Statistic
Real stock returns	77.29**
Volatility of real stock returns	47.54**

** Significant at 5%.

8 Conclusion

This chapter surveyed all the relevant tests in causal analysis provided in the relevant literature, while emphasized inferences on linear time series and panel causality, on asymmetric causality and nonparametric causality. Moreover, the presentation of each test was accompanied by an empirical application that had been already published in the literature to depict the practical application of those testing methodologies in a clearer way. We do hope that readers will appreciate it.

References

Anderson, T.W., Hsiao, C., 1981. Estimation of dynamic models with error components. J. Am. Stat. Assoc., 589–606.

Arellano, M., Bond, S., 1991. Some tests of specification for panel data: Monte Carlo evidence and an application to employment equations. Rev. Econ. Stud. 58, 277–297.

Babajide, F., 2010. The finance-growth nexus in Sub-Saharan Africa: panel cointegration and causality tests. J. Int. Dev. 23, 220–239.

Bekiros, S.D., Diks, C.G.H., 2008. The nonlinear dynamic relationship of exchange rates: parametric and nonparametric causality testing. J. Macroecon. 30, 1641–1650.

Blundell, R., Bond, S., 1998. Initial conditions and moment restrictions in dynamic panel data models. J. Econometr. 87, 115–143.

Bouezmarni, T., Taamouti, A., 2014. Nonparametric tests for conditional independence using conditional distribution. J. Nonparametric Stat. 26, 697–719.

Brock, W., 1991. Causality, chaos, explanation and prediction in economics and finance. In: Casti, J., Karlqvist, A. (Eds.), Beyond Belief: Randomness, Prediction and Explanation in Science. CRC Press, Boca Raton, FL.

Brown, R., Durbin, L., Evans, J., J F.M., 1975. Techniques for testing the constancy of regression relationships over time. J. R. Stat. Soc. Ser. B 37, 149–192.

Diks, C., Panchenko, V., 2006. A new statistic and practical guidelines for nonparametric Granger causality testing. J. Econ. Dyn. Control 30, 1647–1669.

Diks, C.G.H., Degoede, J., 2001. A general non-parametric bootstrap test for Granger causality. In: Broer, H.W., Krauskopf, W., Vegter, G. (Eds.), Global Analysis of Dynamical Systems. Institute of Physics Publishing, UK, Bristol.

Dolado, J., Lutkepohl, H., 1996. Making Wald tests work for cointegrated VAR systems. Econometr. Rev. 15, 369–386.

Dufour, J.M., 2006. Monte Carlo tests with nuisance parameters: a general approach to finite-sample inference and nonstandard asymptotics in econometrics. J. Econometr. 133, 443–477.

Dufour, J.M., Renault, E., 1998. Short- and long-run causality in time series: theory. Econometrica 66, 1099–1125.

Dufour, J.M., Pelletier, D., Renault, E., 2006. Short- and long-run causality in time series: inference. J. Econometr. 132, 337–362.

Dumitrescu, E.I., Hurlin, C., 2012. Testing for Granger non-causality in heterogeneous panels. Econ. Modell. 29, 1450–1460.

Emirmahmutoglu, F., Kose, N., 2011. Testing for Granger causality in heterogeneous mixed panels. Econ. Modell. 28, 870–876.

Enders, W., Siklos, P.L., 2001. Cointegration and threshold adjustment. J. Bus. Econ. Stat. 19, 166–176.

Florens, J., Fougère, D., 1996. Non-causality in continuous time. Econometrica 64, 1195–1212.

Florens, J., Mouchart, M., 1982. A note on non-causality. Econometrica 50, 583–591.

Geweke, J., Meese, R., Dent, W.T., 1983. Comparing alternative tests of causality in temporal systems: analytic results and experimental evidence. J. Econometr. 21, 161–194.

Granger, C.W.J., 1969. Investigating causal relations by econometric models and cross spectral methods. Econometrica 37, 424–438.

Gurgul, H., Lach, L., 2009. Linear versus nonlinear causality of DAX companies. Oper. Res. Decis. 3, 27–46.

Hansen, B.E., 1992. Tests for parameter stability in regressions with I(1) processes. J. Bus. Econ. Stat. 10, 321–335.

Hatemi-J, A., 2012. Asymmetric causality tests with an application. Empir. Econ. 43, 447–456.

Hiemstra, C., Jones, J.D., 1994. Testing for linear and nonlinear Granger causality in the stock price-volume relation. J. Finance 49, 1639–1664.

Hill, J.B., 2007. Efficient tests of long-run causation in triariate VAR processes with a rolling window study of the money-income relationship. J. Appl. Econometr. 22, 747–765.

Holtz-Eakin, D., Newey, W., Rosen, H., 1988. Estimating vector autoregressions with panel data. Econometrica 56, 1371–1395.

Hurlin, C., Venet, B., 2001. Granger Causality Tests in Panel Data Models with Fixed Coefficients. Working Paper, EURISCO NO. Universite Paris Dauphine.

Kònya, L., 2006. Exports and growth: Granger causality analysis on OECD countries with a panel data approach. Econ. Modell. 23, 978–992.

Nishiyama, Y., Hitomi, K., Kawasaki, Y., Jeong, K., 2011. A consistent nonparametric test for nonlinear causality specification in time series regression. J. Econometr. 165, 112–127.

Paparoditis, E., Politis, D., 2000. The local bootstrap for kernel estimators under general dependence conditions. Ann. Inst. Stat. Math. 52, 139–159.

Pesaran, M.H., Pesaran, B., 1997. Working with Microfit 4.0: Interactive Econometric Analysis. Oxford University Press, Oxford.

Pesaran, H., Shin, Y., Smith, R., 1999. Pooled mean group estimation and dynamic heterogeneous panels. J. Am. Stat. Assoc. 94, 621–634.

Rosenblatt, M., 1971. Markov Processes, Structure and Asymptotic Behavior. Springer-Verlag Inc.

Sims, C.A., 1980. Macroeconomics and reality. Econometrica 48, 1–49.

Sims, C.A., 1972. Money, income and causality. Am. Econ. Rev. 62, 540–552.

Skalin, J., Teräsvirta, T., 1996. Another look at Swedish business cycles, 1861–1988. The Economic Research Institute, Stockholm School of Economics. J. Appl. Econ. 130, 359–378.

Teräsvirta, T., 1994. Specification, estimation, and evaluation of Smooth Transition Autoregressive models. J. Am. Stat. Assoc. 89, 208–218.

Toda, H.Y., Phillips, P.C.B., 1994. Vector autoregressions and causality: a theoretical overview and simulation study. Econometr. Rev. 13, 259–285.

Toda, H.Y., Yamamoto, T., 1995. Statistical inference in vector autoregressions with possibly integrated processes. J. Econometr. 66, 225–250.

Further Reading

Engle, R.F., Granger, C.W.J., 1987. Cointegration and error correction: representation, estimation, and testing. Econometrica 55, 251–276.

CHAPTER 10

Simultaneous Equations Modeling in the Energy-Growth Nexus

Heli Arminen

Lappeenranta University of Technology, Lappeenranta, Finland

Chapter Outline

1 Introduction and Background

This chapter contributes to the literature on the causal relationships between economic growth, energy consumption, and air pollution. Although previous energy-growth nexus (EGN) literature has relied heavily on the Granger causality test, some studies have used

The Economics and Econometrics of the Energy-Growth Nexus
http://dx.doi.org/10.1016/B978-0-12-812746-9.00010-9

simultaneous equations modeling (SEM) to capture how the variables are interrelated. This chapter reviews these studies and presents a simplified framework for using simultaneous equations in the nexus. This framework is then utilized to discuss SEM in general and simultaneous equations estimation methods in particular. Special attention is paid to the relevant statistical tests and the reporting of results.

The energy-environment-growth nexus can be arranged into three strands that focus on the causal links between: (1) energy consumption and economic growth (or the level of economic development), (2) (air) pollution and economic growth (or development), and (3) energy consumption and economic growth (or development) and (air) pollution. In particular, the previous literature relies heavily on the Granger causality test (Granger, 1969), but with inconclusive and even conflicting results. Moreover, as Granger causality only implies that changes in a variable take place before changes in another variable, the concept may be weak in finding out whether a variable has a positive or negative impact on another variable. To avoid this pitfall, some studies, though only a few so far, have used SEM to model how energy consumption, economic growth, and air pollution are interrelated.

The previous studies that have applied SEMs in the EGN are reviewed in Table 10.1. The literature search was performed in the Scopus database on 23 March 2017. Table 10.1 summarizes the literature by sample, analysis and estimation methods, and empirical results.

1.1 Energy Consumption and Economic Growth

The EGN has drawn a great deal of attention since the seminal paper by Kraft and Kraft (1978). For example, Omri (2014), Ozturk (2010), and Payne (2010a,b) surveyed the literature, and Menegaki (2014) has performed a meta-analysis on it. In this literature, the term 'economic growth' is generally used to refer to both the level of economic development, often measured by GDP per capita (also called 'income level' here) and the actual growth in the income level (GDP per capita growth). This chapter makes a distinction between income level and its growth, as these are two different concepts from the perspectives of both theoretical and empirical analyses.

Four different hypotheses have been presented concerning the causal relationship between economic growth and energy consumption (e.g., Payne, 2010a,b). First, according to the so-called growth hypothesis, unidirectional causality runs from energy consumption to economic growth, indicating that energy conservation measures hinder economic growth. Second, according to the conservation hypothesis, unidirectional causality runs from economic growth to energy consumption, in which case conservation measures do not hinder economic growth. Third, the feedback hypothesis supports

Table 10.1: Summary of simultaneous equations modeling studies in the energy–environment-growth nexus.

Study	Data Period	Countries	Analysis and Estimation Methods	Main Findings
Adewuyi and Awodumi (2017)	1980–2010	11 West African countries	• Unit root tests for individual countries, panel unit root test for the whole panel • System generalized method of moments (GMM) and three-stage least squares (3SLS); 3SLS results reported	The results for individual countries vary. *For the whole panel:* 1. Biomass consumption → GDP: (+), GDP → biomass consumption: (+) (energy ↔ GDP) 2. CO_2 emissions → biomass consumption: (+), biomass consumption → CO_2 emissions: (+) (energy ↔ CO_2 emissions) 3. CO_2 emissions → GDP: (−), GDP → CO_2 emissions: (−) (GDP ↔ CO_2 emissions)
Amri (2017)	1990–2012	72 countries; 23 developed and 49 developing countries	• Panel unit root test • Two-step difference GMM estimator	Emissions not included in the model. *For the whole sample and the subsamples:* Renewable energy consumption → GDP: (+), GDP → renewable energy consumption: (+) (energy ↔ GDP)
Liu (2005)	1975–90	24 Organisation for Economic Cooperation and Development (OECD) countries	3SLS	No equation for energy consumption, but energy consumption is used to explain CO_2 emissions in Model 2. *Model 1 (without energy):* GDP → CO_2 emissions: (+) (GDP → CO_2 emissions) *Model 2 (with energy):* 1. Energy consumption → CO_2 emissions: (+) 2. CO_2 emissions → GDP: (+), GDP → CO_2 emissions: (−) (GDP ↔ CO_2 emissions)
Omri (2013)	1990–2011	14 Middle Eastern and North African (MENA) countries	• Unit root test • GMM estimation with panel data (details not given), two-stage least squares (2SLS), and 3SLS; GMM results reported	The results for individual countries vary. *All countries (panel):* 1. Energy consumption → GDP: (+), GDP → energy consumption: (+) (energy ↔ GDP) 2. Energy consumption → CO_2 emissions: (+) (energy → CO_2 emissions) 3. CO_2 emissions → GDP: (−), GDP → CO_2 emissions: (+) (GDP ↔ CO_2 emissions)

(Continued)

Table 10.1: Summary of simultaneous equations modeling studies in the energy–environment-growth nexus. (*cont.*)

Study	Data Period	Countries	Analysis and Estimation Methods	Main Findings
Omri and Kahouli (2014)	1990–2011	65 countries; 26 high-income, 26 middle-income, and 13 low-income countries	• Panel unit root test • Difference GMM	Emissions not included in the model. *All countries (panel), high-income, and middle-income countries*: Energy consumption (growth) → GDP (growth): (+), GDP (growth) → energy consumption (growth): (+) (energy ↔ GDP) *Low-income countries*: Energy consumption (growth) → GDP (growth): (+) (energy → GDP)
Omri et al. (2015a)	1990–2011	12 MENA countries	• Panel unit root and cointegration tests • GMM estimation with panel data (details not given)	No equation for energy consumption, but energy consumption is used to explain CO_2 emissions. CO_2 emissions are used to explain GDP due to the perceived linear relationship between energy consumption and CO_2 emissions. The results for *individual countries* vary but in most cases and for the *whole panel*: 1. Energy consumption → CO_2 emissions:(+) 2. CO_2 emissions → GDP: (−) GDP → CO_2 emissions: (+) (GDP ↔ CO_2 emissions)
Omri et al. (2015b)	1990–2011	17 countries	• Unit root test • GMM estimation with panel data (details not given), 2SLS and 3SLS; GMM results reported	Nuclear and renewable energy consumption, no equation for emissions. The results for *individual countries* vary, but for many countries and the *whole panel*: 1. Nuclear energy consumption → GDP: (+), GDP → nuclear and renewable energy consumption: (+) (nuclear energy ↔ GDP & (GDP → renewable energy) 2. CO_2 emissions → nuclear energy consumption: (+)

Table 10.1: Summary of simultaneous equations modeling studies in the energy–environment-growth nexus. (*cont.*)

Study	Data Period	Countries	Analysis and Estimation Methods	Main Findings
Saidi and Hammami (2016a)	1990–2012	58 countries divided into 3 regional subsamples: • Area 1: Europe and North Asia; • Area 2: Latin America and the Caribbean; • Area 3: The Middle East, North Africa and sub-Saharan Africa	Difference GMM	*Global panel and all area panels:* 1. Energy consumption (growth) → GDP (growth): (+), GDP (growth) → energy consumption (growth): (+) (energy ↔ GDP) 2. CO_2 emissions (growth) → energy consumption (growth): (+), energy consumption (growth) → CO_2 emissions (growth): (+) (energy ↔ CO_2 emissions) 3a. CO_2 emissions (growth) → GDP (growth): (−), *Global panel and area 1:* 3b. GDP (growth) → CO_2 emissions (growth): (+) (GDP ↔ CO_2 emissions) *Area 2:* (CO_2 emissions → GDP) *Area 3:* 3b. GDP (growth) → CO_2 emissions (growth): (−) (GDP ↔ CO_2 emissions)
Sinha (2016)	2001–13	139 Indian cities	• Panel unit root and cointegration tests • GMM estimation with panel data (details not given)	Equation for inequality in energy intensity. No equation for energy consumption, but energy consumption is used to explain NO_2 and SO_2 emissions. Emissions are used to explain income due to the perceived linear relationship between energy consumption and emissions. 1. Energy consumption → NO_2 & SO_2 emissions: (+) 2. NO_2 & SO_2 emissions → income: (−) [in 3 out of 4 models], income → NO_2 & SO_2 emissions: (+) [in 3 out of 4 models] (GDP ↔ NO_2 & SO_2 emissions)

(Continued)

Table 10.1: Summary of simultaneous equations modeling studies in the energy–environment-growth nexus. (*cont.*)

Study	Data Period	Countries	Analysis and Estimation Methods	Main Findings
Tiba et al. (2016)	1990–2011	24 countries; 12 middle-income and 12 high-income countries	• Panel unit root test • GMM estimation with panel data (details not given)	The results vary between individual countries. *High-income countries*: 1. Renewable energy consumption → GDP: (−), GDP → renewable energy consumption: (-) (energy ↔ GDP) 2. CO_2 emissions → renewable energy consumption: (+), renewable energy consumption → CO_2 emissions: (+) (energy ↔ CO_2 emissions) *Middle-income countries*: 1. Renewable energy consumption → GDP: (+), GDP → renewable energy consumption: (+) (energy ↔ GDP) 2. CO_2 emissions → renewable energy consumption: (−) (CO_2 emissions → energy) *Both panels*: 3. CO_2 emissions → GDP: (+), GDP → CO_2 emissions: (+) (GDP ↔ CO_2 emissions)
Xia (2012)	2001–08	30 Chinese provinces	• Principal component analysis to form one measure of pollution • Panel unit root test • Cross-section weights regression • Individual fixed cross-section effects method	Pollution not included in the energy consumption equation. 1. Energy consumption → GDP: (+), GDP → energy consumption: (+) (energy ↔ GDP) 2. Energy consumption → wastewater discharge and solid wastes produced: (+) 3. Pollution → GDP: (+), GDP → pollution: (+/−)depending on the pollutant (GDP ↔ pollution)
Xia and Xu (2012)	2001–08	30 Chinese provinces	• Panel unit root test • Cross-section weights regression • Individual fixed cross-section effects method • Three pollutants: waste gas emissions, wastewater discharge and solid wastes produced	Pollution not included in the energy consumption equation. 1. Energy consumption → GDP: (+), GDP → energy consumption: (+) (energy ↔ GDP) 2. No causality from energy consumption to waste gas emissions 3. Waste gas emissions → GDP: (+), GDP → waste gas emissions: (+) (GDP ↔ waste gas emissions)

bidirectional causality between economic growth and energy consumption, while the fourth, the neutrality hypothesis, suggests that there is no causality between energy consumption and economic growth.

Omri and Kahouli (2014) examined the relationship between economic growth and energy consumption growth using an SEM in a panel of 65 countries between 1990 and 2011. Their results for high- and middle-income countries, as well as the entire panel, support the feedback hypothesis, indicating a bidirectional causal relationship between GDP per capita growth and energy consumption per capita growth. In contrast, the results for low-income countries indicate that energy consumption growth positively affects GDP per capita growth, but that GDP per capita growth lacks a statistically significant effect on energy consumption growth, indicating unidirectional causality from energy consumption to GDP in accordance with the growth hypothesis. Low-income countries thus appear to differ from higher-income countries in this regard. Additionally, the magnitude of the effect appears to depend on the income level: the higher the income level, the larger the effects that energy consumption and GDP have on each other.

Some of the previous studies account for the source of energy. For example, the results from Omri et al. (2015b) study of 17 developed and developing countries between 1990 and 2011 support a bidirectional causal relationship between nuclear energy consumption and GDP (i.e., the feedback hypothesis) and a unidirectional causal relationship from GDP to renewable energy consumption (i.e., the conservation hypothesis), indicating that the relationship between the two variables also depends on the source of energy. Conversely, Amri (2017) found evidence for a bidirectional feedback relationship between renewable energy consumption and GDP for 72 countries and all subsamples based on the income level between 1990 and 2012. Thus, just like in the case of the previous EGN literature relying on the Granger causality test, the results of the previous SEM literature are conflicting.

1.2 Air Pollution and Economic Growth

The previous literature on the relationship between air pollution and economic growth[1] is closely associated with the environmental Kuznets curve (EKC) hypothesis (Grossman and Krueger, 1991, 1995), which predicts an inverted U-shaped relationship between income level and air pollution. Dinda (2004), Kijima et al. (2010), and Stern (2004) surveyed the literature on the EKC. Many studies have concentrated on CO_2 emissions because of data availability (e.g., Costantini and Monni, 2008) and the major role that greenhouse gas emissions play in the current environmental debate (Omri, 2013). While a plethora of studies (e.g., Apergis and Payne, 2009a; Pao and Tsai, 2010; Zhang and Cheng, 2009) utilize the Granger causality test

---

[1] Also this literature mostly uses the term 'economic growth' when referring to the income level.

to account for the fact that income and air pollution are jointly determined, a few studies build an SEM (Liu, 2005; Omri, 2013; Omri et al., 2015a; Sinha, 2016).

In one of the rare studies using a system estimation method, Liu (2005) utilized the three-stage least squares (3SLS)[2] to estimate an SEM with GDP per capita and industrial CO_2 emissions per capita as the endogenous variables for 24 OECD countries between 1975 and 1990. Liu also included energy consumption per capita as an explanatory variable in the air pollution equation. The results indicate that if energy consumption per capita is viewed as a proxy for the economy's structure (instead of being endogenously determined within the framework), CO_2 emissions fall with the income level due to cleaner production technologies and changes in the relative importance of manufacturing and service sectors.

In a more recent study, Omri et al. (2015a) defined an SEM with equations for GDP per capita, CO_2 emissions per capita, financial development and trade openness. They used general method of moments (GMM) estimation for individual countries and panel data and found support for bidirectional causality between GDP and CO_2 emissions in 12 Middle Eastern and North African countries between 1990 and 2011. Their results also indicate that CO_2 emissions negatively affect economic growth, which in turn appears to increase CO_2 emissions. Their results also support the EKC hypothesis. In line with Liu (2005), Omri et al. (2015a) included exogenous energy consumption per capita as an explanatory variable in the air pollution equation and found energy consumption to have an increasing effect on air pollution. These results are mainly in line with Sinha (2016), who found that NO_2 and SO_2 emissions per capita detrimentally affected the income level in 139 Indian cities between 2001 and 2013. Moreover, both the income level and energy consumption appeared to result in increased NO_2 and SO_2 emissions, thus implying a bidirectional causal relationship between air pollution and income.

Because it is more conventional to view energy consumption as an endogenous variable in the energy-environment-growth nexus, the next section reviews the literature in which energy consumption, air pollution, and economic growth are all endogenously determined. For example, Ozturk (2010) and Payne (2010a,b) have recommended consideration of the three variables simultaneously.

1.3 Energy Consumption, Economic Growth, and Air Pollution

Among the studies that account for the endogeneity of energy consumption, Omri (2013) examined the causal relationships between energy consumption, economic growth, and CO_2 emissions in the Middle East and North Africa between 1990 and 2011. His results suggest that the causal relationships vary among the countries in the region. For the region as a whole, he found evidence for bidirectional causality between energy consumption and GDP (with the

---

[2] Using 3SLS accounts for endogenous explanatory variables and nonzero covariances between the structural errors in the SEM; see Section 4.2.

expected positive signs) and GDP and CO_2 emissions (with GDP having an increasing effect on emissions and emissions having a decreasing effect on GDP), while there appears to be unidirectional causality from energy consumption to CO_2 emissions.

In a related study, Saidi and Hammami (2016a) utilized panel data for 58 countries over the period of 1990–2012. They divided their panel into three regions: (1) Europe and North Asia; (2) Latin America and the Caribbean; and (3) the Middle East, North Africa and sub-Saharan Africa. According to their results based on growth equations, bidirectional causal relationships run between GDP growth and energy consumption growth and between energy consumption growth and CO_2 emissions growth globally and in all of the areas, respectively. In addition to that, the effect of CO_2 emissions growth on economic growth appears to be negative in all of the areas. The largest differences can be observed in the effect of GDP growth on CO_2 emissions growth, which ranges from positive (global panel and area 1) to statistically insignificant (area 2) and to negative (area 3).

Xia (2012) and Xia and Xu (2012) found evidence for bidirectional causality between energy consumption and GDP in 30 Chinese provinces between 2001 and 2008. In addition, Xia and Xu (2012) used industrial wastewater discharge and industrial solid wastes produced as measures of pollution and found that the causal relationships depends on the pollutant included in the model.

Tiba et al. (2016) concentrated on renewable energy consumption. Their results for 12 high-income and 12 middle-income countries between 1990 and 2011 indicate significant differences in the causal relationships between the groups of countries. These are summarized in Table 10.1. However, the statistically significant differences might be caused by the relatively small sample sizes or the sole focus on renewable energy consumption. In another study on renewable energy consumption, Adewuyi and Awodumi (2017) focus on biomass consumption in a sample covering 11 West African countries from 1980 to 2010. According to their results, the causal relationships between biomass consumption and GDP, biomass consumption and CO_2 emissions, and CO_2 emissions and GDP are all bidirectional with the expected signs.

1.4 A Summary of the Previous Literature

To summarize, although the results for individual countries vary, a clear majority of the previous studies using an SEM support the feedback hypothesis indicating bidirectional causality between economic growth and energy consumption. In addition, in almost all of the models, economic growth appears to increase energy consumption, which in turn seems to positively affect economic growth.

Most of the previous studies employ panel data to assess the relationships between the three variables. From the econometric point of view, the problem with many of the studies is that the descriptions of estimation methods and model specifications are rather vague, which makes it difficult to evaluate the reliability of the results.

2 A Simultaneous Equations Model for Economic Growth (Income Level), Energy Consumption and Air Pollution

The fundamental question in the EGN is the nature of the causal relationship between the two variables. The estimation method should thus account for the fact that it is possible that economic growth and energy consumption are simultaneously determined or that there is reverse causality between them. In addition to that, Payne (2010a,b) has called for examination of both the sign and magnitude of the relationships between economic growth and energy consumption, indicating that something other than the Granger causality test should be used (see also Ozturk, 2010). An alternative that few studies have used so far is SEM. Other benefits of the approach are that it has a strong theoretical background in the production model and that it allows for including control variables to avoid the omitted variables bias, which are also among the features of future research called for by Ozturk (2010) and Payne (2010a,b).

An SEM is a system that consists of at least two equations and two endogenous variables, the values of which are determined within the system. In the EGN, such a model should thus include an equation for economic growth and another for energy consumption. In accordance with the recommendations of Payne (2010a,b) and Tiba and Omri (2017), a third equation for air pollution can also be included in the system, resulting in altogether three equations and three endogenous variables (economic growth, energy consumption, and air pollution). The values of exogenous explanatory variables are determined outside of the system, although depending on the estimation method, some of them could also be defined as endogenous variables to avoid inconsistency.

The important part of building an SEM is to be able to identify the so-called 'structural equations'; each equation in the system has to be specified based on theory. Moreover, each structural equation should include at least one variable that is not included in other equations, so that it can be used as an instrument. The next three subsections discuss the theoretical background and explanatory variables that should be included in (1) an income equation, (2) an energy consumption equation, and (3) an air pollution equation. As the same SEM can be used in cross-sectional, time-series and panel-data analysis, country and time subscripts have been suppressed in the equations presented below.

2.1 Production Function

The previous EGN literature has often included physical capital and labor in the production function (e.g., Amri, 2017; Huang et al., 2008; Hung and Shaw, 2004; Liu, 2005; Omri, 2013; Sharma, 2010; Stern, 1993, 2000), but so far, few studies have accounted for the role of human capital in economic development (e.g., Adewuyi and Awodumi, 2017; Hung and Shaw, 2004; Xia, 2012; Xia and Xu, 2012). Within the nexus, energy consumption is also usually considered to be a factor of production (e.g., Apergis and Payne, 2009b, 2010; Omri, 2013; Sadorsky, 2012; Sharma, 2010; Stern, 1993, 2000), while energy has also been seen as a necessary complement to capital and labor inputs (Saidi and Hammami, 2015). In general, using energy in commercial (e.g., transport) and noncommercial activities links it to the income level (Sharma, 2010). Furthermore, Hung and Shaw (2004), Liu (2005), and Stern et al. (1996), for example, emphasize that pollution and income are jointly determined and that pollution should thus be included as an explanatory variable in the production function. In addition to that, Xia (2012) claimed that since pollution cannot be eliminated from the production process with current technologies, it can be considered to be a factor of production. Among others, Omri (2013), Saidi and Hammami (2016a), Xia (2012), and Xia and Xu (2012) incorporated both energy consumption and pollution into the aggregate production function. The augmented Cobb–Douglas production function can thus be written as:

$$Y = K^{\alpha_1}(AhL)^{\alpha_2} E^{\alpha_3} POL^{\alpha_4}, \tag{10.1}$$

where Y denotes aggregate output, K denotes physical capital stock, A denotes effectiveness of labor, h denotes human capital per person, L denotes labor force (and total population) (meaning that hL equals the effective units of labor), E denotes energy consumption and POL denotes CO_2 emissions. Assuming that the production function exhibits constant returns to scale—that is that $\alpha_1 + \alpha_2 + \alpha_3 + \alpha_4 = 1$—and dividing it by L, the production function can be expressed in per capita terms:

$$\frac{Y}{L} = \left(\frac{K}{L}\right)^{\alpha_1} (Ah)^{\alpha_2} \left(\frac{E}{L}\right)^{\alpha_3} \left(\frac{POL}{L}\right)^{\alpha_4}, \tag{10.2}$$

The log-linearized function, in per capita format, can then be written as:

$$y = gdp = \alpha_0 + \alpha_1 k + \alpha_2 h + \alpha_3 e + \alpha_4 pol + \varepsilon_1, \tag{10.3}$$

where $\alpha_0 = \alpha_2 \ln(A)$ and the variables in lower-case letters are natural logarithms of the aggregate output (measured by real GDP) per capita (y, henceforth denoted by 'gdp'), physical capital stock per capita (k),[3] human capital per capita (h), energy consumption per capita (e), and CO_2 emissions per capita (pol). The random error term ε_1 has been added into the equation.

As factors of production, physical and human capital are expected to have a positive effect on income level, indicating that α_1 and α_2 should be positive. As energy consumption can also be considered to be a factor of production, α_3 should be positive as well. In contrast, α_4 is expected to be negative, because pollution can restrict or reduce production, either directly, through adversely affecting the factors of production, or by raising the costs of reducing emissions (Hung and Shaw, 2004). For example, pollution could cause health problems, thereby reducing labor productivity and increasing absence from work due to illness. Moreover, the more resources are used for reducing emissions, the less resources are left to be used as inputs in actual production.

---

[3] Box 10.1 discusses the measures of physical capital that have been and that should be used in empirical EGN studies.

Box 10.1 Measuring Physical Capital Stock

Several studies (e.g., Omri et al., 2015a) have used gross fixed capital formation from the World Bank (2016b) as a proxy for physical capital stock. A better approach would be to use the perpetual inventory method to calculate the capital stock (see Lorde et al. (2010), Xia (2012), and Xia and Xu (2012) for studies that use the method within the EGN). An alternative approach is to use the existing estimates for the physical capital stock (e.g., the Penn World Table by Feenstra et al. (2015)). If the income equation is estimated in first differences (see Section 4.2 for a discussion on the reason why this might be necessary), gross fixed capital formation can naturally be used as a measure of the absolute change in the physical capital stock.

It is also possible to estimate the income equation using growth rates of the variables (e.g., Saidi and Hammami, 2016a) or in a form where some of the variables are included as growth rates (or first differences) and the other variables are included in levels (e.g., Sharma, 2010). This might be necessary with time-series or panel data (see discussion below), but it should be borne in mind that estimating an equation for economic growth is not equal to estimating an equation for the income level.

2.2 Function for Energy Consumption

The function for energy consumption can be seen as the most problematic structural equation in the energy-growth-pollution SEM in the sense that its theoretical background is less strong than those of production and air pollution functions. Special attention should also be given to explaining why an explanatory variable included in the energy equation should only be included in this equation and not in the income and air pollution equations.

In accordance to the previous EGN literature (e.g., Adewuyi and Awodumi, 2017; Omri, 2013; Saidi and Hammami, 2016a; Shahbaz and Lean, 2012), a simple energy consumption function can be written as:

$$e = \beta_0 + \beta_1 gdp + \beta_2 ind + \beta_3 pol + \varepsilon_2 \qquad (10.4)$$

where the variables in lower-case letters are again natural logarithms of energy consumption per capita (e), real GDP per capita (gdp), industrialization (ind), and CO_2 emissions per capita (pol). Finally, ε_2 is the error term.

First, economic growth is expected to increase energy consumption (e.g., Omri, 2013; Sadorsky, 2010; Zhang and Cheng, 2009), indicating that β_1 should be positive. Second, industrialization is included as an explanatory variable in the energy consumption equation, because higher value added manufacturing tends to use more energy than basic manufacturing, agriculture, or services (Sadorsky, 2013), and because the higher the share of the value added from the industrial sector as part of the economy total value added, the more energy is required to maintain economic growth (Shahbaz and Lean, 2012; see also Xia, 2012). Thus, β_2 should be positive. Third, if the causal relationship between energy consumption and CO_2 emissions is bidirectional, as indicated by the results of Apergis and Payne (2009a), Saidi and Hammami (2016a), Soytas and Sari (2009), and Wang et al. (2011), CO_2 emissions should also be included as an explanatory variable in the energy consumption function. The expected sign of β_3 can, however, be considered not to be known as a priori.

In this chapter, the number of explanatory variables has been kept to minimum to facilitate the educative and illustrative purpose of the chapter. Other variables that have been included in the energy consumption function in the previous literature to avoid the omitted variables bias include, among others, energy (oil) price (e.g., Omri et al., 2015b; Omri and Nguyen, 2014); financial development (e.g., Adewuyi and Awodumi, 2017; Omri, 2013; Omri and Kahouli, 2014; Sadorsky, 2010; Shahbaz and Lean, 2012); foreign direct investment (e.g., Omri and Kahouli, 2014); human capital (e.g., Adewuyi and Awodumi, 2017); physical capital (e.g., Amri, 2017; Omri, 2013; Omri and Kahouli, 2014); population density (e.g., Xia, 2012); trade openness (e.g., Amri, 2017; Omri and Nguyen, 2014; Xia, 2012); and urbanization (e.g., Adewuyi and Awodumi, 2017; Sadorsky, 2013; Shahbaz and Lean, 2012). Also, this equation can be estimated using the growth rates of the variables (e.g., Omri and Kahouli, 2014; Omri and Nguyen, 2014).

2.3 Air Pollution Function

The air pollution function builds on the EKC literature (Grossman and Krueger, 1991, 1995; Stern, 2004). As with the energy consumption function, the number of explanatory variables is restricted here for educative and illustrative purposes, although it might be necessary to include more explanatory variables in the function to avoid the omitted variables bias. A simple air pollution function is:

$$pol = \gamma_0 + \gamma_1 gdp + \gamma_2 gdp^2 + \gamma_3 e + \gamma_4 urb + \varepsilon_3, \tag{10.5}$$

where all the variables have again been log transformed, and where *pol* stands for CO_2 emissions per capita, *gdp* stands for real GDP per capita, gdp^2 stands for real GDP per capita squared, *e* stands for energy consumption per capita, *urb* stands for urbanization, and ε_3 is the error term.

First, GDP per capita and GDP per capita squared capture the inverted U-shape of the EKC hypothesis. Accordingly, γ_1 is expected to be positive and γ_2 is expected to be negative. Second, as indicated by the previous literature (e.g., Apergis and Payne, 2009a; Omri, 2013; Pao and Tsai, 2010; Zhang and Cheng, 2009), higher energy consumption per capita (*e*) is expected to increase CO_2 emissions, indicating that γ_3 should be positive. Third, urbanization (*urb*) can affect emissions in many ways (Martínez-Zarzoso and Maruotti, 2011). It changes the way in which resources are used by accelerating the transition to modern fuels and it promotes efficient use of natural resources and economies of scale in production, thus decreasing the level of emissions. Conversely, urban consumption patterns are often more pollution-intensive than rural patterns, and a more urbanized population thus tends to exert greater pressure on the environment (Martínez-Zarzoso and Maruotti, 2011; Sharma, 2011). The so-called 'compact city theory' suggests, that urbanization results in economies of scale in providing public infrastructure, which should reduce emissions (e.g., Sadorsky, 2014). Altogether, it is not surprising that the previous results on urbanization's effect on pollution are contradictory (e.g., Cracolici et al., 2010; Gangadharan and Valenzuela, 2001; Omri, 2013; Sadorsky, 2014).

Other explanatory variables that have been included in the air pollution equation include, among others: education (Gangadharan and Valenzuela, 2001); foreign direct investment (e.g., Omri et al., 2014); population density (Gangadharan and Valenzuela, 2001; Martínez-Zarzoso and Maruotti, 2011); and trade openness (e.g., Adewuyi and Awodumi, 2017; Omri, 2013; Omri et al., 2014). Also, this equation can be estimated using the growth rates of the variables (e.g., Saidi and Hammami, 2016b).

3 Simultaneous Equations Modeling: Key Concepts

Among others, Gujarati and Porter (2009), Hill et al. (2012), and Wooldridge (2013) discuss SEM at an introductory level, while more advanced treatment is given in Greene (2008) and Wooldridge (2010).

3.1 Structural Equations

An SEM is a system that consists of at least two structural equations that determine the values of at least two endogenous variables. These equations should be based on economic theory. As Wooldridge (2013, pp. 531–534) highlights, each structural equation should have a ceteris paribus causal interpretation in isolation from the other structural equations. Moreover, because only the equilibrium values of the variables are observed in the SEM context, counterfactual questions must be asked to determine the effect of a variable on another, if the value of the explanatory variable was different from its equilibrium value (Wooldridge, 2010, pp. 239–241).

Following Greene (2008, pp. 358–361), a linear SEM consisting of M equations with M endogenous variables ($y_1,...,y_M$) and K exogenous variables ($x_1,..., x_K$) can be written as:

$$\gamma_{11}y_{i1} + \gamma_{21}y_{i2} + \cdots + \gamma_{M1}y_{iM} + \beta_{11}x_{i1} + \beta_{21}x_{i2} + \cdots + \beta_{K1}x_{iK} = \varepsilon_{i1}$$
$$\gamma_{12}y_{i1} + \gamma_{22}y_{i2} + \cdots + \gamma_{M2}y_{iM} + \beta_{12}x_{i1} + \beta_{22}x_{i2} + \cdots + \beta_{K2}x_{iK} = \varepsilon_{i2}$$
$$\vdots$$
$$\gamma_{1M}y_{i1} + \gamma_{2M}y_{i2} + \cdots + \gamma_{MM}y_{iM} + \beta_{1M}x_{i1} + \beta_{2M}x_{i2} + \cdots + \beta_{KM}x_{iK} = \varepsilon_{iM}$$

$$(10.6)$$

or, using matrices, as:

$$\mathbf{y}_i' \Gamma + \mathbf{x}_i' \mathbf{B} = \varepsilon_i' \qquad (10.7)$$

The first element of $\mathbf{x}_i$ is usually the constant 1. As is commonly done with cross-sectional data, the subscript i—with $i = 1,\ldots, N$—is used to index observations. This formulation of the model applies to an observation in a particular cross section (or a point in time with time-series data).

Box 10.2 Example: Structural Equations

First, based on Section 2, the structural equations for a simplified two-equation energy-growth SEM are:

$$gdp = \alpha_0 + \alpha_1 k + \alpha_2 h + \alpha_3 e + \varepsilon_1$$
$$e = \beta_0 + \beta_1 gdp + \beta_2 ind + \varepsilon_2$$

Air pollution has been excluded from this SEM for the sake of simplicity. This system will be called 'energy-growth SEM' in the examples below (see Boxes 10.3–10.9).
Second, 'energy-growth-pollution SEM' also used as an example looks like:

$$gdp = \alpha_0 + \alpha_1 k + \alpha_2 h + \alpha_3 e + \alpha_4 pol + \varepsilon_1$$
$$gy = \beta_0 + \beta_1 gdp + \beta_2 ind + \beta_3 pol + \varepsilon_2$$
$$pol = \gamma_0 + \gamma_1 gdp + \gamma_2 gdp^2 + \gamma_3 e + \gamma_4 urb + \varepsilon_3$$

The coefficients α, β, and γ are known as 'structural parameters' or 'structural coefficients', while the error terms ε_1, ε_2, and ε_3 are known as 'structural errors' or 'structural disturbances'. Because the dependent variables are included as explanatory variables in the other structural equations of the SEMs, both systems are interdependent. The variables gdp, e, and pol are endogenous and variables k, h, ind, and urb are exogenous because their values are determined outside the SEMs.[4]
The number of explanatory variables has been kept as low as possible here for the sake of simplicity, as this chapter aims to discuss simultaneous equations modeling using these SEMs as examples. It is, however, important to note that all the structural equations include one exogenous explanatory variable that is not included in the other equations of the SEM in question. An additional issue related to the energy-growth-pollution SEM is that it includes gdp^2 as an endogenous explanatory variable. Because this nonlinearity makes it more complicated to analyse the full system (see Greene, 2008, p. 380; and Wooldridge, 2010, pp. 262–271), some examples discussed below only use the energy-growth SEM. The issue of nonlinearity will be discussed in more detail in Section 3.4.

---

[4] It can be disputed whether all of the variables defined as exogenous are truly exogenous (i.e., uncorrelated with the error terms). Luckily, many of the estimation methods covered below allow for more explanatory variables to be defined as endogenous. The problem with this approach is that more so-called 'instrumental variables' are then needed.

It is assumed that the structural errors ε are randomly drawn from an M-variate distribution with $\left[\varepsilon_i \mid x_i\right] = 0$ and $E\left[\varepsilon_i \varepsilon_i' \mid x_i\right] = \Sigma$. For the full set of N observations, the model can be written as:

$$\mathbf{Y\Gamma + XB = E} \tag{10.8}$$

with $E[\mathbf{E}|\mathbf{X}] = \mathbf{0}$ and $E\left[(1/N)\mathbf{E}'\mathbf{E}|\mathbf{X}\right] = \Sigma$.

Two examples of the structural equations in the EGN context are given in Box 10.2.

3.2 Reduced-Form Equations

The structural equations can be used to solve the so-called 'reduced-form equations', which express all the endogenous variables as linear functions of all the exogenous variables (including possible lagged endogenous variables) and stochastic disturbances. There is one reduced-form equation for every endogenous variable, and every reduced-form equation includes all of the system's exogenous variables as explanatory variables. Because the reduced-form equations do not include any endogenous explanatory variables, the ordinary least squares (OLS) estimator can be used to estimate them. Using the notation of the previous section, and again in accordance with Greene (2008, pp. 360–361), the reduced form of the SEM is:

$$y_i' = [x_1 \ x_2 \ \cdots \ x_K]_t \begin{bmatrix} \pi_{11} & \pi_{12} & \cdots & \pi_{1M} \\ \pi_{21} & \pi_{22} & \cdots & \pi_{2M} \\ \vdots & \vdots & \vdots & \vdots \\ \pi_{K1} & \pi_{K2} & \cdots & \pi_{KM} \end{bmatrix} + [v_1 \ \ v_2 \ \cdots \ v_K]$$

$$y_i' = -x_i'\mathbf{B\Gamma}^{-1} + \varepsilon_i' \ \mathbf{\Gamma}^{-1}$$

$$y_i' = -x_i'\mathbf{\Pi} + v_i' \tag{10.9}$$

The coefficients π are known as the 'reduced-form parameters' or 'reduced-form coefficients', while the error terms v are known as the 'reduced-form errors' or 'reduced-form disturbances'.

Box 10.3 Energy-Growth SEM: Reduced-Form Equations

The reduced-form equations for the energy-growth SEM can be written as:

$$gdp = \pi_{10} + \pi_{11}k + \pi_{12}h + \pi_{13}ind + v_1$$
$$e = \pi_{20} + \pi_{21}k + \pi_{22}h + \pi_{23}ind + v_2$$

The coefficients π thus represent the reduced-form coefficients, while error terms v represent the reduced-form errors.

The nonlinearity in endogenous variables imposed by the air pollution equation in the energy-growth-pollution SEM, makes it considerably more difficult or even impossible to derive the reduced-form equations (see Greene [2008, p. 380] and Wooldridge [2010, pp. 262–271] for discussion). This issue will be discussed in Section 3.4.

Thus, the reduced form for the full set of N observations is:

$$\mathbf{Y} = \mathbf{X\Pi} + \mathbf{V} \tag{10.10}$$

where $\mathbf{V} = \mathbf{E\Gamma}^{-1}$.

An example of the reduced-form equations in the EGN context is presented in Box 10.3.

3.3 The Identification Problem

A question that arises in the SEM framework is whether the numerical estimates of the structural parameters can be obtained, that is, whether the structural equation is identified. For example, Greene (2008, pp. 361–370) discusses the concept thoroughly. If an equation is identified, its parameters can be estimated consistently, and if it is unidentified (or underidentified), its parameters cannot be consistently estimated. Another way of posing this, is that 'if you know that your estimate of a structural parameter is in fact an estimate of that parameter and not an estimate of something else, then that parameter is said to be identified' (Kennedy, 2008, p. 173). In the context of the EGN, it can be considered particularly important for a researcher, to justify why an explanatory variable is expected to appear in one structural equation but not in the other. The answer to this question has been somewhat neglected in the previous literature.

The necessary (but not sufficient) condition for identification is called 'the order condition'. In a two-equation system, it requires that at least one exogenous explanatory variable included in the other structural equation be excluded from a structural equation in order for the latter structural equation to be identified. The so-called 'rank condition' for identification requires, additionally, that the coefficient of the exogenous variable, in the structural equation where it is included, is nonzero. The rank condition is both necessary and sufficient for an equation to be identified (e.g., Gujarati and Porter, 2009, pp. 699–703; Wooldridge, 2013, pp. 538–539).

In a system with more than two simultaneous equations, it is more difficult to show that an equation is identified because, as mentioned, identification depends on the unknown parameter values in the other equations. The order condition for identification in this case is that the number of exogenous explanatory variables excluded from an equation has to be at least as large as the number of endogenous explanatory variables. In practice, an equation satisfying the order condition is often assumed to be identified (Gujarati and Porter, 2009, p. 703; Wooldridge, 2013, p. 543). If every structural equation in a system of simultaneous equations is identified, the whole system is said to be identified.

An identified equation can be either just (or exactly or fully) identified or over-identified. If unique numerical values can be obtained for the structural parameters, the equation is just identified. If more than one numerical value can be obtained for some of the structural parameters, the equation is over-identified. In such a case, there is too much information to identify the structural equation. In practice, over-identification is more common than just identification (Kennedy, 2008, p. 176). Gujarati and Porter (2009, pp. 689–698) provide examples for both cases. In general, the rank condition tells whether an equation is identified, and the order condition tells whether the equation is just identified or over-identified (Gujarati and Porter, 2009, p. 702). As long as an equation is just identified or over-identified, its parameters can be estimated, but the identification status has to be taken into account when choosing the estimator.

A simple way to put the necessary condition for identification is: 'In a system of M simultaneous equations, which jointly determine the values of M endogenous variables, at least $M - 1$ variables must be absent from an equation for the estimation of its parameters to be possible' (Hill et al., 2012, p. 451). If exactly $M - 1$ variables are absent, the equation is just identified, and if more than $M - 1$ variables are absent, the equation is over-identified.

The identification of the energy-growth SEM defined in Box 10.2 is discussed in Box 10.4.

Box 10.4 Energy-Growth SEM: Identification

In the energy-growth SEM, (1) the income equation is just identified because industrialization (*ind*) is not included in it (and it is assumed that the coefficient of *ind* in the energy consumption equation is nonzero) and (2) the energy consumption equation is over-identified because physical capital (*k*) and human capital (*h*) are not included in it (and it is assumed that the coefficients of *k* and *h* in the income equation are nonzero).

3.4 Nonlinearity in Endogenous Variables

The issue of nonlinearity in endogenous variables arises in the energy-growth-pollution SEM (defined in Box 10.2), where the third equation includes the squared income level as an explanatory variable:

$$gdp = \alpha_0 + \alpha_1 k + \alpha_2 h + \alpha_3 e + \alpha_4 pol + \varepsilon_1$$
$$e = \beta_0 + \beta_1 gdp + \beta_2 ind + \beta_3 pol + \varepsilon_2 \qquad (10.11)$$
$$pol = \gamma_0 + \gamma_1 gdp + \gamma_2 gdp^2 + \gamma_3 e + \gamma_4 urb + \varepsilon_3$$

In general, such nonlinearity makes it impossible to obtain linear reduced-form equations for all the endogenous variables in the SEM. However, because this SEM is still linear in parameters, the estimation methods covered in the next chapter remain applicable. Two topics that require particular attention, due to the observed nonlinearity, are identification and the choice of instrumental variables (IV) (see Greene, 2008, p. 380; Wooldridge, 2010, pp. 262–271).

One solution is to define a new endogenous variable, so that the SEM is linear in endogenous variables (Wooldridge, 2010, pp. 263–267). In the energy-growth-pollution SEM, a new endogenous variable $gdp_{new} = gdp^2$ can thus be defined. Using this, the air pollution equation can be written as:

$$pol = \gamma_0 + \gamma_1 gdp + \gamma_2 gdp_{new} + \gamma_3 e + \gamma_4 urb + \varepsilon_3 \qquad (10.12)$$

The key challenge is then to define the linear projection(s)—that is, the new structural equation(s)—for the newly defined endogenous variable(s). Relying only on the existing exogenous variables of the SEM is too restrictive, but gdp^2 is likely to be correlated with the cross-products of the existing exogenous variables as well as their squares, which can thus be used as instruments (see also Greene, 2008, p. 380). Although exact guidelines for choosing and building additional instruments for nonlinear SEMs do not exist, all the exogenous variables appearing in the entire SEM must at least be included in every linear projection.

The energy-growth-pollution SEM can now be augmented with the following equation for gdp_{new}:

$$gdp_{new} = \pi_{400} + \pi_{401}k + \pi_{402}h + \pi_{403}ind + \pi_{404}urb + \pi_{405}k^2 + \pi_{406}h^2 + \pi_{407}ind^2 + \pi_{408}urb^2$$
$$+ \pi_{409}kh + \pi_{410}kind + \pi_{411}kurb + \pi_{412}hind + \pi_{413}hurb + \pi_{414}indurb + v_4, \qquad (10.13)$$

where $k^2 = k^2$, $h^2 = h^2$, $ind^2 = ind^2$, $urb^2 = urb^2$, $kh = k*h$, $kind = k*ind$, $kurb = k*urb$, $hind = h*ind$, $hurb = h*urb$, and, finally, $indurb = ind*urb$. Using these, the SEM can be written as:

$$gdp = \alpha_0 + \alpha_1 k + \alpha_2 h + \alpha_3 e + \alpha_4 pol + \varepsilon_1$$
$$e = \beta_0 + \beta_1 gdp + \beta_2 ind + \beta_3 pol + \varepsilon_2$$
$$pol = \gamma_0 + \gamma_1 gdp + \gamma_2 gdp_{new} + \gamma_3 e + \gamma_1 urb + \varepsilon_3$$
$$gdp_{new} = \pi_{400} + \pi_{401}k + \pi_{402}h + \pi_{403}ind + \pi_{404}urb + \pi_{405}k^2 + \pi_{406}h^2 + \pi_{407}ind^2 + \pi_{408}urb^2$$
$$+ \pi_{409}kh + \pi_{410}kind + \pi_{411}kurb + \pi_{412}hind + \pi_{413}hurb + \pi_{414}indurb + v_4 \qquad (10.14)$$

The fourth equation given above illustrates the problem with this approach, particularly with small samples. Namely, the number of parameters to be estimated increases quickly. It might, however, be possible to use $(\hat{gdp})^2$ as an instrument for gdp^2, instead of the long list of additional nonlinear functions (Wooldridge, 2010, p. 268) although Greene (2008, p. 380) warned that such an approach could be inconsistent. Another way to deal with the endogenous gdp^2 is to use the so-called 'control function approach', which in this case involves including the OLS residuals (and perhaps their squares) from the first-stage regression, as extra regressor(s) in the air pollution function (Wooldridge, 2010, pp. 126–129, 268–271).

In general, if the most general linear version of the SEM is identified, the nonlinear version of the SEM will likely be identified too. As for estimation with the presence of this kind of nonlinearity, it is important to apply IV estimators directly to the structural equations, as trying to substitute the fitted values of the endogenous variables into nonlinear functions, does not usually produce consistent estimators.

4 Simultaneous Equations Modeling: Estimation Methods

After checking that an equation is identified, the estimation method must be selected. It is possible to use either a single-equation method or a system estimation method. Both approaches rely on using IV. In this context, they consist of the exogenous variables that appear in the SEM. With panel data, it is also possible to use lags of the instrumented variables as instruments (see Section 4.3.1), in which case it is not necessary to use any external instruments.

4.1 Cross-Sectional Analysis

4.1.1 Single-equation estimation methods

Single-equation estimation methods can be used to estimate each (identified) equation of an SEM separately, meaning that one endogenous variable is examined at a time. Because single-equation estimation methods do not utilize knowledge of the restrictions in the other equations of the SEM, they are also called 'limited information methods'.

The OLS estimator is biased and inconsistent in the presence of an endogenous explanatory variable that correlates with the error term, meaning that the OLS estimators do not converge to their true population values even when the sample size approaches infinity (see e.g., Gujarati and Porter [2009, pp. 679–683] and Wooldridge [2013, pp. 354–356] for proofs of the inconsistency). In the context of SEM, this bias is referred to as simultaneity bias or simultaneous equations bias. As a result, OLS should not be used to estimate a structural equation of an SEM. There are various alternative estimation methods that could be used in this context, but 2SLS estimation has been the most popular one.

4.1.1.1 Two-stage least squares

The most commonly used approach to overcome the simultaneity bias is the 2SLS estimator, which is sometimes referred to as the generalized instrumental variables estimator (GIVE). The 2SLS method can be considered to be a special case of the more general IV technique. As the name suggests, the 2SLS estimation consists of two stages. For example, suppose that the first (identified) structural equation of an SEM is:

$$y_1 = \delta_0 + \delta_1 y_2 + \delta_2 y_3 + \delta_3 x_1 + \delta_4 x_2 + \varepsilon_1, \tag{10.15}$$

where instances of y represent the endogenous variables and instances of x represent the exogenous variables of the SEM.

In the first stage, OLS estimation is used to regress each endogenous explanatory variable on all of the exogenous variables in the SEM, which is equivalent to estimating the reduced-form equations. For example, assuming that there are K exogenous variables in the SEM, we get the following reduced-form equations for y_2 and y_3:

$$y_2 = \pi_{21}x_1 + \pi_{22}x_2 + \cdots + \pi_{2K}x_K + v_2$$
$$y_3 = \pi_{31}x_1 + \pi_{32}x_2 + \cdots + \pi_{3K}x_K + v_3 \quad (10.16)$$

The estimated reduced-form equations can then be used to calculate the predicted values for the endogenous explanatory variables:

$$\hat{y}_2 = \hat{\pi}_{21}x_1 + \hat{\pi}_{22}x_2 + \cdots + \hat{\pi}_{2K}x_K$$
$$\hat{y}_3 = \hat{\pi}_{31}x_1 + \hat{\pi}_{32}x_2 + \cdots + \hat{\pi}_{3K}x_K \quad (10.17)$$

In the second stage, the predicted values of the endogenous explanatory variables are plugged into the structural equations of the SEM to acquire consistent estimates for the structural parameters:

$$y_1 = \delta_0 + \delta_1\hat{y}_2 + \delta_2\hat{y}_3 + \delta_3 x_1 + \delta_4 x_2 + \varepsilon_1 \quad (10.18)$$

The second-stage regressions for the energy-growth SEM are presented in Box 10.5.

In practice, the 2SLS estimators built in statistical software packages should be used, as otherwise the second-stage standard errors are incorrect.

Box 10.5 Energy-Growth SEM: Second-Stage Regressions

As an example, the second-stage regressions for the energy-growth SEM can now be written as:

$$gdp = \alpha_0 + \alpha_1 k + \alpha_2 h + \alpha_3 \hat{e} + \varepsilon_1$$
$$e = \beta_0 + \beta_1 g\hat{d}p + \beta_2 ind + \varepsilon_2$$

More generally, using matrix notation based on Greene (2008, pp. 371–375), the nonzero terms in the jth structural equation are for all N observations:

$$\mathbf{y_j} = \mathbf{Y_j}\mathbf{\gamma_j} + \mathbf{X_j}\mathbf{\beta_j} + \mathbf{\varepsilon_j}$$
$$\mathbf{y_j} = \mathbf{Z_j}\mathbf{\delta_j} + \mathbf{\varepsilon_j} \quad (10.19)$$

while the reduced-form equations for the included endogenous variables $\mathbf{Y}_j$ are:

$$\mathbf{Y_j} = \mathbf{X\Pi_j} + \mathbf{V_j} \quad (10.20)$$

The first-stage estimation results can be used to derive the predicted values for the endogenous explanatory variables:

$$\hat{\mathbf{Y}}_j = \mathbf{X}\left[(\mathbf{X}'\mathbf{X})^{-1}\mathbf{X}'\mathbf{Y}_j\right] = \mathbf{X}\mathbf{P}_j \tag{10.21}$$

and the 2SLS estimator is then:

$$\hat{\delta}_{j,2SLS} = \begin{bmatrix} \hat{\mathbf{Y}}_j'\mathbf{Y}_j & \hat{\mathbf{Y}}_j'\mathbf{X}_j \\ \mathbf{X}_j'\mathbf{Y}_j & \mathbf{X}_j'\mathbf{X}_j \end{bmatrix}^{-1} \begin{bmatrix} \hat{\mathbf{Y}}_j'\mathbf{Y}_j \\ \mathbf{X}_j'\mathbf{Y}_j \end{bmatrix} \tag{10.22}$$

Based on some useful simplifications (see Greene, 2008, pp. 374–375), the 2SLS estimator can also be written as:

$$\hat{\delta}_{j,2SLS} = [(\mathbf{Z}_j'\mathbf{X})(\mathbf{X}'\mathbf{X})^{-1}(\mathbf{X}'\mathbf{Z}_j)]^{-1}(\mathbf{Z}_j'\mathbf{X})(\mathbf{X}'\mathbf{X})^{-1}\mathbf{X}'\mathbf{y}_j \tag{10.23}$$

A problem associated with 2SLS and IV estimators, in general, is that the standard errors tend to be large, therefore making the 2SLS estimator relatively inefficient (e.g., Wooldridge, 2010, pp. 108–111).

4.1.1.2 Other estimation methods

Reduced-form parameters can also be used to estimate the structural parameters. The process described in Gujarati and Porter (2009, pp. 715–718), for instance, is called 'indirect least squares' (ILS). The idea is that, the OLS estimates of the reduced-form parameters can be used to acquire the ILS estimates of the structural parameters, because the structural parameters can be expressed as functions of the reduced-form parameters. The ILS can only be applied to just identified equations, because it is only then that there is only one manner in which to calculate the structural parameters based on the reduced-form parameters. The ILS estimates are consistent, but not unbiased. The ILS is also equivalent to 2SLS with the just identified structural equations.

Another alternative is the 'limited information maximum likelihood' (LIML) estimator, which uses maximum likelihood estimation to estimate the reduced-form parameters (see e.g., Greene, 2008, pp. 375–377). The estimated reduced-form parameters are then used to calculate the estimates of the structural parameters like in ILS. The procedure can also be viewed as a case of IV estimation, because the estimated reduced-form equations can be used to calculate the estimated values of the endogenous explanatory variables in the structural equations, which can then be used as instruments in the 2SLS estimation. For just identified equations, 2SLS and LIML give the same results. The recognition of the problems that weak instruments cause for the 2SLS estimator (see Section 4.1.3) has resulted in increased interest in the LIML estimator.

Box 10.6 Energy-Growth SEM: 2SLS and LIML Estimation Results With Cross-Sectional Data

Data description

The sample is restricted to high-income and upper-middle-income countries according to the World Bank's classification for the year 2011 (World Bank, 2016a), which is the last year for which data for some of the variables are available. The sample used in the cross-sectional analysis consists of 69 countries in 2011. The detailed descriptions of the variables are provided in Table A.1. Of the key variables, energy consumption is measured by energy use (kilogram of oil equivalent per capita) obtained from the World Bank (2016b), income level by real GDP from the Penn World Table (Feenstra et al., 2015), and air pollution by CO_2 emissions from the World Bank (2016b). The exogenous variables comprise physical capital stock and human capital from the Penn World Table (Feenstra et al., 2015) and industrialization and urbanization from the World Bank (2016b). The variables not measured in percentage, or originally in per capita terms, were divided by the population size, also from the World Bank (2016b). Finally, all of the variables were log transformed.

Estimation results

The 2SLS and LIML estimation results for the energy-growth SEM are presented in Table 10.2. The dependent variable in Models 1 and 3 is GDP per capita, and in Models 2 and 4, it is energy consumption per capita. Stata/IC version 14 and the module ivreg2 by Baum et al. (2010) were used in the estimations. Option *robust* was used to get statistics that are robust to the presence of arbitrary heteroscedasticity, and option *small* was used to report small-sample statistics. With the just identified income equation, 2SLS and LIML estimates are identical, as they should be. With the over-identified energy equation, 2SLS and LIML estimates differ, but only very slightly in this case. Overall, the results support unidirectional causality from GDP per capita to energy consumption per capita. In addition, Models 1 and 3 indicate that physical capital per capita is the most significant determinant of GDP per capita, while Models 2 and 4 suggest that higher industrialization increases energy consumption per capita, as expected.

Table 10.2: 2SLS and LIML estimation results with cross-sectional data.

	2SLS		LIML	
	(1)	(2)	(3)	(4)
	GDP	Energy	GDP	Energy
GDP		0.922***		0.925***
		(9.86)		(9.67)
Energy consumption	0.195		0.195	
	(1.44)		(1.44)	
Physical capital	0.728***		0.728***	
	(6.75)		(6.75)	
Human capital	−0.219		−0.219	
	(−0.88)		(−0.88)	
Industrialization		0.435***		0.435***
		(2.83)		(2.83)
Constant	0.574	−2.678***	0.574	−2.709***
	(1.24)	(−2.88)	(1.24)	(−2.85)
Observations	69	69	69	69

t-statistics are shown in parentheses (***significant at 1%).

> **Note**
> Using SEM in the EGN thus requires that the income level, energy consumption, and air pollution (if included in the model) are allowed to be endogenous. However, some of the other explanatory variables could also be endogenous. Although it is possible to define them as endogenous in 2SLS estimations, the problem is that new instruments are then needed. Therefore it might be necessary to use internal rather than external instruments. This is possible with panel data (see Section 4.3.1).

2SLS and LIML estimation results for the energy-growth example are compared in Box 10.6.

4.1.2 System estimation methods

System methods estimate all the identified equations of an SEM simultaneously. Because they utilize every zero restriction of the entire SEM to estimate all of the structural parameters, they are also called 'full information methods'. When the SEM has been specified correctly, system estimation methods are generally more efficient than estimating each structural equation with 2SLS.

4.1.2.1 Three-stage least squares

The most commonly used system estimation method is the 3SLS estimator, which differs from 2SLS by allowing for nonzero covariances between the structural errors. As the name implies, 3SLS estimation can be thought to consist of three stages, which are described below. Again using matrix notation, the full SEM can be formulated as (Greene, 2008, pp. 380–383):

$$\begin{bmatrix} \mathbf{y}_1 \\ \mathbf{y}_2 \\ \vdots \\ \mathbf{y}_M \end{bmatrix} = \begin{bmatrix} \mathbf{Z}_1 & \mathbf{0} & \cdots & \mathbf{0} \\ \mathbf{0} & \mathbf{Z}_2 & \cdots & \mathbf{0} \\ \vdots & \vdots & \vdots & \vdots \\ \mathbf{0} & \mathbf{0} & \cdots & \mathbf{Z}_M \end{bmatrix} \begin{bmatrix} \delta_1 \\ \delta_2 \\ \vdots \\ \delta_M \end{bmatrix} + \begin{bmatrix} \varepsilon_1 \\ \varepsilon_2 \\ \vdots \\ \varepsilon_M \end{bmatrix} \tag{10.24}$$

or

$$\mathbf{y} = \mathbf{Z}\delta + \varepsilon \tag{10.25}$$

with $E[\varepsilon \mid \mathbf{X}] = \mathbf{0}$ and $E[(\varepsilon'\varepsilon|\mathbf{X}] = \overline{\Sigma} = \Sigma \otimes \mathbf{I}$.

The matrix of IV is denoted by $\overline{\mathbf{W}}$, which is assumed to satisfy the requirements for an IV estimator.

$$\overline{\mathbf{W}} = \hat{\mathbf{Z}} = \text{diag}[\mathbf{X}(\mathbf{X}'\mathbf{X})^{-1}\mathbf{X}'\mathbf{Z}_1,\ldots,\mathbf{X}(\mathbf{X}'\mathbf{X})^{-1}\mathbf{X}'\mathbf{Z}_M] = \begin{bmatrix} \hat{\mathbf{Z}}_1 & \mathbf{0} & \cdots & \mathbf{0} \\ \mathbf{0} & \hat{\mathbf{Z}}_2 & \cdots & \mathbf{0} \\ \vdots & \vdots & \vdots & \vdots \\ \mathbf{0} & \mathbf{0} & \cdots & \hat{\mathbf{Z}}_M \end{bmatrix} \tag{10.26}$$

Using this, the 3SLS estimator can be formulated as a generalized least squares (GLS) estimator:

$$\hat{\delta}_{3SLS} = [\hat{Z}'(\Sigma^{-1} \otimes I)\hat{Z}]^{-1} \hat{Z}'(\Sigma^{-1} \otimes I)y \qquad (10.27)$$

Box 10.7 Energy-Growth SEM: 3SLS Estimation Results With Cross-Sectional Data

The 3SLS estimation results for the energy-growth SEM are presented in Table 10.3 using the same data as in Box 10.6. Instruments that are exogenous in one equation are assumed to be exogenous in both equations, although this assumption is often violated in practice (see the discussion in Section 4.4). Another potential problem is that the 3SLS procedure that is built in Stata does not account for system heteroscedasticity.

Table 10.3: 3SLS estimation results with cross-sectional data.

	3SLS	
	GDP	**Energy**
GDP		0.922***
		(10.74)
Energy consumption	0.230**	
	(2.21)	
Physical capital	0.684***	
	(8.35)	
Human capital	0.038	
	(0.15)	
Industrialization		0.435***
		(3.20)
Constant	0.510	−2.678***
	(1.22)	(−2.69)
Observations	69	69

z-statistics are shown in parentheses (**significant at 5%; ***significant at 1%).

If both equations were just identified, 2SLS and 3SLS would produce identical parameter estimates (see also Wooldridge, 2010, pp. 254–255). Now that the income equation is just identified and the energy consumption equation is over-identified, the 3SLS parameter estimates for the energy consumption equation are equivalent with the 2SLS parameter estimates, but the parameter estimates for the income equation differ.

In contrast to the 2SLS results, which supported unidirectional causality from GDP per capita to energy consumption per capita, the 3SLS results support a bidirectional causal relationship between the income level and energy consumption and are thus more in line with the previous EGN literature using SEM. The 3SLS results also imply (in accordance with the 2SLS results) that physical capital per capita is a significant determinant of GDP per capita and that higher industrialization increases energy consumption per capita.

with the following asymptotic covariance matrix:

$$\text{Asy. Var}[\hat{\boldsymbol{\delta}}_{3SLS}] = [\bar{\mathbf{Z}}'(\boldsymbol{\Sigma}^{-1} \otimes \mathbf{I})\bar{\mathbf{Z}}]^{-1} \tag{10.28}$$

where $\bar{\mathbf{Z}} = \text{diag}[\mathbf{X}\boldsymbol{\Pi}_j, \mathbf{X}_j]$

In the first stage of 3SLS estimation, OLS is used to estimate $\boldsymbol{\Pi}$, which is then used to compute $\hat{\mathbf{Y}}_j$ for each equation. In the second stage, $\hat{\boldsymbol{\delta}}_{j,2SLS}$ is calculated for each equation and then used to estimate the contemporaneous variance-covariance matrix of the structural errors. Finally, in the third stage, the GLS estimator is computed using equation (10.27) and the estimated asymptotic covariance using equation (10.28), where $\bar{\mathbf{Z}}$ may be estimated with $\hat{\mathbf{Z}}$ and $\boldsymbol{\Sigma}$ with $\hat{\boldsymbol{\Sigma}}$.

The 3SLS estimation results for the energy-growth SEM are presented in Box 10.7.

4.1.2.2 Other methods

'Full information maximum likelihood' (see Greene, 2008, pp. 383–384) is the system estimation version of LIML. It involves estimating all structural equations jointly using maximum likelihood estimation.

4.1.3 Specification tests

With SEM, at least one explanatory variable is endogenous by definition. If an explanatory variable is endogenous, the OLS estimators are inconsistent. 'The Hausman specification test' (Hausman, 1978; described in Wooldridge, 2010, pp. 129–134; Hill et al., 2012, pp. 420–421) can be used to test whether this is the case. The test is also called 'the Durbin–Wu–Hausman (DWH) test' (Durbin, 1954; Wu, 1973; Hausman, 1978). Among others, Adewuyi and Awodumi (2017), Hung and Shaw (2004), and Omri (2013) used the test within the EGN. The idea behind the Hausman test is that, if there is no simultaneity bias (which is the null hypothesis of the test), OLS and 2SLS estimators are both consistent, and the estimated parameters should not differ significantly between the two estimation methods. If the null hypothesis is rejected, indicating that at least one of the explanatory variables is endogenous, the consistent 2SLS estimator should be used. However, due to the properties of the 2SLS estimator, it can sometimes perform worse than the OLS estimator even in this case (see discussion below). If, however, the Hausman test implies that the explanatory variables are exogenous, the 2SLS estimator will be consistent, but not efficient. In such a case, the efficient OLS estimator should be used.

A good IV is highly correlated with the endogenous explanatory variable for which it is used as an instrument. If, the so-called 'instrument relevance', is violated and the IV only weakly correlates with the endogenous explanatory variable, the 2SLS estimator tends to have very poor properties and can be inconsistent even in large samples (see e.g., Wooldridge, 2010, pp. 108, 111). Testing for weak instruments is thus an essential part of IV estimation (see e.g., Greene, 2008, pp. 350–352 for discussion). A standard approach has been to rely on the first-stage results of the 2SLS procedure. A rule-of-thumb, which only works with a single endogenous explanatory variable, is that the F-statistic on the joint significance of the excluded instruments in the estimated first-stage equation should exceed 10. Thus, it is necessary to compute and report the F-statistic in the context of all 2SLS estimation results.

When the number of endogenous explanatory variables exceeds one, testing each of them separately is not sufficient, because collinearity could affect the test result (Greene, 2008, pp. 350–351). A solution to identify the weak instruments with several endogenous explanatory variables is using canonical correlations, which describe how two sets of variables are associated (Baum et al., 2007; Hill et al., 2012, pp. 434–440). 'The Cragg–Donald F-test' uses the so-called 'minimum canonical correlation'. The null hypothesis of this test is that the instruments are only weakly correlated with the endogenous explanatory variables.

A valid IV, denoted by 'z' for now, is also uncorrelated with the structural error, so that $cov(z,\varepsilon) = 0$. This instrument validity is something that is often difficult to test in a reliable way. For example, 'the over-identifying restrictions test', also called 'the Sargan test', (Sargan, 1958; described in e.g., Wooldridge, 2010, pp. 134–137) can only be used if an equation is over-identified, meaning that there are more instruments than would be needed to identify the equation. A simple way to calculate the test statistic is described in Hill et al. (2012, pp. 421–422) and Verbeek (2012, p. 164): the statistic can be calculated by multiplying the R^2 of an auxiliary regression of IV residuals upon the full set of instruments, by the sample size. If the null hypothesis of instrument validity gets rejected, the model specification should be reconsidered. The Sargan test does not, however, tell which of the instruments are invalid.

The Sargan test is only valid if the errors are homoskedastic. If the errors are heteroskedastic, a more general test called 'the Hansen's J-test' should be used. Among others, Adewuyi and Awodumi (2017), Omri (2013), and Omri et al. (2015a) have used the Hansen test within the EGN.

The specification tests are demonstrated in Box 10.8.

4.2 Time-Series Analysis

An SEM can also be used with time-series data. It is then essential to take the special features of the data into account. In particular, one should be aware that using nonstationary time series might result in spurious regression results (see Hill et al. [2012, pp. 482–483] for an example). It is thus necessary to test the time series for unit roots (i.e., nonstationarity) (see also Huang et al., 2008). This applies to many of the time series used in the EGN. A way to avoid problems with trends and high persistence is to specify the SEM in first differences or growth rates, which are more likely to be stationary. In fact, Wooldridge (2013, p. 548) found that growth rates of variables that are integrated of order 1, I(1),[5] or that are trend stationary are fairly commonly used in time-series SEM applications. It should, however,

---

[5] That is that are nonstationary as levels, but stationary in first differences.

Box 10.8 Energy-Growth SEM: Specification Tests With Cross-Sectional Data

The first-stage regression results for the energy-growth SEM are presented in Table 10.4. Thus, Model 1 is the reduced-form equation for GDP per capita, and Model 2 is the reduced-form equation for energy consumption per capita. The module ivreg2 by Baum et al. (2010) was again used.

Table 10.4: The reduced-form equations for the 2SLS estimations with cross-sectional data.

	(1)	(2)
	GDP	Energy
Physical capital	0.865***	0.704***
	(16.30)	(6.89)
Human capital	0.097	1.622***
	(0.33)	(3.23)
Industrialization	0.131	0.673***
	(1.24)	(3.67)
Constant	−0.168	−3.806***
	(−0.27)	(−4.30)
F-statistic	192.58	13.45
Cragg–Donald F-statistic	217.61	19.87

t-statistics are shown in parentheses (***significant at 1%).

First, instrument relevance can be tested using the F-test of excluded instruments and the Cragg–Donald F-test. The F-statistic depicted at the bottom of Table 10.4 indicates that the excluded instruments are not weak, as the statistic is above 10 in both first-stage regression models. Moreover, using the critical values provided by Stock and Yogo (2005) and reproduced, for example in Hill et al. (2012, p. 436), the null hypothesis of instruments being weak can also be rejected based on the Cragg–Donald F-statistics. Therefore it can be concluded that the instruments do not appear to be too weak in this case, which is something good.

Table 10.5: Hansen J-test results (2SLS).

	GDP	Energy
Hansen J-statistic	0.000	10.379
Hansen J-test P-value		0.0013

Second, instrument validity can only be tested with the over-identified energy consumption (structural) equation. The Hansen J-test is used instead of the Sargan test to account for heteroscedasticity. The results depicted in Table 10.5 imply that the null hypothesis of instrument validity is rejected even at the 1% level, indicating that the instruments are not valid and that the model should thus be reconsidered.

As for the 3SLS results, the Hansen–Sargan over-identification statistic of 8.461 with a P-value of 0.0036 indicates again that the instruments are not valid (i.e., are correlated with the structural error). This problem is likely to be common in the EGN SEM, suggesting that equation-by-equation estimation with different instruments for different equations might have to be used (see Section 4.4 for a discussion on using different instruments for different equations and Section 4.3.1 for a solution in the panel data context).

be recognized that the first-differenced or growth SEM differs from the original SEM in levels (Wooldridge, 2013, p. 546). An additional point of interest related to SEM with time series data, is that if a structural equation includes a time trend, the trend acts as its own IV (Wooldridge, 2013, p. 548).

Another issue that arises more often with time-series than cross-sectional data is serial correlation, also called autocorrelation. It is thus necessary to test for autocorrelation and, when necessary, to calculate the (heteroscedasticity and) autocorrelation consistent standard errors, often called robust standard errors (see e.g., Hill et al., 2012, pp. 347–358).

With time-series (and panel) data, it is also possible to estimate dynamic models that include lagged explanatory variables. Lagged endogenous variables that are included as explanatory variables are predetermined variables.[6] Therefore they are independent of the contemporaneous and future errors in the relevant equation. Because of this property, they can be treated as exogenous and consistent estimators can still be derived.

Accounting for potential nonstationarity and autocorrelation and applying advanced econometric methods should, in principle, reduce the previously observed discrepancies in the results of the EGN. The problem, however, is that when it comes to the country-level annual time-series data most often used in the nexus, the time dimension tends to be relatively short for these methods. The insufficient number of observations leads to the tests having low statistical testing power, which leads to inconsistent results (Huang et al., 2008). Using the panel data alleviates this problem. Applying SEM in the panel data context is thus discussed in the next section.

4.3 Panel Data Analysis

Using panel data can thus alleviate the problems related to the relatively short time series available for the variables often used in the EGN. In addition, using panel data makes it easier to control for heterogeneity between countries because it allows for controlling for country-specific effects. Following Greene (2008, p. 182), a structural equation can be written in the panel data format as:

$$y_{it} = x'_{it}\beta + c_i + \varepsilon_{it},\qquad(10.29)$$

where the subscript i indexes cross-sectional units (here, countries) and t time. The heterogeneity between units is captured by the individual effect $c_i = z'_i\alpha$, where z_i contains the constant and the time-invariant observed and unobserved individual characteristics. If z_i

---

[6] As defined in Bond (2002), a variable x_t is endogenous if it is correlated with ε_t and earlier shocks but uncorrelated with ε_{t+1} and subsequent shocks. It is called 'predetermined' if it is uncorrelated with ε_t but correlated with ε_{t-1} and the earlier shocks. If x_t is exogenous, it is uncorrelated with all past, present, and future realizations of ε_s.

contains unobserved characteristics, which is likely to be the case in practice, the error term (which naturally includes the unobserved individual effect) is likely to correlate with many of the explanatory variables in **x**, thus leading to the endogeneity problem.

According to Wooldridge (2013, p. 548), estimating SEMs with panel data consists of two steps. The first step, which becomes necessary due to the endogeneity problem, is to eliminate the unobserved individual effects from the structural equations with the fixed-effects transformation, or first-differencing. The second step is to find instruments for the endogenous explanatory variables (to be more specific, the first differences or changes, of the endogenous variables in the SEM) in these transformed equations. Possible instruments can again be found from the other structural equation(s) of the SEM. Because an instrument is needed for the change in the endogenous explanatory variable, which is unlikely to be highly correlated with a level of an IV, instruments that vary over time are needed. Baltagi (2013, pp. 129–150) discusses simultaneous equations in the panel data context and covers both single-equation and system estimation.

4.3.1 Single-equation estimation methods
4.3.1.1 The 2SLS

The 2SLS estimation can also be used with panel data. In line with the panel data estimation methods in general, there are several different 2SLS estimators for panel data. Baltagi (2013, pp. 129–137) discusses the so-called 'within', 'between', 'error-component', and 'generalized 2SLS' estimators in more detail. However, the problem with using these estimators in the EGN is that the instruments are unlikely to be valid. Hung and Shaw (2004) used the within (i.e., fixed effects) 2SLS estimator to analyze the relationship between the income level and several air pollutants.

4.3.1.2 Difference and system GMM estimation

The problem of finding valid external instruments can be solved by using the dynamic panel GMM estimators by Arellano and Bond (1991) and Blundell and Bond (1998), also called the difference and system GMM estimators. Baltagi (2013, pp. 155–183) and Roodman (2009) discussed the estimators and their properties. In the SEM context, these estimators have also been called 'equation-by-equation estimators' due to their focus on only one dependent variable at a time (e.g., Carrión-Flores and Innes, 2010). Within the EGN, for example, Amri (2017) and Omri and Kahouli (2014) have used the difference GMM estimator, while Adewuyi and Awodumi (2017), Huang et al. (2008), and Sadorsky (2010) have used the system GMM estimator.

Starting with a dynamic panel data model (and following the notation of Greene, 2008, pp. 340–349), a structural equation to be estimated can be written as:

$$y_{it} = \boldsymbol{x}_{it}'\beta + \delta y_{i,t-1} + c_i + \varepsilon_{it}, \tag{10.30}$$

where the subscript i indexes again cross-sectional units and t indexes time. The error term is assumed to consist of two orthogonal components: the fixed individual effects c_i and the idiosyncratic shocks ε_{it}. It is, moreover, assumed that (see Roodman, 2009):

$$E[c_i] = E[\varepsilon_{it}] = E[c_i \varepsilon_{it}] = 0 \tag{10.31}$$

Differencing equation y_{it} eliminates the individual effects c_i, thus resulting in:

$$\Delta y_{it} = (\Delta x_{it})' \beta + \delta(\Delta y_{i,t-1}) + \Delta \varepsilon_{it}, \tag{10.32}$$

where Δ is the first-difference operator.

Although predetermined variables become endogenous in first differences, longer lags of the explanatory variables remain available as potential instruments. The difference GMM estimator (Arellano and Bond, 1991) thus uses the lagged predetermined and endogenous variables as instruments for the same variables in first differences. The system GMM estimator (Arellano and Bover, 1995; Blundell and Bond, 1998) adds the original level equations into the system of first-differenced equations. It continues to instrument the endogenous and predetermined variables in the first differences with lagged endogenous and predetermined variables, while the endogenous and predetermined variables in levels are additionally instrumented with their lagged first differences.[7] The main assumption is that $E[c_i \Delta \varepsilon_{it}] = 0$, implying that the individual effects should not be correlated with disturbance-term changes. The Hansen test for over-identifying restrictions can be used to test the validity of the instruments in the presence of heteroscedasticity.

As Roodman (2009) highlighted, difference and system GMM estimators are designed for panel datasets with many units and few time periods. However, the EGN literature has not always taken this into account. The first problem with ignoring this prerequisite is that the cluster-robust standard errors and the Arellano–Bond test for serial correlation are unreliable if the number of units is too small. The second issue is that dynamic panel bias becomes insignificant if the number of time periods is high. In such a case, the less complex fixed-effects estimator can be used. Moreover, the number of instruments is quadratic in the number of time periods in the sample, which might result in overfitting the endogenous variables by including too many instruments relative to the sample size. A rule-of-thumb presented by Roodman (2009) is that the number of instruments should be kept below the number of units. To achieve this goal, it might be necessary to limit the number of lags of the instruments used and collapse the instrument matrix (see Roodman (2009) for more detailed discussion). Importantly, it is necessary to mind and report the instrument count relative to the sample size, together with the difference and the system GMM estimation results.

---

[7] Although using external instruments would be the best remedy for reverse causality, valid exogenous instruments varying over time and across units are not easily found, making the internal instruments used by the difference and system GMM estimators a good alternative (Farhadi et al., 2015). However, it is also possible to include external instruments when using these estimators (Roodman, 2009).

Box 10.9 Energy-Growth-Pollution SEM: Difference and System GMM Estimation Results With Panel Data

---

Data description

The variables used are described in Box 10.6 and Table A.1. Due to gaps in the original data, some of the variables were interpolated to preserve the sample size. Four-year averages of the variables are used to reduce the effect of the business cycle fluctuations. The final sample thus consists of 8 periods starting with 1980–83 and ending with 2008–11. A total of 40 countries are included in the analysis.

Estimation results

Stata/IC version 14 and the module *xtabond2* by Roodman (2009) were used. Because the one-step GMM estimator requires homoscedasticity and uncorrelated error terms to be efficient, the asymptotically more efficient two-step estimator was used. The potentially downward-biased standard errors (see Arellano and Bond, 1991; Blundell and Bond, 1998) were corrected with Windmeijer (2005) finite-sample correction for the asymptotic variance of the two-step GMM estimator. The number of lags of the IVs in the transformed equation was restricted to lags 2–5 for the endogenous variables and to lags 1–4 for the predetermined variables. Small-sample statistics were calculated and period dummies were included in the models to make the assumption of no correlation across countries in the idiosyncratic error terms more likely to hold.

The results for the difference and system GMM estimations are presented in Panels A–C of Table 10.6. Models 1, 3, and 5 have been estimated with the difference GMM estimator and Models 2, 4 and 6 with the system GMM estimator. Although it is necessary to test for the stationarity of the series before conducting panel data estimations (See Chapters 8 and 9 in this book for a discussion on the panel unit root tests), three different models are presented here for illustrative purposes: variables have been included in levels in Models 1 and 2, in first differences in Models 3 and 4, and as growth rates in Models 5 and 6. Although the results of Models 1 and 2 could be spurious due to nonstationarity, they have been included in Table 10.6 to highlight that the implications of growth equations might differ from those of the equations in levels and that the same model specification might not be suitable for both approaches. For example, it might be necessary to control for the initial income level in an equation for GDP growth (Panel A, Models 5 and 6).

The results for the income equation presented in Panel A of Table 10.6 indicate that the model specification could probably be improved, as physical capital and human capital are statistically insignificant in all of the models. The Hansen test of over-identifying restrictions, however, implies that the instruments are valid, while the Arellano-Bond test for second-order serial correlation in first differences indicates that the results of Model 2 should be taken with a grain of salt, as the estimator consistency relies upon the assumption that there should not be second-order serial correlation between the first-differenced errors (Baltagi, 2013, p. 159). The number of instruments is clearly below the number of groups in all of the models, which is the rule-of-thumb presented by Roodman (2009). The positive effect of energy consumption on income appears relatively robust, thus affirming the role of energy as a factor of production.

Table 10.6: Difference and system GMM estimation results for the income, energy consumption, and air pollution equations.

	(1) diff GMM	(2) sys GMM	(3) diff GMM	(4) sys GMM	(5) diff GMM	(6) sys GMM
Panel A (dependent variable: GDP)						
Lagged GDP	0.637**	0.654***	0.185	0.137	0.156	0.119
	(2.57)	(5.53)	(1.29)	(1.51)	(0.98)	(1.00)
Physical capital	0.035	0.118	−0.016	0.095	0.123	0.204
	(0.10)	(1.14)	(−0.07)	(0.63)	(0.42)	(1.27)
Human capital	−0.139	0.192	−0.115	0.317	−0.020	0.017
	(−0.31)	(1.02)	(−0.17)	(0.71)	(−0.37)	(0.43)
Energy consumption	0.497**	0.103	0.273*	0.492***	0.292**	0.453***
	(2.37)	(1.38)	(1.81)	(4.47)	(2.31)	(4.20)
CO_2 emissions	−0.076	−0.036	−0.095	−0.085	−0.034	−0.034
	(−0.59)	(−0.32)	(−0.97)	(−1.17)	(−1.12)	(−1.32)
Observations	162	202	146	186	146	186
Number of groups	39	40	39	40	39	40
Number of instruments	26	32	25	31	25	31
Arellano-Bond test for second-order serial correlation in first differences (*P*-value)						
	0.179	0.077	0.625	0.654	0.587	0.577
Hansen test of over-identifying restrictions (*P*-value)						
	0.253	0.186	0.342	0.340	0.233	0.287
Panel B (dependent variable: energy consumption)						
Lagged energy consumption	0.105	0.791***	0.099	0.098	0.049	0.075
	(0.72)	(9.12)	(1.60)	(1.38)	(0.62)	(1.18)
GDP	0.546**	−0.032	0.536*	0.374**	0.495	0.482**
	(2.23)	(−0.32)	(1.76)	(2.08)	(1.12)	(2.37)
Industrialization	0.403	0.239*	0.055	0.176	0.074	−0.003
	(1.53)	(1.84)	(0.14)	(1.33)	(0.37)	(−0.07)
CO_2 emissions	0.165	0.054	0.242***	0.164**	0.120***	0.092***
	(0.86)	(0.44)	(3.06)	(2.22)	(2.78)	(3.82)
Observations	162	202	146	186	146	186
Number of groups	39	40	39	40	39	40
Number of instruments	22	27	21	26	21	26
Arellano-Bond test for second-order serial correlation in first differences (*P*-value)						
	0.195	0.027	0.105	0.084	0.126	0.110
Hansen test of over-identifying restrictions (*P*-value)						
	0.702	0.210	0.267	0.141	0.438	0.247

(Continued)

Panel C (dependent variable: CO_2 emissions)					
Lagged CO_2 emissions	0.271	0.746**	−0.294*	−0.115	−0.090	−0.007
	(1.22)	(2.63)	(−1.86)	(−0.67)	(−0.39)	(−0.03)
GDP	5.734	0.619	1.171	6.333	19.166	−88.333
	(1.10)	(0.18)	(0.11)	(1.02)	(0.10)	(−0.53)
GDP squared	−0.322	−0.028	−0.088	−0.362	−9.947	42.927
	(−1.13)	(−0.17)	(−0.16)	(−1.04)	(−0.10)	(0.52)
Energy consumption	0.869***	0.171	1.542***	1.548***	4.110**	4.434**
	(8.89)	(0.31)	(2.83)	(3.65)	(2.41)	(2.66)
Urbanization	−0.684	−0.316	−0.081	−0.544	0.378	0.186
	(−0.45)	(−0.50)	(−0.09)	(−1.11)	(0.36)	(0.46)
Observations	162	202	140	180	140	180
Number of groups	39	40	39	40	39	40
Number of instruments	26	32	25	31	25	31
Arellano-Bond test for second-order serial correlation in first differences (*P*-value)						
	0.536	0.916	0.131	0.188	0.302	0.299
Hansen test of over-identifying restrictions (*P*-value)						
	0.129	0.116	0.027	0.302	0.103	0.281

t-statistics are reported in parentheses (*significant at 10%; **significant at 5%; ***significant at 1%). Period dummies and constant excluded from the table. The null hypotheses of the tests are as follows: (1) The Arellano-Bond test for autocorrelation: null = no autocorrelation; (2) The Hansen test: null = the set of instruments is valid.

The results for the energy consumption equation presented in Panel B imply that increases in GDP per capita increase energy consumption per capita. Together with the results for the income equation, the results thus indicate that the causal relationship between energy consumption and GDP is bidirectional. Otherwise, the results for the energy consumption equation in levels (Models 1 and 2) and the equation in first differences or as growth rates (Models 3–6) differ. Somewhat surprisingly, the association between CO_2 emissions and energy consumption (Models 3–6) appears stronger than that observed between industrialization and energy consumption (Model 2). The results for the Hansen test indicate again that the instruments are valid, while the Arellano-Bond test for second-order serial correlation in first differences implies that the system GMM results of Models 2 and 4 might not be reliable due to serial correlation. The number of instruments is again clearly below the number of groups in all of the models, as it should be.

According to the results for the air pollution equation (Panel C), energy consumption is clearly the most important determinant of CO_2 emissions with the expected increasing effect. In contrast, the results do not support the EKC hypothesis. The results for the Hansen test imply that the instruments are not valid in Model 3.

Box 10.9 presents and discusses the difference and system GMM estimation results for the energy-growth-pollution SEM.

4.3.2 System estimation methods

Baltagi (2013, pp. 138–140) discussed 3SLS estimation methods for panel data. As with single-equation methods, 'within', 'between', 'error-component', and 'generalized 3SLS' estimators are available. Because the problem with using these estimators in the EGN is, again, that the instruments are unlikely to be valid, the methods are not covered here in more detail.

4.4 Single-Equation Versus System Estimation Methods

Wooldridge (2010, p. 252) discussed the pros and cons of single-equation and system estimation methods. If an SEM is correctly specified, system estimation is asymptotically more efficient than using a single-equation estimation method. Single-equation estimation methods, however, are more robust. A major disadvantage of system estimation is that, if any of the structural equations are mis-specified, all structural parameters in the entire SEM are affected, instead of only the structural parameters of the equation suffering from the mis-specification. Mis-specifying one of the structural equations can thus contaminate the results of the other structural equations in system estimation.

To avoid such contamination, difference and system GMM have sometimes been used (e.g., by Carrión-Flores and Innes (2010) in a related context) and can also be recommended in the EGN where the theoretical background for the energy consumption equation, in particular, could be stronger.

4.5 Different Instruments for Different Equations

Wooldridge (2010, pp. 271–273) discussed the cases where it is necessary to use different instruments for different structural equations. In general, if a structural equation contains other endogenous variables apart from the variables with values determined within the SEM, different instruments need to be used. This is highly relevant from the perspective of the EGN, where model misspecification or data availability issues often result in the endogeneity problem.

According to Wooldridge, if there are more endogenous variables, their linear projections on all of the exogenous variables need to be added to the system of equations. A problem with this approach is the poor performance of the IV estimators with weak instruments. Finding new instruments might thus be necessary. In principle, however, having different instruments for different equations is not a problem in single-equation estimation, where valid instruments can be searched for each of the endogenous variables before estimating the equations separately. With system methods, the matter is more complicated, and the GMM 3SLS version should be used (Wooldridge, 2010, p. 270).

5 Conclusions

Some recommendations for future research can be made based on this chapter. First, regarding data, it is important to pay attention to the sample size and use appropriate variables in the analysis. For example, gross fixed capital formation is not the same as physical capital stock. Second, because the risk of model mis-specifications is relatively large in EGN SEM, and because instrument validity is also likely to be an issue, using single-equation estimation, instead of system estimation, is a safer choice. Related to this, finding exogenous, perhaps geographical, variables that would affect energy consumption but not GDP could be essential for the system estimation methods to work. Third, special attention should be paid to reporting estimation methods and all specification choices in detail. All the results of the relevant statistical tests should be reported. Finally, regarding estimation results, the practice of performing and reporting robustness checks is highly recommended.

Appendix

Table A.1: Variables.

Variables	Descriptions	Sources	Calculations by the Author
Income level	Real GDP at constant 2005 national prices (in mil. 2005 USD)	Penn World Table (Feenstra et al., 2015)	Divided by population
Energy consumption	Energy use (kilogram of oil equivalent per capita)	World Bank (2016b)	Interpolated
Air pollution	CO_2 emissions (metric tons per capita)	World Bank (2016b)	
Physical capital	Physical capital stock at constant 2005 national prices (in mil. 2005 USD)	Penn World Table (Feenstra et al., 2015)	Divided by population
Human capital	Index of human capital per person	Penn World Table (Feenstra et al., 2015)	
Industrialization	Industry, value added (% of GDP)	World Bank (2016b)	
Urbanization	Urban population (% of total)	World Bank (2016b)	
Population	Population, total	World Bank (2016b)	

Acknowledgments

I would like to thank Mahmut Yasar and the NAERE workshop (Helsinki, 2017) participants for their valuable comments. I am responsible for all remaining errors.

References

Adewuyi, A.O., Awodumi, O.B., 2017. Biomass energy consumption, economic growth and carbon emissions: fresh evidence from West Africa using a simultaneous equation model. Energy 119, 453–471.

Amri, F., 2017. Intercourse across economic growth, trade and renewable energy consumption in developing and developed countries. Renew. Sustainable Energy Rev. 69, 527–534.

Apergis, N., Payne, J.E., 2009a. CO_2 emissions, energy usage, and output in Central America. Energy Policy 37 (8), 3282–3286.

Apergis, N., Payne, J.E., 2009b. Energy consumption and economic growth in Central America: evidence from a panel cointegration and error correction model. Energy Econ. 31 (2), 211–216.

Apergis, N., Payne, J.E., 2010. Energy consumption and growth in South America: evidence from a panel error correction model. Energy Econ. 32 (6), 1421–1426.

Arellano, M., Bond, S., 1991. Some tests of specification for panel data: Monte Carlo evidence and an application to employment equations. Rev. Econ. Stud. 58 (2), 277–297.

Arellano, M., Bover, O., 1995. Another look at the instrumental variable estimation of error-components models. J. Econometr. 68 (1), 29–51.

Baltagi, B.H., 2013. Econometric Analysis of Panel Data, fifth ed. John Wiley & Sons Ltd, Chichester, United Kingdom.

Baum, C.F., Schaffer, M.E., Stillman, S., 2007. Enhanced routines for instrumental variables/GMM estimation and testing. Stata J. 7 (4), 465–506.

Baum, C.F., Schaffer, M.E., Stillman, S., 2010. ivreg2: Stata module for extended instrumental variables/2SLS, GMM and AC/HAC, LIML and k-class regression. http://ideas.repec.org/c/boc/bocode/s425401.html.

Blundell, R., Bond, S., 1998. Initial conditions and moment restrictions in dynamic panel data models. J. Econometr. 87 (1), 115–143.

Bond, S., 2002. Dynamic panel data models: a guide to micro data methods and practice. Portuguese Econ. J. 1 (2), 141–162.

Carrión-Flores, C.E., Innes, R., 2010. Environmental innovation and environmental performance. J. Environ. Econ. Manage. 59 (1), 27–42.

Costantini, V., Monni, S., 2008. Environment, human development and economic growth. Ecol. Econ. 64 (4), 867–880.

Cracolici, M.F., Cuffaro, M., Nijkamp, P., 2010. The measurement of economic, social and environmental performance of countries: a novel approach. Social Indic. Res. 95 (2), 339–356.

Dinda, S., 2004. Environmental Kuznets curve hypothesis: a survey. Ecol. Econ. 49 (4), 431–455.

Durbin, J., 1954. Errors in variables. Rev. Int. Stat. Inst. 22 (1/3), 23–32.

Farhadi, M., Islam, M.R., Moslehi, S., 2015. Economic freedom and productivity growth in resource-rich economies. World Dev. 72, 109–126.

Feenstra, R.C., Inklaar, R., Timmer, M.P., 2015. The next generation of the Penn World Table. Am. Econ. Rev. 105 (10), 3150–3182.

Gangadharan, L., Valenzuela, M.R., 2001. Interrelationships between income, health and the environment: extending the environmental Kuznets curve hypothesis. Ecol. Econ. 36 (3), 513–531.

Granger, C.W., 1969. Investigating causal relations by econometric models and cross-spectral methods. Econometrica, 424–438.

Greene, W.H., 2008. Econometric Analysis, sixth ed. Pearson Prentice Hall, Upper Saddle River, NJ.

Grossman, G.M., Krueger, A.B., 1991. Environmental Impacts of a North American Free Trade Agreement (No. w3914). National Bureau of Economic Research.

Grossman, G.M., Krueger, A.B., 1995. Economic growth and the environment. Q. J. Econ. 110 (2), 353–377.

Gujarati, D.N., Porter, D.C., 2009. Basic Econometrics, fifth ed. McGraw-Hill Education, Singapore.

Hausman, J.A., 1978. Specification tests in econometrics. Econometrica 46 (6), 1251–1271.

Hill, R.C., Griffiths, W.E., Lim, G.C., 2012. Principles of Econometrics, fourth ed. (international student version) John Wiley & Sons, Asia.

Huang, B.N., Hwang, M.J., Yang, C.W., 2008. Causal relationship between energy consumption and GDP growth revisited: a dynamic panel data approach. Ecol. Econ. 67 (1), 41–54.

Hung, M.F., Shaw, D., 2004. Economic growth and the environmental Kuznets curve in Taiwan: a simultaneity model analysis. In: Boldrin, M., Chen, B.L., Wang, P. (Eds.), Human Capital, Trade and Public Policy in Rapidly Growing Economies: From Theory to Empirics. Edward Elgar, UK, pp. 269–290.

Kennedy, P., 2008. A Guide to Econometrics, sixth ed. Blackwell Publishing Ltd, Malden, MA, USA.

Kijima, M., Nishide, K., Ohyama, A., 2010. Economic models for the environmental Kuznets curve: a survey. J. Econ. Dyn. Control 34 (7), 1187–1201.

Kraft, J., Kraft, A., 1978. Relationship between energy and GNP. J. Energy Dev. 3, 401–403.

Liu, X., 2005. Explaining the relationship between CO_2 emissions and national income—the role of energy consumption. Econ. Lett. 87 (3), 325–328.

Lorde, T., Waithe, K., Francis, B., 2010. The importance of electrical energy for economic growth in Barbados. Energy Econ. 32 (6), 1411–1420.

Martínez-Zarzoso, I., Maruotti, A., 2011. The impact of urbanization on CO_2 emissions: evidence from developing countries. Ecol. Econ. 70 (7), 1344–1353.

Menegaki, A.N., 2014. On energy consumption and GDP studies; a meta-analysis of the last two decades. Renew. Sustainable Energy Rev. 29, 31–36.

Omri, A., 2013. CO_2 emissions, energy consumption and economic growth nexus in MENA countries: evidence from simultaneous equations models. Energy Econ. 40, 657–664.

Omri, A., 2014. An international literature survey on energy-economic growth nexus: evidence from country-specific studies. Renew. Sustainable Energy Rev. 38, 951–959.

Omri, A., Daly, S., Rault, C., Chaibi, A., 2015a. Financial development, environmental quality, trade and economic growth: what causes what in MENA countries. Energy Econ. 48, 242–252.

Omri, A., Kahouli, B., 2014. Causal relationships between energy consumption, foreign direct investment and economic growth: fresh evidence from dynamic simultaneous-equations models. Energy Policy 67, 913–922.

Omri, A., Mabrouk, N.B., Sassi-Tmar, A., 2015b. Modeling the causal linkages between nuclear energy, renewable energy and economic growth in developed and developing countries. Renew. Sustainable Energy Rev. 42, 1012–1022.

Omri, A., Nguyen, D.K., 2014. On the determinants of renewable energy consumption: international evidence. Energy 72, 554–560.

Omri, A., Nguyen, D.K., Rault, C., 2014. Causal interactions between CO_2 emissions, FDI, and economic growth: evidence from dynamic simultaneous-equation models. Econ. Modell. 42, 382–389.

Ozturk, I., 2010. A literature survey on energy–growth nexus. Energy Policy 38 (1), 340–349.

Pao, H.T., Tsai, C.M., 2010. CO_2 emissions, energy consumption and economic growth in BRIC countries. Energy Policy 38 (12), 7850–7860.

Payne, J.E., 2010a. Survey of the international evidence on the causal relationship between energy consumption and growth. J. Econ. Stud. 37 (1), 53–95.

Payne, J.E., 2010b. A survey of the electricity consumption-growth literature. Appl. Energy 87 (3), 723–731.

Roodman, D., 2009. How to do xtabond2: an introduction to difference and system GMM in Stata. Stata J. 9 (1), 86–136.

Sadorsky, P., 2010. The impact of financial development on energy consumption in emerging economies. Energy Policy 38 (5), 2528–2535.

Sadorsky, P., 2012. Energy consumption, output and trade in South America. Energy Econ. 34 (2), 476–488.

Sadorsky, P., 2013. Do urbanization and industrialization affect energy intensity in developing countries? Energy Econ. 37, 52–59.

Sadorsky, P., 2014. The effect of urbanization on CO_2 emissions in emerging economies. Energy Econ. 41, 147–215.

Saidi, K., Hammami, S., 2015. The impact of energy consumption and CO_2 emissions on economic growth: fresh evidence from dynamic simultaneous-equations models. Sustainable Cities Soc. 14, 178–186.

Saidi, K., Hammami, S., 2016a. Economic growth, energy consumption and carbon dioxide emissions: recent evidence from panel data analysis for 58 countries. Qual. Quantity 50 (1), 361–383.

Saidi, K., Hammami, S., 2016b. An econometric study of the impact of economic growth and energy use on carbon emissions: panel data evidence from fifty eight countries. Renew. Sustainable Energy Rev. 59, 1101–1110.

Sargan, J.D., 1958. The Estimation of economic relationships using instrumental variables. Econometrica 26 (3), 393–415.

Shahbaz, M., Lean, H.H., 2012. Does financial development increase energy consumption? The role of industrialization and urbanization in Tunisia. Energy Policy 40, 473–479.

Sharma, S.S., 2010. The relationship between energy and economic growth: empirical evidence from 66 countries. Appl. Energy 87 (11), 3565–3574.

Sharma, S.S., 2011. Determinants of carbon dioxide emissions: empirical evidence from 69 countries. Appl. Energy 88 (1), 376–382.

Sinha, A., 2016. Trilateral association between SO_2/NO_2 emission, inequality in energy intensity, and economic growth: a case of Indian cities. Atmos. Pollut. Res. 7 (4), 647–658.

Soytas, U., Sari, R., 2009. Energy consumption, economic growth, and carbon emissions: challenges faced by an EU candidate member. Ecol. Econ. 68 (6), 1667–1675.

Stern, D.I., 1993. Energy and economic growth in the USA: a multivariate approach. Energy Econ. 15 (2), 137–150.

Stern, D.I., 2000. A multivariate cointegration analysis of the role of energy in the US macroeconomy. Energy Econ. 22 (2), 267–283.

Stern, D.I., 2004. The rise and fall of the environmental Kuznets curve. World Dev. 32 (8), 1419–1439.

Stern, D.I., Common, M.S., Barbier, E.B., 1996. Economic growth and environmental degradation: the environmental Kuznets curve and sustainable development. World Dev. 24 (7), 1151–1160.

Stock, J.H., Yogo, M., 2005. Testing for weak instruments in linear IV regression. In: Andrews, D.W.K., Stock, J.H. (Eds.), Identification and Inference for Econometric Models: Essays in Honor of Thomas Rothenberg. Cambridge University Press, Cambridge, pp. 80–108.

Tiba, S., Omri, A., 2017. Literature survey on the relationships between energy, environment and economic growth. Renew. Sustainable Energy Rev. 69, 1129–1146.

Tiba, S., Omri, A., Frikha, M., 2016. The four-way linkages between renewable energy, environmental quality, trade and economic growth: a comparative analysis between high and middle-income countries. Energy Syst. 7 (1), 103–144.

Verbeek, M., 2012. A Guide to Modern Econometrics, fourth ed. John Wiley & Sons Ltd.

Wang, S.S., Zhou, D.Q., Zhou, P., Wang, Q.W., 2011. CO_2 emissions, energy consumption and economic growth in China: a panel data analysis. Energy Policy 39 (9), 4870–4875.

Windmeijer, F., 2005. A finite sample correction for the variance of linear efficient two-step GMM estimators. J. Econometr. 126 (1), 25–51.

Wooldridge, J.M., 2010. Econometric Analysis of Cross Section and Panel Data, second ed. MIT Press, Cambridge, Massachusetts.

Wooldridge, J.M., 2013. Introductory Econometrics: A Modern Approach, fifth ed. Cengage Learning, Canada, South-Western.

World Bank, 2016a. How are the income group thresholds determined? Available from: https://datahelpdesk.worldbank.org/knowledgebase/articles/378833-how-are-the-income-group-thresholds-determined.

World Bank, 2016b. World DataBank. Available from: http://databank.worldbank.org/data/home.aspx.

Wu, D., 1973. Alternative tests of independence between stochastic regressors and disturbances. Econometrica 41 (4), 733–750.

Xia, Y., 2012. An empirical research on the interactions of China's energy consumption, pollution emissions and economic growth. Int. J. Global Energy Issues 35 (5), 411–425.

Xia, Y., Xu, M., 2012. A 3E model on energy consumption, environment pollution and economic growth – an empirical research based on panel data. Energy Procedia 16, 2011–2018.

Zhang, X.P., Cheng, X.M., 2009. Energy consumption, carbon emissions, and economic growth in China. Ecol. Econ. 68 (10), 2706–2712.

CHAPTER 11

The Energy-Growth Nexus (EGN) Checklist for Authors

Stella Tsani*,**, Angeliki N. Menegaki[†,‡]

**International Centre for Research on the Environment and the Economy, Athens, Greece;
**Athens University of Economics and Business, Athens, Greece; [†]TEI STEREAS ELLADAS, University of Applied Sciences, Lamia, Greece; [‡]Hellenic Open University, Patras, Greece*

Chapter Outline

1 Introduction

The study of the EGN is of ongoing interest. Understanding and analyzing the energy-growth links are important and of high academic and policy interest for several reasons. First, the need to understand the relationship between energy and economic growth becomes imminent in the light of the unpredictable developments with regard to energy prices. In-depth understanding of the energy-growth links is essential for the analysis of the impact of energy cost fluctuations in the growth process of the nations, their effects, and their magnitude. Understanding the energy-growth links also appears important to assess the viability and the usefulness of energy efficiency policies, which are much advocated by major energy consumers, such as the European Union (EU). Moreover, developments with international initiatives and commitments to address climate change issues have intensified the debate on the implementation of energy conservation policies and switching to alternative forms of energy production, moving away from carbon-emitting fossil fuels and highlighting the need to understand the implications of such actions for economic growth.

Empirical research dating back to the 1970s provides an extensive volume of research on the energy-growth links touching mainly upon causality issues. The results are categorized in four main groups. The first group of findings identifies the presence of bidirectional causality between energy consumption and economic growth, confirming the so-called feedback hypothesis. The feedback hypothesis identifies a joint effect between energy consumption and economic growth, wherein each one affects the other in such a way that an increase (decrease) in energy consumption increases (decreases) GDP levels and vice versa. The second group of results shows the existence of a unidirectional causality running from energy consumption to economic growth. These results confirm what is known as the growth hypothesis, which postulates that an increase (decrease) in energy consumption can cause an increase (decrease) in GDP level. According to the growth hypothesis, energy consumption plays a vital part in economic growth. The third group of results suggests the existence of unidirectional causality running from GDP to energy consumption. The findings in this group of results confirm the so-called conservation hypothesis. According to the conservation hypothesis, economic growth increases energy consumption. Last, the fourth group of studies provides evidence in support of the so-called neutrality hypothesis. According to this hypothesis, no causal relationship is identified between energy consumption and economic growth, suggesting that no effect runs from energy consumption to economic growth and vice versa.

The identification, understanding, and quantification of the relationship between energy consumption and economic growth have profound implications for energy-related policies and actions aiming at boosting the economic growth of the nations. For instance, in case there is a causal relationship running from energy consumption to economic growth, energy consumption can be regarded as a stimulus to economic growth. When such causal relationships are the case, then international binding initiatives, energy efficiency, or conservation policies might negatively impact economic growth. If no causal relationships are identified between energy consumption and economic growth, then energy policies can be disentangled from economic performance, and economic policies can be expected not to impact energy consumption.

Given the importance of this matter, the economic literature provides an extensive list of articles studying the relationship between energy consumption and economic growth. Studies develop different methodologies and examine different countries or different groups of countries as case studies, often distinguishing between developed and/or developing ones. In addition, studies also vary in their time dimensions. Some studies examine recent evidence covering a time span of several years. Other studies make use of data that cover several decades looking well back into the past.

A plethora of papers on this subject renders it important to warn researchers and prospective authors interested in this field that not much is left to be done. As a matter of fact, one of the leading peer-reviewed journals in the field of energy economics once advised authors wishing to submit their papers for publication to the journal not to opt for papers studying the causality between energy consumption and economic growth. The journal suggested that the subject of causality has been examined in detail, and not much room has been left for innovative studies on the subject. Yet, it appears that the literature has not yet come to an agreement on the existence and the direction of the causality between energy consumption and economic growth. In fact, what almost all papers on this subject use in their introductory part as an argument in support of their work is that, despite the volume of literature on this matter, no consensus has been reached on the energy-growth causal links.

Thus, it appears that until the scientific community comes up with clear evidence and a unanimous view on the existence and the direction of causality between energy consumption and economic growth, researchers will keep producing papers on the subject and will keep submitting them to peer-reviewed journals to be considered for publication. Keeping pace with developments in energy policy and targets, researchers need to consider energy efficiency and conservation, climate change mitigation actions, penetration of renewables, and other energy-related targets that nations have set for the immediate future. From a methodological point of view, researchers need to identify recent developments in methodologies and to what extent they can be implemented for the study of the EGN.

This chapter unfolds with the intention to assist aspiring authors dealing with the EGN. We intend to provide a comprehensive review of the energy-growth literature with the aim to identify the missing pieces in the puzzle, as the title of the respective section suggests. With this, we aim to offer a clear and condensed summary of what has been done and what yet remains to be examined in the EGN literature, indicating possible research paths for the future. To do so, we review the existing literature in the field in terms of the questions that have already been addressed. We extend our discussion in terms of the methodologies and the data used with the goal to identify not only recent methodological advances, but also methodological and data considerations that authors need to take into account.

Next, we turn our attention to more practical questions, which are associated with transferability issues of research findings and the comments commonly raised by reviewers. The last part of this chapter reviews the subjects discussed in the previous parts of the chapter from a reviewer's point of view. This is done with the aim to identify and discuss common reviewer comments and indicate the checklist that potential authors in this research area should keep in mind. The discussion is wrapped up with a comprehensive author's checklist.

The remainder of the chapter develops as follows: Section 2 provides a snapshot overview of the literature with the aim to identify recent developments and missing pieces from the discussion on EGN. Section 3 discusses issues and advancements associated with data and methodologies employed. Section 4 discusses ways and directions by which the findings of studies on the EGN can be of use, and their applications can become transferable to other research fields. Section 5 reviews the most common questions raised by reviewers and the checklist that authors need to keep in mind when preparing manuscripts on the EGN. Section 6 provides the concluding remarks for the chapter.

2 Snapshot of the Literature on the Energy-Growth Nexus (EGN): Missing Pieces from the Puzzle

Theoretical approaches to the energy-growth relationship suggest different models. In the traditional neoclassical growth model, energy is introduced as an intermediate input next to the basic factors of land, labor, and capital. Energy contributes to economic growth both in a direct and/or indirect manner (Akarka and Long, 1979; Stern, 1993, 2000; Yuan et al., 2008). Under the biophysical approach, energy becomes an important determinant of income, with economies depending significantly on energy and being affected heavily by changes in energy consumption (Cleveland et al., 1984). In general, most analyses recognize that adequate and reliable access to energy is important to economic growth.

Aiming at the identification of the existence of links between energy consumption and economic growth, the literature offers abundant studies looking at the existence of causality, and if so, at its direction. Consequently, an overwhelming number of studies in the EGN are concerned with whether higher energy consumption leads to GDP growth, whether GDP growth leads to more energy consumption, whether the bidirectional causality is the case, or whether no causal relationship exists between the two variables.

The pioneer study by Kraft and Kraft (1978) on this topic dealt with applying a standard Granger (1969) causality test. They found a unidirectional long-run relationship running from GDP to energy consumption in the United States for the period 1947–1974. Following the same methodology, Yu and Hwang (1984) found no causality in the case of the United States for the period 1947–1979. These initial studies were followed by an explosion of interest on this matter and application of different methods and data drawing on the evidence from developed and developing countries. Following the seminal survey of Kraft and Kraft (1978), Granger causality tests have been the most common approach used for the investigation of the EGN. The results are mixed with regard to the nature and the direction of causality.

Yu and Choi (1985) employed the standard Granger causality test for a set of countries for the period 1954–1976. The authors found that causality runs from GDP to energy consumption for Korea; it runs in the opposite direction for Philippines, while no causal relationship is identified in the case of United States, Poland, and the United Kingdom. Erol and Yu (1987) employed the same test and found unidirectional causality running from energy consumption to growth for Japan for the period 1950–1982. Following a similar methodology with data from Taiwan for the period 1955–1993, Hwang and Gum (1992) and Yang (2000) found a bidirectional causal relationship between energy consumption and economic growth. Similarly, Cheng (1997) and Wolde-Rufael (2004) found the long-run causal relationship running from energy consumption to GDP in the case of Brazil for the period 1963–1993. Wolde-Rufael (2004) provided evidence of a long-run causal relationship running from energy consumption to GDP based on data from Shanghai for the period 1952–1999.

Cheng and Lai (1997) employed Hsiao (1981) Granger causality test for Taiwan for the period 1955–1993. Their results, in contrast to the study of Hwang and Gum (1992) and Yang (2000), suggest the causal relationship as being unidirectional and running from GDP to energy consumption. Chiou-Wei et al. (2008) applied both linear and nonlinear Granger causality tests to examine the causal relationship between energy consumption and economic growth for a sample of newly industrialized Asian countries as well as the United States for the period 1954–2006. Their study found evidence supporting the neutrality hypothesis for the United States, Thailand, and South Korea. Moreover, they found the existence of a unidirectional causality running from economic growth to energy consumption for Philippines and Singapore, while energy consumption may have affected economic growth for Taiwan, Hong Kong, Malaysia, and Indonesia.

Chontanawat et al. (2008) tested for determining causality between energy and GDP using a consistent data set and Granger test methodology for 30 OECD countries and 78 non-OECD countries. They found that causality from energy to GDP was more prevalent in the developed OECD countries compared to the developing non-OECD countries. More recently, Bozoklu and Yilanci (2013) provided evidence in support of the conservation hypothesis, suggesting that income Granger causes energy consumption, thus providing evidence in support of the conservation hypothesis. They also showed that energy consumption Granger causes the income level for the case of 20 OECD countries.

Pao and Fu (2013) applied cointegration tests, and their findings confirm the presence of unidirectional causality in the case of Brazil, running from GDP to energy consumption and based on data that cover the period 1980–2010. Mandal and Madheswaran (2010) made use of data for the period 1979–2005 for India. With the application of cointegration methods, they found that there existed bidirectional causality between energy consumption and economic growth. Ozturk and Acaravci (2010) used data for Turkey for the period 1968–2005 and applied the autoregressive distributed lag (ARDL) bounds test. Their findings suggest that there exists no causality running in either direction between energy consumption and economic growth. Ahamad and Islam (2011) applied a Vector Error Correction Model (VECM) for Bangladesh, with data running from 1971 to 2008 and showed that causality runs from energy consumption to economic growth and vice versa.

In a different methodological path, Fallahi (2011) employed Markov-switching Vector Autoregressive Models on data from the United States covering the period 1960–2005. The results indicated the presence of bidirectional causality. Al-Iriani (2006) applied panel cointegration method for six Gulf countries for data for the period 1970–2002 and showed that in this case causality exists and runs from income to energy consumption. In a larger sample of data and countries (82 countries, 1972–2002), Huang et al. (2008) found that income Granger causes energy consumption. Payne (2009) undertook Pedroni Panel cointegration and Granger causality tests for six Central American countries and data covering the period 1980–2004. His findings showed causality running from energy consumption to economic growth.

Despite a significant number of studies on the causality between energy consumption and economic growth, there is a lack of agreement in the literature about the existence and, if so, the direction of the causal links between energy consumption and economic growth. This point has been formalized by Ozturk (2010) in a survey of literature published in a leading energy journal. The point was further reinforced by a recent meta-analysis of 158 articles by Kalimeris et al. (2014) and 51 studies by Menegaki (2014). The authors showed that each of the four possible patterns of causal links had been identified on a roughly equal number of occasions. They also showed that there was no systematic correlation between the causal pattern identified and the methodological approach adopted. Similar findings have been reported by Payne (2010a,b).

Thus, it appears that the core question of the EGN studies remains at large unanswered. Studies contradict each other in findings even in cases when the same countries and data or similar methodologies are employed. Therefore, it may be argued that the core piece of the puzzle in the EGN literature is still missing, that is we are still unaware of a clear and definitive description of the links and the relationship between energy consumption and economic growth. Thus, authors who intend to contribute to this debate or to reach a clear answer have still ample room for improvements (see Table 11.1 for a summary). In doing so, authors need to identify what has been already been done at large, and what can be suggested as an innovative contribution to the ongoing studies in this field.

In a recent review of the energy, economic growth, and poverty reduction-related literature, Bacon and Kojima (2016) identified and discussed some core methodological issues in the study of energy-growth links. According to the authors, a simple correlation between energy consumption and GDP cannot be taken to necessarily support the view that increasing energy use will increase GDP or energy consumption. The causation may be entirely in the reverse direction: increased GDP may lead to increased energy use. It is only the full specification and estimation of both possible links that can establish their relative importance. On a closely related issue, in the majority of the existing studies to date, the significance of and the explicit inclusion in the analysis of other major determinants of energy consumption and economic growth have been at large ignored. Thus, more can be done to complement existing knowledge with regard to the correct specification of the relationship between energy consumption and economic growth.

In methodological terms, this limitation is associated to the problem of omitted variables. Studies that test the simultaneous presence of a production function relation, in which higher energy use contributes to GDP growth and a demand function relation in which GDP growth results in higher energy consumption, should include all possible major determinants of both relations. Under this assumption, the production function would have to include capital and labor variables as well as energy as inputs for the determination of GDP. In addition, the demand function would include the relative energy price as well as GDP in the determinants of the use of energy, because omission of an important variable leads to bias and possible misidentification of the causal pattern.

To evaluate the prevalence of the possible omitted variables bias, Bacon and Kojima (2016) surveyed the studies included in the meta-analysis of Ozturk (2010) and subsequent publications in the leading energy journals. Their analysis showed that out of 136 studies, 126 applied some form of testing for the direction of causality, and 116 applied testing and estimation techniques that allowed for nonstationarity of the data. In this group of studies surveyed, only in 3 of them were tests on the direction of causality applied, allowed for nonstationarity, and included possible major explanatory variables for both the production and the demand function relationships.[1]

Moving further, several studies have extended the topic of the EGN so as to account also for the environment in the development process. The extension into the analysis of the energy-environment-growth nexus follows the Environmental Kuznets Curve hypothesis. The curve was first conceived as depicting the relationship between income and inequality as Kuznets (1955) deducted. Subsequently, this pattern has been used so as to describe the long-run environment-growth nexus (Dinda, 2004; Friedl and Getzner, 2003; Grossman and Krueger, 1991; Managi and Jena, 2008; Stern, 2000). The Environmental Kuznets Curve hypothesis postulates that the nexus between the environment and economic growth is of an inverted U-shape (Omri, 2014; Saboori et al., 2012). This means that environmental damage increases with output until a threshold is reached, after which it begins to decline.

Literature studying the extended energy-growth-environment nexus also show significant variation in methodology, data and results. Indicatively, Apergis and Payne (2014) studied 11 countries of Commonwealth of Independent States with data for the period 1992–2004, with the use of a panel VECM methodology. Their findings indicate the existence of unidirectional causality running from energy use to environmental quality, and from economic growth to environmental quality. The authors also found the presence of a bidirectional relationship

---

[1] The authors conclude that only the study of Stern and Enflo (2013) using Swedish data can be seen as a reliable guide to the energy–GDP relationship. Their study used a significant long time span of data (150 years) and it indicated that energy use affects economic output over the full sample period. They also showed that economic output affected energy use in recent smaller samples. Relative energy prices have been found to have a significant negative link to both energy use and GDP.

Table 11.1: Indicative parts of the energy-growth nexus puzzle and related questions yet to be answered.

Existence and direction of causality in the energy-growth links	• Are there any causal links between energy and growth? • What is the direction of causality in the energy-growth relationship?
EGN and the importance of the environment	• Is there an environmental Kuznets curve? • What are the interactions between energy, growth, and environmental degradation? • What is the direction of causal links (if any) in the energy-growth-environment nexus?
RES and energy efficiency policies	• What are the implications of the RES/energy efficiency targets? • Does growth Granger cause RES deployment? • How does energy efficiency relate to economic growth? • What are the implications of energy efficiency and RES deployment policies for economic growth?

EGN, Energy–growth nexus; *RES*, renewable energy sources.
Source: The authors.

between economic growth and energy use, and a long-run bidirectional relationship among environment and energy use. Pao and Tsai (2010) studied the energy-growth-environment status using data from the BRIC countries for the period 1971–2005. With the employment of panel causality tests, their study showed that there exists bidirectional causality between environment and energy use, and between energy use and output. Their results also showed that there exists a short-run unidirectional causality running from environment and energy use, respectively, to output. Working along the same lines, Omri (2014) investigated the environment-energy-growth nexus for a sample of 14 MENA countries with data covering the period 1990–2011 and the application of simultaneous equations models. The results indicated the presence of bidirectional causality among output and energy use.

Introducing the environment into the investigation of the EGN allows for greater completeness of the analysis. It also introduces a multidisciplinary view in the study and lends the dynamics of analysis of contemporary topics of high policy priority and relevance. The dilemma between economic growth and environmental degradation is an emerging distress. A growing number of studies in the field of ecological economics confirm the existence of links between environmental degradation and economic growth (Coondoo and Dinda, 2008; Soytas et al., 2007). In addition, a growing number of studies show the impact of environmental degradation on human activity (indicative are the studies of Lee and Lee, 2009; Liddle, 2013a).

Similar to the study of the EGN, the investigation on the energy-growth-environment nexus is characterized by an abundance of methodologies, data, and usage of country samples. However, studies fail to reach to a conclusion on the interplay between the three variables. Ozturk (2010), Payne (2010a), and Dinda (2004) argue that no substantial findings are provided by the majority of previous literature devoted to the study of the causality relationship among energy, the environment, and economic growth. The same authors suggest that the problem should be overcome with the use of new economic modeling, and with more attention being paid to new approaches and refreshing perspectives as opposed to being limited to standard methods being applied to a set of common variables for various countries, a given region, and different sample periods. In this regard, prospective authors need to identify the methodological advancements that can be employed in an innovative manner rather than applying well-used methodologies in different samples and data.

Another stream of closely related literature has been looking at the links between energy efficiency and economic growth. This is associated at large to the policy targets set in major energy-consuming countries or to the implications of energy efficiency policies for developing countries. The literature looking at the EGN, from an efficiency perspective, provides again mixed results. While energy efficiency might be the key to reducing emissions and achieving the climate change targets set by international agreements, studies argue that environmentally friendly growth policies might challenge the growth opportunities of developing countries (Dercon, 2012). Other studies provide evidence that countries that make progress in energy efficiency have seen growth accelerating (Hu and Wang, 2006; Miketa and Mulder, 2005). However, what is important to note here is the importance of the measures of energy efficiency used. These studies consider the reduction in carbon intensity, which however is not necessarily associated with greater energy efficiency. In this respect, what appears to be supportive to the existing literature is the identification of the appropriate measures of energy efficiency.

Recent literature on the energy-growth links has been looking at the relationship between renewable energy sources (RES) and economic growth. Domac et al. suggested that RES may positively impact the macroeconomic performance of the countries through the creation of employment and other economic gains. Awerbuch and Sauter argue in support of RES, as they can have a positive effect on economic growth by reducing the exposure of the economies to the volatility of energy prices linked to the unpredictable ups and downs of oil and gas prices. RES may have a further positive impact to economic growth through their contribution to the security of energy supply.

Moving on with the empirical investigation of the relationship between RES and the economic growth relationship, the existing literature provides several attempts in this regard. Indicatively, Ewing et al. (2007) employed a generalized forecast error variance decomposition analysis to investigate the links between disaggregated energy consumption (coal, oil, natural gas, hydropower, wind power, solar power, wood, and waste) and industrial output in the United States. The findings of their study showed that non-renewable energy shocks (coal, gas, and oil) had a higher effect on output variation as compared to other forms of energy sources. Chien and Hu (2007) looked at the relationship between RES and GDP for 116 economies with the employment of Structural Equation Modeling. Their study showed that RES are not linked to trade balance improvements and have no import substitution effect. Sari et al. (2008) used the ARDL approach to examine the relationship among disaggregated energy consumption (coal, fossil fuels, natural gas, hydro, solar and wind power, wood, and waste), industrial output, and employment in the United States. Their results provide evidence in support of the view that in the long run, industrial production and employment are important determinants of fossil fuel, hydro, solar, waste, and wind energy consumption. Chang et al. (2009) used a panel threshold regression model to examine the impact of energy prices on RES development under different economic growth rates for the OECD countries over the period 1997–2006. Their analysis showed that there is no direct and simple relationship between GDP and the contribution of RES to energy supply. They also found that changes in economic growth were related to past levels of renewable energy use. Chang et al. (2009) argued that high-economic growth countries deploy and use RES so as to minimize the effects of adverse price shocks, hence recording a substitution effect toward RES, while low-economic growth countries may not be able to do so.

Payne (2009) examined the causal relationship between renewable and non-renewable energy consumption and real GDP for the United States with the use of annual data for the period 1949–2006 and the employment of the Toda–Yamamoto causality tests. His findings suggest that no Granger causality exists between renewable energy consumption and real GDP. Apergis and Payne (2014) investigated the relationship between renewable energy consumption and economic growth for 20 OECD countries over the period 1985–2005 within a multivariate framework. The authors included capital formation and labor in their analysis. The Granger causality test results suggest the presence of bidirectional causality between RES consumption and economic growth, both in the short and in the long run. Their investigation concluded that there exists a long-run equilibrium relationship between real GDP and renewable energy. RES is found to be indirectly impacted by GDP through capital formation.

Further research on the relationship between RES and economic growth could result in significant contributions to the EGN assessment. The research in the field is relatively nascent and provides significant room for complementary studies. Moreover, the RES-economic growth interaction is characterized by significant complexity and provides a plethora of directions into which the research on this relationship could look at. The research on the links between RES and economic growth should consider the different economic effects of RES deployment, which include costs and benefits not only at the micro level, but also at the macro level of the economies. In terms of complexity, RES deployment can bring about profound changes in the energy sector. Indicatively, RES penetration necessitates or implies changes in generation technologies, energy supply and transport modes, and appropriate adjustments in the energy marketing activities.

From a pricing perspective, RES deployment can have profound implications for the market prices of energy. Shifts in energy supply or demand result in price changes, which however happen at the wholesale energy market. The extent to which these changes pass to the final consumers depends on the market structure. Besides price changes on the wholesale market, energy consumers might be subject to policy-induced charges for RES that are usually applied so as to cover the additional costs of RES penetration and deployment. The way RES deployment costs are covered or to which economic actors these costs are applied to (for instance, it can be considered an additional fee applied to final consumers, the case of industries being exempted from paying the additional costs for RES deployment, or RES deployment being financed by the public budget via subsidies or tax credits) may impact in different ways on the micro and the macro levels with significant implications for economic growth.

The deployment and use of RES can impact the available technology and the technological progress. This impact might result through learning by doing and/or learning by research and may differ in terms of the induced effects or the time these effects take place in the economy. Such effects also impact production, the costs associated to technology, energy efficiency, and trade. As these effects take place over time, it is important to examine the RES deployment-economic growth relationship under a dynamic perspective.

3 Data and Methodologies: Done Those, Been There, so What's New?

A review of the literature on the EGN shows that research in the field is characterized by the abundance of methodologies used to address related questions as well as of a large sample of countries and time periods examined. A nonexhaustive list of popular methodologies that have been used frequently in the studies on the EGN includes the following:

- Granger causality
  - Vector Autoregressive Model (VAR)
  - Vector Error Correction (VEC)
  - Autoregressive Distributed Lag Bounds Test (ARDL)
  - Bootstrapped
  - Pair wise
  - Toda–Yamamoto
  - Dolado–Lutkepohl
- Sim causality
- Hsiao causality
- Cointegration
  - Johansen–Juselius cointegration
  - Engle–Granger
- Dynamic panel estimation

- Dynamic simultaneous equation panel data models
- Panel causality
- Panel data with structural breaks
- Bootstrap panel unit root tests
- OLS regression
- Dynamic panel causality
- Panel cointegration
  - Pedroni panel cointegration
- Vector Error Correction Model
- Error Correction Model
- Parametric and nonparametric test
- Forecast Error Variance Decomposition
- Dynamic Modeling
- Computable General Equilibrium Modeling

Several methodological issues need to be considered when estimating the relationship between energy consumption and economic growth, and indicate where future methodological efforts should concentrate. Following Bacon and Kojima (2016), we discuss here several methodological limitations to take into account. A first point to consider regards simultaneity. The vast majority of the studies in the field recognize the issue of simultaneity, that is the possible presence of reverse relationships between energy consumption and economic growth. In general, it is assumed that overall an increase in income (output) will lead to an increase in the demand for energy, although shifts in relative demand between sectors (e.g., from manufactures to services) may cause the energy intensity of an economy to fall. This is equivalent to recognizing that the income elasticity of demand for energy may be less than unity—but it would be rare for the elasticity to become negative at the level of an aggregate economy. Thus, a positive correlation between energy and output would result either from the causal link from energy to growth, or from growth to energy, or both. This indicates the need to employ a methodology that can appropriately disentangle the two directions of causation and evaluate their relative importance. Disregarding the possibility of other-way round links, means that ignoring the possible reverse link can lead to biased results.

An additional methodological shortcoming regards nonstationarity in time series data. Energy and output time series data can be subject to strongly increasing trends. If both are determined by a third factor, also trend-dominated, they could appear to be significantly correlated even if there is no actual link between them. Studies failing to address issues of nonstationarity of the main data series or those working through a known cointegrated relation are at risk of producing spurious correlations between the variables investigated. Thus, using specific econometric techniques that can correct the nonstationarity in data emerges as highly important to ensure that what is being captured in the empirical investigation is not a spurious correlation. To address this issue, data need to be checked for whether they are stationary. If stationarity in data is not the case, data need to be checked for cointegration.

The nonstationarity and cointegration issues indicate the importance to discuss another methodological consideration of high relevance to the energy-growth studies. Such studies may often be forced to use short time series. In cases with few observations, the statistical tests for the presence of non-stationary series and their cointegration can be characterized by low power. In an attempt to address the few observation limitations, it has become a common practice in recent studies to employ panel estimations, that is combining data from different countries (indicative studies include Akkemik and Göksal, 2012; Apergis and Tang, 2013; Narayan and Popp 2012; Yoo and Lee, 2010). This approach comes with several advantages such as enabling for large samples that are required to obtain reliable estimates, as well as ensuring variation in data, because differences among countries are larger than the overtime changes within a country. On the downside, the employment of panel data implies the adoption of a quite strong assumption that the estimated coefficient is the same for all countries and constant over time.

Panel estimations employed in multicountry studies use fixed effects, which correspond to allowing the intercept of the relationship being different for each country. This is done by creating a dummy variable, which is 1 for the country in question and 0 for all others. The dummy variable controls any variable that explains differences among countries but is constant over time for each country. This approach assumes that the slopes (coefficients on the explanatory variables) are the same across all countries. This assumption implying that the coefficient of the effect of a unit change in the energy/output of the economy is the same for every country may be too restrictive. Imposing such a restriction can result in unreliable estimates of the effects under investigation. To address this issue, some studies have disaggregated their panels into groups differentiating by region (Narayan and Popp, 2012) or income (Kahsai et al., 2012; Liddle, 2013b). Other approaches attempt to address this restriction with the use of econometric techniques that allow for heterogeneity (Akkemik and Göksal, 2012).

Equally important remains the appropriate formulation of the relationship between energy consumption and economic growth. The testing of the EGN often takes place in the absence of the specification of other key explanatory variables that may be affecting the relationship. If energy is linked to output by a production function, it is to be expected that output will be affected even at a minimum capital and labor. If these factors are left out of the statistical estimation, then the estimation results may be biased to the extent of the possible correlation between the variable of interest and the omitted variables. Similarly, in a demand-type function with links running from output to energy consumption, the price of energy could also be expected to determine energy consumption. Thus, failure to include the energy price as an explanatory variable can lead to biased results.

Many studies acknowledge the possibility of bidirectional links between energy and output and carry out appropriate tests. However, many of them include only production function variables (capital and labor) or only demand variables (energy prices), but not all of them. Some recent studies discuss possible additional explanatory variables that should be included in the analysis when testing the energy-growth links. For instance, Apergis and Tang (2013) and Liddle (2013b) included urbanization as a shift factor in the production function through spillover effects and economies of scale. Altintas and Kum (2013) included exports as a factor determining GDP.

Moving forward, appropriate methodologies need to be identified and put in place with regard to the way energy and output are measured. Energy use and output measurement might be fairly straightforward, and the majority of the existing studies use similar measurements with very small variations. Nevertheless, if other explanatory variables are considered to be included in the econometric specification or in the statistical testing carried out with regard to the variables of interest, then possible measurement errors can add to the estimation bias.

Bacon and Kojima (2016) identified two measurement problems commonly encountered in the literature. First, the capital stock is rarely available on a time series basis, particularly with regard to developing countries. To address this issue, most studies use Gross Fixed Capital Formation (investment) as a proxy for capital (Akkemik and Göksal, 2012; Eggoh et al., 2011). Nevertheless, the use of investment series may be problematic. If capital is cointegrated with output, it is unlikely that its difference over time (investment) will not be. This can result in the rejection of an important variable in the estimation of the production function and hence to a biased estimation of the energy-growth link.

Second, the price of energy is not available for most countries. Some studies have used the consumer price index (CPI) as a proxy (Altintas and Kum, 2013; Eggoh et al., 2011), while others have used the international price of crude oil expressed in US dollars (Ouedraogo, 2013). Despite their advantages, such approaches may be problematic. As Stern and Enflo (2013) note, the price of energy should be measured relative to a general price index within the country in order to allow for substitution in demand between energy and other commodities. Proxying this price ratio by the CPI or GDP price deflator introduces a measurement error into the variable that can result in biased estimation.

In studies using panel data, it is often the case that output is measured in constant US dollars. In such a case, energy prices should also be measured in US dollars. To do so, studies use the CPI; nevertheless, this measure is not corrected for exchange rate movements. The international price of crude oil, as it is used in a number of studies, is measured in US dollars but it does not reflect a relative price movement. In addition, the crude oil price does not represent the general movement of all domestic energy prices, even allowing for exchange rate movements.

Turning next to the importance of country traits, the methodological approaches to the investigation of the EGN should take into consideration the specific characteristics of the economies in which the investigation tools are applied to. Efforts toward methodological advancements should focus on developing approaches that capture in a sufficient manner the specific characteristics of the economies under investigation and have a clear view of the anatomy of the countries under study (for instance, it is important to set up in a correct manner the cointegrating function that takes into consideration all the necessary aspects). In a related issue, it appears useful to assess the links between energy consumption and economic growth looking at both not only at the aggregate level, but also at disaggregated levels of the economy, allowing for the appropriate capturing of the relationships at the different levels.

The study on the EGN should also advance in a clear manner with the measurement of the flip sides of energy consumption such as energy efficiency, energy conservation, carbon emission reduction, and so on. For instance, it should make clear that the link between energy efficiency and energy demand (or emissions) is not straightforward. The first does not necessarily lead to the second, and both can be measured in several ways. Energy efficiency is defined as the ratio of useful outputs to energy inputs, while energy intensity is the inverse of this measure. In any case, measures of energy efficiency depend on the definition and the measurement of inputs and outputs. These can be measured in energy, in physical or in economic terms. Different approaches might respond better to different systems or models, and it might be often the case that not all of them capture the same values or effects.

Improvements in energy efficiency might not always lead to lower energy demand, while lower energy demand might not always be related to improved energy efficiency. In order to claim "energy savings" or "demand reduction," it is important to specify the reference against which those savings are measured or estimated (Sorrell, 2015). That involves specifying the relevant spatial and temporal boundaries and unit of measure, as well as adopting *ceteris paribus* assumptions. Data on energy consumption are not always available and/or accurate. Counterfactuals may be difficult to model and countervailing variables are difficult to control; thus, the causal link between specific changes and energy savings may be hard to establish. The relationship between energy efficiency and energy demand is further complicated by the presence of multiple rebound effects. Consider, for instance, cost-effective energy efficiency improvements in industry that enable output expansion, lower product prices, and higher market demand that in its turn stimulates economic growth and aggregates energy consumption. Overall, it can be misleading to associate energy efficiency with lower energy consumption. The definition and measurement of the latter term deserves a thorough consideration, as does the complex relationship between them.

Last, as discussed in the previous section, the relationship between energy consumption and economic growth might be complex and entails the generation of a set of interrelated effects.

The links between energy consumption and economic growth are identified in terms of improved technology, technological innovation, market price changes and demand functions, higher investments and creation of employment opportunities, and so on. Thus, it appears that the development of appropriate models that can capture these effects and provide an integrated picture might contribute to the energy-growth debate.

Recent developments in the literature in this direction include the use of complex integrated models such as Computable General Equilibrium (CGE) models. Such models allow the handling of complex interconnections and extend the analysis in a longer time horizon. They also enable a detailed representation of the different sectors of the economy and of the different economic actors (consumers, producers, public sector, financial institutions, etc.). CGE models can provide detailed representations of not only the energy markets, but also of the sectors or other markets of interest, such as the labor market.

Despite their advantages, CGE models are often criticized on the grounds of consisting a black box, which due to their complexity provide results that are difficult to comprehend. Another point of criticism is regarding the rather simple, in many cases, representation of the energy-growth process. Early developments of CGE models relied on the assumption that economic growth is purely exogenous. Recent developments provide modeling approaches to endogenous economic growth. The research on the development and the use of complex models for the assessment of the relationship between energy consumption and economic growth is ongoing. In this direction, efforts to appropriately capture energy and growth dynamics in the modeling process and provide integrated results could be significant contribution in the field.

4 Findings and Transferability

The research on the EGN can provide valuable insights and methodological suggestions that can extend beyond the specific countries examined or the specific field of interest. Transferability of the results can extend into several directions. Results based on the empirical evidence from specific countries or from groups of countries with similar characteristics may prove useful for research and/or policy making in other similar contexts or countries with similar energy and economic traits. As an example, evidence based on data from small- and medium-sized economies (Rapanos and Polemis, 2005; Samouilidis and Mitropoulos, 1984; Santamouris et al., 2007; Sardianou, 2007; Tsani, 2010) provide useful insights for other economies of similar size and structural characteristics. Additionally, the investigation of evidence from case studies can be of transferable use to other countries faced with similar energy efficiency and conservation policies, as resulting from commitments to international initiatives and high levels of dependence on energy imports. Keeping these points in mind, authors need to formulate their argumentation on the transferability of their results in a clear and coherent manner, considering not only the similarities, but also the differentials between countries and contexts.

From a conceptual point of view, the research in the EGN has considerable implications for other areas of research looking into the links between growth and infrastructure or the natural resources-economic growth nexus. Infrastructure can be an important determinant of growth. Infrastructure shortages can impede growth in developing countries. As the energy sector is an important component of infrastructure, a demonstrated link between infrastructure and growth supports the hypothesis that power sector infrastructure is linked to growth (Calderón et al., 2011; Straub, 2008a,b).

In formulating the links between infrastructure and growth, it is important to treat infrastructure in an integrated manner. As an example, the impact of electricity supply to economic growth could be higher if an adequate road or port system is in place. The benefits of infrastructure are associated with the existence of complementarities. As a result of these complementarities, it would be easier to demonstrate a link between a package of interventions regarding, for instance, the provision of reliable infrastructure including electricity, telecommunications, transport, and water to the level of economic output.

Similar to the investigation of the relationship between energy and economic growth, an abundant volume of research looks into the links between infrastructure and the level of GDP of an economy. This is usually done via a production function relationship, in which infrastructure enters alongside capital and employment in the determination of GDP, or with the investigation of empirical relations between the level of infrastructure and GDP growth. In a similar manner to the measurement challenges associated with energy, in the studies examining the infrastructure-growth nexus, the definition and measurement of infrastructure emerges as primary importance. Furthermore, the identification of additional explanatory variables entering the models employed remains equally important, which in many cases can be the same as those employed in the energy-growth modeling.

Popular approaches here include measuring infrastructure by public capital. This is a straightforward approach, which however extends the definition of infrastructure beyond its traditional concept of accounting, also for schools, hospitals, and so on. Moreover, this approach fails to account for the inclusion of private stocks of infrastructure, which can be significant in the case of energy and telecommunications. Herein, the approaches on the measurement of energy in the energy-growth literature could provide useful insights and methodological alternatives (Fig. 11.1).

The research on the EGN could further provide useful insights on the infrastructure-growth nexus research with regard to the identification and the tackling of methodological drawbacks associated with causality and heterogeneity. Reverse causation and heterogeneity might be some of the caveats present in the analysis of the relationship between infrastructure and growth. Although intuitively we might argue that infrastructure supply affects growth, it might also be the case that higher incomes may have a positive impact on the demand for infrastructure services. Failure to take this reverse causality effect into account might lead to overestimation of the impact of infrastructure on economic growth. Studies in the field deal

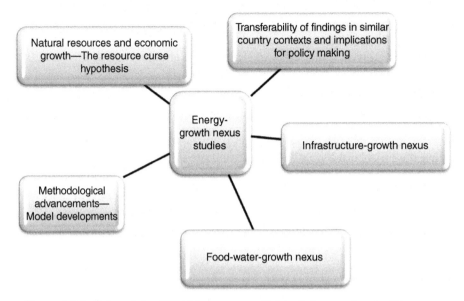

Figure 11.1: Links of the EGN Literature to Other Relevant Areas of Interest.

with this problem mainly with the use of instrumental variables and in few cases with the explicit modeling of infrastructure demand (Khandker et al. 2012; Van de Walle et al., 2013).

Additional limitations are with regard to heterogeneity across countries. The output contribution of infrastructure may differ across countries and time periods, and this may depend on many factors such as quality differentials in infrastructure. Similar to the energy-growth studies that are faced with the same problem, this caveat is associated to measurement problems. In this regard, approaches to overcome these limitations in the energy-growth literature can be applicable or transferable to the infrastructure-growth literature.

Extending beyond energy, the research in the EGN can add to the wider resources-economic growth debate. Researchers and policy makers recognize that one of the greatest challenges that humankind is faced with is natural resources, water, and food security. This challenge is expected to intensify in the future and more pressure will be the case as a result of population growth, economic developments, and climate change. This increasing pressure on natural resources raises concerns on the availability, accessibility, and sustainable use of natural resources. Research on the importance of natural resources to economic growth can benefit from the debate on the EGN, the methodologies employed and the findings.

Since the seminal paper of Sachs and Warner (1995), many studies have shown that natural resource abundance impacts negatively on economic growth. This phenomenon labeled as the "resource curse" has received much attention in the recent literature. However, in a similar manner to the energy-growth or the infrastructure-growth nexus debate, this stream of literature has underestimated the possible reverse causation or heterogeneity limitations in the analysis. Moreover, it has not been able to capture in full the interrelationships between natural resources and economic growth.

It might be well the case that higher incomes increase demand for natural resources (and make profitable the exploitation of natural resources that would otherwise be costly to extract), food, or water, or that countries differ in terms of institutions and quality of policy formulation and implementation with regard to natural resources. On the other hand, policies regarding the management of natural resources may impact on economic growth through several channels. Indicatively, these could include investments, changes in market prices and demand, technological innovation, productivity effects, and so on. Given that other natural resources resemble in nature and characteristics to energy resources (e.g., they are exhaustible, concentrated in specific geographic areas, etc.) models developed for the integrated assessment of the relationship between energy consumption and economic growth (e.g., CGE models) could be used, after appropriate updates are made for the investigation of the natural resources-growth nexus.

5 Reviewer's Turn: Some Common Questions and Comments Raised

We now turn to the reviewing and publishing process of studies and research results in the EGN field. We proceed with the discussion here on the assumption that the authors are well aware of the developments in the field and of their contribution. Thus, we focus on preparation matters and technical comments that may be raised by the reviewers.

Every author wishing to publish his/her study in a peer-reviewed journal will be faced, at some point, with comments raised by reviewers. These may vary in substance, nature, depth of analysis and enquiry, and on the difficulty in addressing them. The latter depend at large not only on the substance and completeness of the submitted paper, but also on the expertise of the reviewer in the subject. Despite the fact that editors put effort in ensuring a good match between the subject area of the submitted study and the expertise of the reviewer, this might not always be the case. Thus, authors need to make sure that their study is presented in a clear, comprehensive, and straightforward manner, and in a way that it appears interesting to not only the expert, but also to the non-expert readers.

To help authors in their submission process, journals provide guidelines, hints, and tips. Although most of them concern technical and presentation aspects, some provide indications on the aspects of the submitted study that make it worthy of publication. These general guidelines are also provided to reviewers with the intention to help them in the reviewing process. In fact, in many cases, reviewers are given a reviewer report to fill, in which very often the specific guidance to authors is used as a checklist for the reviewers. Indicative parts of a reviewer's report include the following:

- Reviewer recommendations (accept/accept with major revisions/accept with minor revisions/decline)
- Manuscript rating in a predefined scale (for instance in a 0—poor to 5—excellent scale)
- Manuscript rating question on whether the (graphical) abstract is a meaningful and an accurate representation of the article
- Manuscript rating question on whether the highlights are a meaningful and accurate representation of the article
- Comments on subject matter (within or not the scope of the journal)
- Comments on originality (similar papers published by the journal or elsewhere)
- Comments on technical quality
- Comments on clarity of presentation
- Comments on importance in field
- Comments on title (title is an accurate reflection of the study or it needs revision)
- Comments on language (grammar and spelling)
- Comments on abstract (clear and adequate, or in need of condensing)
- Comments on quantity and quality of illustrations and tables
- Comments on abbreviations, formulae, and units (whether they respond to acceptable standards)
- Comments on references (incorrect, insufficient, or too extensive)

So let us start from the basics of reviewing. Reviewers will be asked to identify and they will also want to see to what extent the papers provide novel insights to important issues of contemporary research. Moving further, it is not only the novelty but also the impact of the results to policy and decision-making, and ultimately to the society, that reviewers will be asked to comment on or look for in a paper. Thus, authors need to make sure to list out in a clear manner where the novelty of their study lies and what are the implications for decision and policy makers. On the novelty aspects, authors need to keep in mind that this part of contribution may extend into several dimensions including developments in theoretical explanations and model developments, novel advancements in methodological approaches, or novel empirical evidence. In addition, the emerging novelties that are equally important are in terms of questions addressed or raised.

Moving further, reviewers (are often asked to) comment on the methodologies employed. With regard to the methodological approaches used, authors need to make sure (and also present in a clear manner) that the employed methods and tools are appropriate so as to address the specific questions, and that they have been employed in a correct way. At this point, we should acknowledge the ever-increasing importance that publishers and the scientific community are putting on the replicability of the results. Studies showing good

---

BOX 11.1 General Publishing Guidelines[2]

8 Reasons I Accepted Your Article

1. It provides insight into an important issue—for example, by explaining a wide variance when numbers are spread out from the mean or expected value, or by shedding light on an unsolved problem that affects a lot of people.
2. The insight is useful to people who make decisions, particularly long-term organizational decisions or, in our particular field, family decisions.
3. The insight is used to develop a framework or theory, either a new theory or advancing an existing one.
4. The insight stimulates new, important questions.
5. The methods used to explore the issue are appropriate (e.g., data collection and analysis of data).
6. The methods used are applied rigorously and explain why and how the data support the conclusions.
7. Connections to prior work in the field or from other fields are made and serve to make the article's arguments clear.
8. The article tells a good story, meaning it is well written and easy to understand, the arguments are logical and not internally contradictory.

---

[2] The reasons, provided as publishing tips by Elsevier, are summed up by Dr. Torsten Pieper, Assistant Editor of the *Journal of Family Business Strategy* and Assistant Professor at the Cox Family Enterprise Center, Coles College of Business, at Kennesaw State University in Georgia, and his colleague, Dr. Joseph Astrachan, Editor-in-Chief of the journal and Executive Director of the Cox Family Enterprise Center and Professor of Management and Entrepreneurship. For more information see: https://www.elsevier.com/connect/8-reasons-i-accepted-your-article.

---

replicability can have important effects on future research. To ensure reproduction, authors need to provide as much information as possible, as well as all the necessary data and methodological clarifications, so as to allow any interested party and researchers in the field to employ the same methodology(ies) and reproduce the reported result(s).

An additional task for the reviewers is to identify the contribution of the studies under review to the ongoing debate on the specific subject (Box 11.1). This reviewing task coincides with one of the core driving forces of research, that is complementing and advancement of existing knowledge. Thus, authors need to show, in their study, in a clear manner where their study stands with regard to existing studies, prior knowledge, and state-of-the-art standards. Authors should also be able to demonstrate how their work fits with the existing studies and how it links to the accumulated stock of knowledge.

Keeping these general guidelines in mind, an author's checklist with a set of specific questions and points that authors should keep in mind can be developed as follows:

- Is there a particular policy question that drives the need for implementation of this study?
- Am I conducting global analysis for the sake of theory development?
- Have I presented some stylized facts to motivate the paper?

- Have I made sure I have not mixed the introduction with literature review?
- Shall I conduct a specific country focus or adopt a global perspective based on data availability?
- Does my country grouping make sense or is it a convenience grouping?
- What variables should I use and why?
- Does my selection of variables reflect the mechanics of the economy(-ies) I am studying?
- How do I dig for big issues in an economy that should be taken into account in my estimated function?
- Where do I source my data from?
- How and when do I use proxies?
- Do I use current or constant prices in my selected variables?
- What is my search yardstick in my literature review?
- Do I make a convenience literature review?
- Is my literature review searching reproducible?
- Is my literature review meta-analytic?
- Do I know exactly which gap I am trying to fill in literature?
- Why do I use this methodology?
- How many tests should I employ?
- How do I opt for the right estimation method for cointegration analysis?
- How do I opt for the right estimation method for the causality analysis?
- Have I explained adequately the choice of my estimation strategies?
- Are my models significant and pass all relevant tests?
- Have I checked heteroskedasticity and multicollinearity?
- Do my models have significant variables and the expected signs?
- Have I interpreted the estimated coefficients?
- Are my estimated models stable?
- Have I described causality and speed of adjustment in sufficient detail?
- How much forecasting can be done with my models?
- What do my results say about policy making?
- How does my research contribute to theory development and advancement?
- Does my abstract contain a sufficient description of the paper together with the most important result and novelty of the paper?
- Do I know why a reader would read my paper?
- How do I choose a journal for submission?
- How do I choose suggested reviewers?
- How do I set up my answers to reviewers?
- How do I set up an author team?
- What does the corresponding author do?
- Does the description of the methodology provide in a clear manner all the necessary information that allow for the reproduction of the steps followed?

- Do I provide detailed information on data definition, sources, and management?

To assist prospective authors in terms of what to expect from the reviewing process, we provide next some real-life examples of (verbatim) comments raised by reviewers on articles submitted for publication in leading energy journals on the topic of EGN:

- It could be further useful for the author(s) to bring out some example countries faced with similar energy consumption patterns and energy conservation policies as country X.
- The author(s) argue that contradictive previous results could be attributed to a diverse set of samples studied with different characteristics in each case, a variety of variables included, as well as of the implementation of different econometric approaches developed. However, the author(s) seem not to deal with those aspects with exception to the econometric approach. The author should elaborate more on their choice of variables, and why they are appropriate in comparison to previous studies cited in the literature.
- I would suggest that the author(s) provide the correlations of different sectors with growth.
- My suggestion is that the paper is short of a balance on background and policy analysis, though the empirical results seem robust. If the author is ready to revise the paper as suggested, I will be glad to re-consider my suggestion.
- To spare space, table 1 and 2 can be summarized together.
- Last paragraph of P. 5, the Wald statistics converge to distribution "with m degrees." I think it is a typing error, for m should be k.
- The results should be compared to the literature.
- In the present literature review, the author(s) use a lot of ECM results, which may cause some confusion.
- The aim of this paper is to assess the existence of the "long run" relationships between variables; however, from the representations of the eq.(1) and eq.(2), I could not identify a long-run causality term.
- It seems that the authors extend tremendous efforts on testing the potential unit roots. There are five of them altogether and they take up over half of the empirical results section! I would suggest picking a couple of the appropriate.
- What is the economic sense or policy implication of the findings? This question may come from the rationale not offered in the beginning of this paper: why is the residential energy consumption connected to energy growth, and what is the meaning behind it?
- It is a bit repetitive and could do with some editing to remove the repetition
- My recommendation is: ACCEPT with very minor revision. The only aspect that should be worked out more clearly is in the conclusions: Please re-write this text in a way that allows also policy makers—who are addressed—to understand it.
- My major point of criticism is that the conclusions are no real conclusions but rather a summary of the results. I suggest a more careful discussion of the results.
- The model is likely to be known by those working in the field, but it is not described in detail here.

- There is no clear error analysis and, although to produce one might be difficult for such modelling activities, the text speaks about "sensitivity variants" but does not show these.
- The text could be improved by giving clearer and if possible more quantitative information on the sensitivity of the results to key parameters.
- Scenario assumptions are complex and "heavy" to read. Moreover, it is quite difficult to understand what scenario assumptions are and what results from running the model are.
- The conclusions are certainly credible and likely to be valid, but how exactly they have been reached on the basis of modelling results is not very clear.
- The text is not very easy to read, but is grammatically correct. In order to understand the text, a reader would need to have quite a detailed knowledge of the field. It would be helpful to include some definitions.
- The tables are very important to this text, but are not very well discussed. This should be improved.
- Not all of the links in the references actually work—these should be checked and corrected.
- It feels quite long, with a lot to get through before getting to the main topic, i.e. the estimation results. I would reduce the length of section 2, possibly putting into an annex. This should make the paper overall more accessible, especially for non-technical readers.
- The results are highly dependent on two sets of parameters—energy price elasticities and trade price elasticities (which look high for the energy-intensive industries). As the choice of parameters can be arbitrary these need to be documented in order to properly interpret the results.
- There is no discussion of uncertainty—what are the key assumptions and what are the results most sensitive to (e.g. parameter choices)? It would also be good to include a short discussion about how results might change if some of the underlying assumptions about optimal energy use were relaxed.
- The paper would benefit from professional editing as there are a few areas where the language makes it difficult to interpret the method/findings.
- The text suggests that price differentials cannot persist in the long run but in reality there are surely many examples where price differentials are maintained? This could do with some reworking.
- Table 13 (and possibly others)—it would be good to have totals for quick comparison.
- Be more concrete in the abstract. It is no real surprise that higher energy prices have a negative impact on GDP. Mention concrete impacts. Also the term "minimizes" could be reconsidered.
- The impact on GDP of the lower energy prices is so much lower than that of the higher prices. The authors should relate this first of all to the energy prices differences: is the price difference the same in both cases?
- Small things: page 3 gives kind of the impression that unconventional energy is cheaper than conventional—which is not the case.

- There is a lonesome b on page 28 that should be "by".
- In fact the authors apply standards methods to explore the causal interplays between economic variables for Saudi Arabia, the results and methodology are not innovative as more paper considered more flexibe approach, in addition it will be interesting to make a wider analysis by considering GCC countries instead of single country. Overall wavelet based causality was applied and I do not recommend a publication.
- The study neither is carefully written nor methodologically innovative. In brief, it is not good fit to the journal's level of standard which looks for more methodologically innovative papers and also theoretical contribution to the literature is marginal.
- Author(s) needs to motivate the reader more by why this study is relevant and how their contribution is important to the literature. The author(s) fails to establish a link between the existing literature and the topic of the paper.
- The contribution of this paper must be presented to the reader in a much more convincing manner, clearly stating how it adds something new to the field. The way it is presented, it is not clear at all how this paper is different from the existing literature. In other words the originality of research must be established.
- Hypotheses are not "confirmed" by your analysis. Hypotheses may be supported but they cannot be confirmed, particularly when so many unanswered problems with the statistical method exist.
- Your results are not firm enough to be used by governments.
- This sentence is too long and the meaning gets lost.
- This seems like an exaggeration. Not all policy makes care about energy conservation.
- It is essential that you say that Causality or Granger causality, as used in this section, means nothing other than that a statistical criterion has been satisfied for a given set of data. Real-life causality is not so simple. Your summary of literature shows that the statistical criterion of Granger causality may support of the four hypotheses in one case and another hypothesis in another case, even though both cases are for the same country. That demonstrates that what really causes the GDP to grow or contract is not completely captured and may be completely missed or disguised, by the statistical procedure.
- When the Granger method shows uni-directional causality running from GDP growth to energy consumption growth, the conservation hypothesis is suggested. Namely, the statistical analysis may be used to support the hypothesis that efforts to conserve energy will not lead to a decrease in GDP and may even lead to an increase in GDP. However, just as energy may be used to support GDP growth or to not support it, differing energy conservation measures may differ in their effects on the economy. One may reasonably expect that certain energy conservation measures may promote (or at least not hinder) economic growth even in an economy for which Granger causality supports the growth hypothesis. In all cases, analysis beyond the Granger method may either support or weaken the hypothesis.

- In various places of the paper, it would help to have the meaning of the hypothesis re-emphasized.
- With the exception of GDP and Energy use, it is not clear how the other parameters here are used. What does labor and trade have to do with anything? If you use them in an equation as controls, you should state this and show it.
- Do not re-state the Table in the text like this. You should pull out specific interesting facts or patterns shown in the table and discuss them.
- Numbers should be appropriately and homogenously formatted.
- In Figure X, you should use a bar graph rather that a line graph. The lines between various data points do not indicate anything.
- If you modify your conclusion, the abstract and the highlights will have to be modified to be consistent with them.
- Explain qualitatively what the unit root test and cointegration are and why they are significant.
- For unit root testing: Authors might yield explicit inferences about the output of unit root tests as "stationary" or "not stationary" in the last column.
- Why do authors follow short run causality analyses to measure the effects of variables on GDP? What is the purpose of conducting short run causality between the variables? The main motivation of this manuscript is to observe the growth within the framework of energy use. The short run causality analyses consider how fast deviations from the long-run equilibrium are corrected to reach long-run equilibrium. Hypothetically, the growth occurs in the long run and not in the short run. Authors to this end might conduct alternative block exogeneity tests to confirm/disconfirm these outcomes.
- What follows really focuses on the technical details of the exercise. The technical part should be there of course but my view is that there are the details without the general frame. Some equations and tables should be reported as annex or supplementary material.
- I think that this paper has a potential but in its current structure it shows as a quantitative exercise for insiders. The meaning and usefulness of the whole analysis must be made understandable to readers and practitioners.
- On a more general note, although the authors use the dynamic Driscoll–Kraay estimator methodology, which seems adequate, the paper appears to be in an initial phase. Some relevant results are not sufficiently explored, namely important ones (such as effects of conservation policies), and others that contradict existing research (such as non-significance of typical energy variables).
- To sum up, the paper shows potential interest for a study more relevant because of its results, and less because of the methodology. It appears to be in an initial development phase and needs further groundings, in line with previous comments.
- The novelty/originality shall be further justified by highlighting that the manuscript contains sufficient contributions to the new body of knowledge. The knowledge gap needs to be clearly addressed in Introduction.

- There is no discussion of results or empirical analysis but only a presentation of the results and statistics tests. The authors must write the analysis and discussion of results. What are the implications of their results?
- The simultaneous use of both nominal and real variables seems inadequate.
- A large concern comes from the simultaneous use of variables such as labor and dLabor. Most likely there is a severe problem of collinearity. Unfortunately neither the individual VIF nor the correlation matrix are provided.
- A wide range of both specification and diagnostic testst are absent. They are crucial to understand the quality of the estimations.
- Please indicate the order of the lags in the ARDL model.

Reading through the reviewers' comments, it becomes apparent that these may vary in substance and the level of difficulty in addressing them. In any case, reviewers and editors are those having a final say on whether the study under consideration should be published or not. It is thus essential for the authors to convince the reviewers on the merits of their study and to make sure that they address each comment raised in detail and in a straightforward manner. In some instances, it might also be the case that authors feel that they disagree with the comments raised by the reviewers or the suggested revisions to their work. It is up to the authors to choose how they are going to treat their study and research. In every case though, it should be made clear to the peer reviewers why authors disagree with specific comments or disagree in implementing the proposed changes. In doing so, authors need to remember that the reviewing process is one more form of scientific debate and interaction of peers interested in the same subject matters, and as such, argumentation, agreement, and disagreement are vital parts of it.

6 Concluding Remarks

Despite the plethora of literature on the EGN, many questions remain yet to be addressed. Given the importance of energy and economic growth to mankind and societies and the significance of the research results on the EGN for policy making, research in this field is expected to continue intensely. Prospective authors in this field of research need to be well aware of the research results and directions explored to date, as well as of not only the methodological abundance that characterizes the existing efforts, but also of the limitations present and possible paths for advancement. In an interconnected world with many imminent challenges that need to be immediately addressed such as climate change, food, water and natural resource sustainable management, environmental resilience, and economic recession, researchers need to conduct their research on the basis of a wider perspective and identify connections between their studies with other relevant areas of research.

On a technical level, authors need to place their study in a well-defined context of existing studies and contributions to future research, which may include novel theoretical approaches, fresh empirical insights, or new questions that the scientific community needs to be looking at. The links to the existing literature and the way that the studies in question advance the established knowledge need to be documented in a clear manner for the community of interest. As a first reaction to their research publication, authors should consider thoroughly the peer-reviewing process associated with their work. Authors need to keep in mind a set of qualitative requirements that their work should meet in addition to the substantial content and contribution. In addition, authors are not to forget that the reviewers of their papers are on their side and not their enemies.

References

Ahamad, M.G., Islam, A., 2011. Electricity consumption and economic growth nexus in Bangladesh: revisited evidences. Energy Policy 39, 6145–6150.

Akarka, A.T., Long, T., 1979. On the relationship between energy consumption and GNP: a re-examination. J. Energy Dev. 5, 326–331.

Akkemik, A., Göksal, K., 2012. Energy consumption-GDP nexus: heterogeneous panel causality analysis. Energy Econ. 34, 865–873.

Al-Iriani, M.A., 2006. Energy–GDP relationship revisited: an example from GCC countries using panel causality. Energy Policy 34, 3342–3350.

Altintas, H., Kum, M., 2013. Multivariate Granger causality between electricity generation, exports prices and economic growth in Turkey. IJEEP 3, 41–51.

Apergis, N., Payne, J.E., 2014. Renewable energy, output, CO_2 emissions, and fossil fuel prices in Central America: evidence from a non linear panel smooth transition vector error correction model. Energy Econ. 42, 226–232.

Apergis, N., Tang., C., 2013. Is the energy-led growth hypothesis valid? New evidence from a sample of 85 countries. Energy Econ. 38, 24–31.

Bacon, R., Kojima, M., 2016. Energy, economic growth and poverty reduction. World Bank Working Paper 104866.

Bozoklu, S., Yilanci, V., 2013. Energy consumption and economic growth for selected OECD countries: further evidence from the Granger causality test in the frequency domain. Energy Policy 63, 877–881.

Calderón, C., Moral-Benito, E., Servén, L., 2011. Is infrastructure capital productive? A dynamic heterogeneous approach. Policy Research Working Paper 5682. World Bank.

Chang, T.H., Huang, C.M., Lee, M.C., 2009. Threshold effect of the economic growth rate on the renewable energy development from a change in energy price: evidence from OECD countries. Energy Policy 37, 5796–5802.

Cheng, B.S., 1997. Energy consumption and economic growth in Brazil Mexico and Venezuela: a time series analysis. Appl. Econ. Lett. 4, 671–674.

Cheng, B.S., Lai, T.W., 1997. An investigation of co-integration and causality between energy consumption and economic activity in Taiwan. Energy Econ. 19, 435–444.

Chien, T., Hu, J.L., 2007. Renewable energy and macroeconomic efficiency of OECD and non OECD economies. Energy Policy 35, 3606–3615.

Chiou-Wei, S.Z., Chen, C., Zhu, Z., 2008. Economic growth and energy consumption revised: evidence from linear and nonlinear Granger causality. Energy Econ. 30, 3063–3076.

Chontanawat, J., Hunt, L.-C., Pierse, R., 2008. Does energy consumption cause economic growth? Evidence from a systematic study of over 100 countries. J. Policy Model. 30, 209–220.

Cleveland, C.J., Costanza, R., Hall, C.A.S., Kaufmann, R.K., 1984. Energy and the US economy: a biophysical perspective. Science 225, 890–897.

Coondoo, D., Dinda, S., 2008. The carbon dioxide emission and income: a temporal analysis of cross-country distributional patterns. Ecol. Econ. 65, 375–385.

Dercon, S., 2012. Is green growth good for the poor? Policy Research Working Paper 6231, World Bank, Washington, D.C.

Dinda, S., 2004. Environmental Kuznets curve hypothesis: a survey. Ecol. Econ. 49, 431–455.

Eggoh, J., Bangake, E., Rault, C., 2011. Energy consumption and economic growth revisited in Africa countries. Energy Policy 39, 7408–7421.

Erol, U., Yu, E.S.H., 1987. On the relationship between electricity and income for industrialized countries. J. Electr. Employ. 13, 113–122.

Ewing, B.T., Sari, R., Soytas, U., 2007. Dissagregate energy consumption and industrial output in the United States. Energy Policy 35, 1274–1281.

Fallahi, F., 2011. Causal relationship between energy consumption (EC) and GDP: a Markov-switching (MS) causality. Energy 36, 4165–4170.

Friedl, B., Getzner, M., 2003. Determinants of CO_2 emissions in a small open economy. Ecol. Econ. 45, 133–148.

Granger, C.W.J., 1969. Investigating causal relations by econometric models and cross-spectral models. Econometrica 37, 424–438.

Grossman, G., Krueger, A., 1991. Environmental impacts of a North American free trade agreement. National Bureau of Economics Research Working Paper, No. 3194.

Hsiao, C., 1981. Autoregressive modelling and money income causality detection. J. Monetary Econ. 7, 85–106.

Hu, J., Wang, S., 2006. Total factor energy efficiency of regions in China. Energy Policy 34 (17), 3206–3217.

Huang, B.N., Hwang, M.J., Yang, C.W., 2008. Causal relationship between energy consumption and GDP growth revisited: a dynamic panel approach. Ecol. Econ. 67, 41–54.

Hwang, D.B.K., Gum, B., 1992. The causal relationship between energy and GNP: the case of Taiwan. J. Energy Dev. 16, 219–226.

Kahsai, M., Nondo, C., Schaeffer, P., Gebremedhin, T., 2012. Income level and the energy consumption-GDP nexus: evidence from Sub-Saharan Africa. Energy Econ. 34, 739–746.

Kalimeris, P., Richardson, K., Bithas, K., 2014. A meta-analysis investigation of the direct of the energy-GDP causal relationship: implications for the growth-degrowth dialogue. J. Cleaner Prod. 67, 1–13.

Khandker, S., Samad, H., Ali, R., Barnes, D., 2012. Who benefits most from rural electrification? Evidence from India. Policy Research Working Paper 6095, World Bank, Washington, D.C.

Kraft, A., Kraft, J., 1978. On the relationship between energy and GNP. J. Energy Dev. 3, 401–403.

Kuznets, S, 1955. Economic growth and income inequality. Am. Econ. Rev. 45, 1–28.

Lee, C., Lee, J.D., 2009. Income and CO_2 emissions: evidence from panel unit root and cointegration tests. Energy Policy 37, 413–423.

Liddle, B., 2013a. Population, affluence, and environmental impact across development: evidence from panel cointegration modeling. Environ Model. Softw. 40, 255–266.

Liddle, B., 2013b. The energy, economic growth urbanization nexus across development: evidence from heterogeneous panel estimates robust to cross-sectional dependence. Energy J. 34, 223–244.

Managi, S., Jena, P.R., 2008. Environmental productivity and Kuznets curve in India. Ecol. Econ. 65, 432–440.

Mandal, S., Madheswaran, S., 2010. Causality between energy consumption and output growth in the Indian cement industry: an application of the panel vector error correction model (VECM). Energy Policy 38, 6560–6565.

Menegaki, A.N., 2014. On energy consumption and GDP studies: a meta-analysis of the last two decades. [Review]. Renew. Sustainable Energy Rev. 29, 31–36.

Miketa, A., Mulder, P., 2005. Energy productivity across developed and developing countries in 10 manufacturing sectors: patterns of growth and convergence. Energy Econ. 27 (3), 429–453.

Narayan, P., Stephan, P., 2012. The energy consumption-real GDP nexus revisited: empirical evidence from 93 countries. Econ. Model. 29, 303–308.

Omri, A., 2014. An international literature survey on energy-economic growth nexus: evidence from country-specific studies. Renew. Sustainable Energy Rev. 38, 951–959.

Ouedraogo, N., 2013. Energy consumption and human development: evidence from a panel cointegration and error correction model. Energy 63, 28–41.

Ozturk, I., 2010. A literature survey on energy-growth nexus. Energy Policy 38, 340–349.

Ozturk, I., Acaravci, A., 2010. CO_2 emissions, energy consumption, and economic growth in Turkey. Renew. Sustainable Energy Rev. 14, 3220–3225.

Pao, H.T., Fu, H.C., 2013. Renewable energy, non-renewable energy and economic growth in Brazil. Renew. Sustainable Energy Rev. 25, 381–392.

Pao, H.T., Tsai, C.M., 2010. CO_2 emissions, energy consumption and economic growth in BRIC countries. Energy Policy 38, 7850–7860.

Payne, J.E., 2009. On the dynamics of energy consumption and output in the US. Appl. Energy 86, 575–577.

Payne, J., 2010a. Survey of the international evidence on the causal relationship between energy consumption and growth. J. Econ. Stud. 37, 53–95.

Payne, J., 2010b. A survey of the electricity consumption-growth literature. Appl. Energy 87, 723–731.

Rapanos, V.T., Polemis, M.L., 2005. Energy demand and environmental taxes: the case of Greece. Energy Policy 33, 1781–1788.

Saboori, B., Sulaiman, J., Mohd, S., 2012. Economic growth and CO_2 emissions in Malaysia: a co-integration analysis of the Environmental Kuznets Curve. Energy Policy 51, 184–191.

Sachs, J., Warner, A., 1995. Natural resource abundance and economic growth. Development Discussion Paper No 517a. Harvard Institute for International Development.

Samouilidis, J., Mitropoulos, C., 1984. Energy and economic growth in industrializing countries. Energy Econ. 6, 191–201.

Santamouris, M., Kapsis, K., Korres, D., Livada, I., Pavlou, C., Assimakopoulos, M.N., 2007. On the relation between the energy and social characteristics of the residential sector. Energy Build. 39 (8), 893–905.

Sardianou, E., 2007. Estimating energy conservation patterns of Greek households. Energy Policy 35 (7), 3778–3791.

Sari, R., Ewing, B.T., Soytas, U., 2008. The relationship between disaggregate energy consumption and industrial production in the United States: an ARDL approach. Energy Econ. 30, 2302–2313.

Sorrell, S., 2015. Reducing energy demand: a review of issues, challenges and approaches. Renew. Sustainable Energy Rev. 47, 74–82.

Soytas, U., Sari, R., Ewing, B.T., 2007. Energy consumption, income, and carbon emissions in the United States. Ecol. Econ. 62, 482–489.

Stern, D., Enflo, E., 2013. Causality between energy and output in the long-run. Energy Econ. 39, 135–146.

Stern, D.I., 1993. Energy and economic growth in USA: a multivariate approach. Energy Econ. 15, 137–150.

Stern, D.I., 2000. A multivariate cointegration analysis of the role of energy in the US macroeconomy. Energy Econ. 22, 267–283.

Straub, S., 2008a. Infrastructure and growth in developing countries: recent advances and research challenges. Policy Research Working Paper 4460, World Bank, Washington, D.C.

Straub, S., 2008b. Infrastructure and development: a critical appraisal of the macro level literature. Policy Research Working Paper 4590, World Bank, Washington, D.C.

Tsani, S., 2010. Energy consumption and economic growth: a causality analysis for Greece. Energy Econ. 32 (3), 582–590.

Van de Walle, D., Ravallion, M., Mendiratta, V., Koolwal, G., 2013. Long-term impacts of household electrification in rural India. Policy Research Working Paper 6527, World Bank, Washington, D.C.

Wolde-Rufael, Y., 2004. Disaggregated energy consumption and GDP: the case of Shanghai, 1952–1999. Energy Econ. 26, 69–75.

Yang, H.Y., 2000. A note on the causal relationship between energy and GDP in Taiwan. Energy Economics 22, (309–317).

Yoo, S.H., Lee, J., 2010. Electricity consumption and economic growth: a cross country analysis. Energy Policy 38, 622–625.

Yu, E.S.H., Choi, J.Y., 1985. The causal relationship between energy and GNP: an international comparison. J. Energy Dev. 10, 249–272.

Yu, E.S.H., Hwang, B.K., 1984. The relationship between energy and GNP: further results. Energy Econ. 6, 186–190.

Yuan, J., Kang, J., Zhao, C., Hu, Z., 2008. Energy consumption and economic growth: evidence from China at both aggregate and disaggregated levels. Energy Econ. 30, 307–3094.

Conclusions

Angeliki N. Menegaki*,**

*Hellenic Open University, Patras, Greece; **TEI STEREAS ELLADAS, University of Applied Sciences, Lamia, Greece

The energy-growth nexus (EGN) is a research field that has attracted, keeps attracting, and will be attracting attention by numerous academics and researchers, because it is a very timely topic that causes vivid debates. Despite the many years since the first paper by Kraft and Kraft in 1978, little consensus has been reached since then in the field. This is because the EGN field has not only been deepening (containing new methods and data), but also widening (using more variables and various countries). If one keeps doing both actions at the same time, results would not be as clear, as it would have been if one had concentrated only in the activity of deepening. Given that environmental protection, together with energy security and energy efficiency are priority issues in the agendas of relevant top international meetings and fora, it is no wonder why there is ample work left to be done in this field, both vertically (deepening) and horizontally (widening).

However, up to date, several survey studies or meta-analyses, as well as individual studies have concluded that research in the EGN has produced conflicting results, which cannot inform policy making in a definite and concrete way. The usage of various econometric methods, data spans, and variables causes chaos and confusion, which can be of little help in policy making, which would require speaking with a degree of certainty. These worries are particularly and eloquently described within Chapters 1, 5, and 11.

The book not only summarizes what has been done up to date in the EGN field, with particular emphasis on what is being done right now, containing the most up to date trends and techniques, but also paves the way for the future with the new variables that need to be introduced, how they must be selected, treated, and so on. The comprehensive description and discussion of all theoretical, empirical, and practical facets in the EGN render this book a valuable guide for EGN researchers of various experience levels. Irrespective of whether the reader is a new or an experienced researcher, the book covers all needs. To recapitulate about the structure of the book: it consists of two fundamental parts—Part 1 is the economics of the EGN, and Part 2 is the econometrics of the EGN.

The Economics and Econometrics of the Energy-Growth Nexus
http://dx.doi.org/10.1016/B978-0-12-812746-9.00012-2

Part 1: The Economics of the Energy-Growth Nexus (Chapters 1–5)

This part started with Chapter 1, which offered a comprehensive overview of the EGN knowledge, history, and development. Particular emphasis was laid on the environmental Kuznets curve, which is a special research field that was considered to be a close relative to the EGN. Readers have also been introduced to the new challenges faced in the field, particularly the water-energy-food and the water-energy nexuses. Most importantly, Chapter 1 familiarized the readers with the "Hajko critique," which introduced a new perspective in the field, hopefully causing a new, decisive turn in the field very soon.

The need for disaggregation in the EGN was presented in Chapter 2. This chapter described various levels and types of disaggregation in the EGN and provided selected, up-to-date, chronologically ordered, literature examples from main types of disaggregation: energy type and sectoral. The former included: nuclear energy, natural gas, various fossil, fuel energy types renewable, and electricity. The latter encompassed: transport, commercial, industry, and residential. The ideal but most difficult type of analysis in the EGN is the disaggregated, both at sectoral level and energy level. Dearth on data can be a hindrance for this analysis, particularly in the underdeveloped countries. The chapter ended with a case study on the United States over the period 1962–2015, using the ARDL bound test and emphasized the importance of utilization of disaggregated data, once available, by referring to the aggregation bias in the energy economics.

Given that renewable energy economics is a novel ascending subfield in the broader field of energy economics, this book could not go without a chapter on the renewable EGN topic. Thus, Chapter 3 focused on the recomposition of the electricity mix, by incorporating the intermittent renewable sources, and the consequences of this diversification in the economic growth. It examined the interactions between sources and the relationship between the latter and economic growth, and concluded on the conditions necessary to make the energy transition suitable. To make a more succinct demonstration, the chapter used Germany as a case study, as this is a country with greater commitment in this energy transition.

One of the most difficult problems in the EGN is to decide which variables to select to best depict the equation that characterized the production function in one economy. Recent trends show that a wide selection of new variables is gradually hosted. Thus, Chapter 4 dealt with this battery of factors that can influence and determine the relationship, such as production factors, international trade, financial development, population growth, and urbanization; but also some less popular ones in the energy economics literature, such as militarization and tourism development. Furthermore, the chapter made a brief introduction into two approaches that attracted attention recently, in the EGN: regime switching modeling and time-varying estimations.

As aforementioned, Chapter 5 is one of the chapters that present the critical aspects faced in the EGN today. For example, the chapter revealed that EGN studies have not been particularly verbose to whether international energy strategies and targets can be fulfilled, although the studies are often motivated by them and they derive their usefulness by the promise to answer such questions. Another issue posed in this chapter is that EGN studies do not make distant future projections and forecasting. Moreover, the EGN studies involving large numbers of countries have appeared to be useful, mostly, for the discovery of global relationships that aim to build and evolve theory, but have not been able to provide single countries with insightful policy making results. Foremost, Chapter 5 has suggested several new directions to which the EGN could be shifted and enriched accordingly, such as the food-energy-water nexus, the accommodation of energy efficiency, and sustainability in the EGN.

Part 2: The Econometrics of the Energy-Growth Nexus (Chapters 6–11)

The discussion of critical aspects in the EGN, but at a more technical level, was continued in Chapter 6. The chapter collected crucial information regarding data sources, variable selection, transformation, analysis, and interpretation in the EGN. It gave readers a lot of condensed information that is usually collected after many years of work in the EGN. While the first part of Chapter 6 had a more practical advice orientation, the second part of the chapter focused on Bayesian data analysis and it was more technical. The introduction to Bayesian analysis tools was complemented with an application that offered readers with a valuable experience.

Chapter 7 offered one of the most up-to date frameworks in time series analysis to work with—that is the asymmetries and nonlinearities framework—and demonstrated their functioning onto a case study for Pakistan. More specifically, the chapter used an ARDL approach in the presence of both symmetric and asymmetric cointegration between energy consumption, oil prices, capital, labor, and economic growth over the period of 1985QI-2016QIV in Pakistan. Besides many interesting findings, the chapter concludes for energy, that a rise in energy consumption (positive shock) adds to economic growth via stimulating economic activity; whereas an energy consumption (negative shock) retards economic growth insignificantly. The chapter suggests a serious number of policy recommendations for Pakistan to be able to develop in a sustainable way.

After devoting the previous chapter on time series analysis, Chapter 8 was devoted on panel data analysis. The chapter sheds light on the ordinary challenges faced, while performing a panel data analysis in the field of the EGN and shows the way without getting stuck in complex statistics and formulae. The chapter concluded with an application on G7 countries.

Besides devoting separate chapters for time series analysis and panel data analysis, Chapter 9 is devoted to the presentation of the most up-to-date causality tests, which is the strongest part of analysis in the EGN. Particular emphasis is placed on asymmetric causality and nonparametric causality. The presentation of each test was accompanied by a relevant case study in order to improve the educational experience of the readers.

Specific attention is paid to simultaneous equations modeling in Chapter 10. This is done with a focus on EGN and a case study.

Last but not least, Chapter 11 provides readers will a collection of advice placed in the form of a checklist that will enable researchers to do innovative research and prepare their future papers in the best way possible for review. In an interconnected world with many imminent challenges, such as climate change, food, water, and natural resource sustainable management, environmental resilience, and economic recession, researchers need to place their new projects into a wider perspective. The same applies for the technical perspective that employs the most up-to-date techniques.

After the brief presentation of the conclusions of the book, I would like to say a few words about the contributors of this book. All of them are established scholars in the EGN who have contributed not only with their particular chapter contribution, but also with their ideas and suggestions to the initial set up of the book. I would like to thank them as much as Elsevier that showed us trust and accommodated our contributions into this single book. Foremost, I would like to thank the researchers who read our book. We look forward to hear comments and suggestions for further improvements and future editions of the book.

Index

Printed in the United States
By Bookmasters